METHODS in
MICROBIOLOGY

METHODS in MICROBIOLOGY

Edited by

J. R. NORRIS
Borden Microbiological Laboratory,
Shell Research Limited,
Sittingbourne, Kent, England

D. W. RIBBONS
Department of Biochemistry,
University of Miami School of Medicine,
and Howard Hughes Medical Institute,
Miami, Florida, U.S.A.

Volume 5A

1971

ACADEMIC PRESS
London and New York

ACADEMIC PRESS INC. (LONDON) LTD
Berkeley Square House
Berkeley Square
London, W1X 6BA

U.S. Edition published by
ACADEMIC PRESS INC.
111 Fifth Avenue
New York, New York 10003

Library of Congress Catalog Card Number: 68–57745
ISBN: 0–12–521505–3

PRINTED IN GREAT BRITAIN BY
ADLARD AND SON LIMITED
BARTHOLOMEW PRESS, DORKING

LIST OF CONTRIBUTORS

IRENE BATTY, *Wellcome Research Laboratories, Beckenham, Kent, England*

J. DE LEY, *Laboratory for Microbiology, Faculty of Sciences, State University, Gent, Belgium*

M. ENOMOTO, *National Institute of Genetics, Mishima, Shizuoka-ken, Japan*

T. IINO, *National Institute of Genetics, Mishima, Shizuoka-ken, Japan*

SUSAN M. JONES, *Division of Agricultural Bacteriology, University of Aberdeen, Scotland*

W. H. KINGHAM, *Shell Research Limited, Milstead Laboratory of Chemical Enzymology, Sittingbourne, Kent, England*

H. KOFFLER, *Department of Biological Sciences, Purdue University, Lafayette, Indiana, U.S.A.*

J. E. M. MIDGLEY, *Department of Biochemistry, University of Newcastle-upon-Tyne, England*

J. R. NORRIS, *Shell Research Limited, Borden Microbiological Laboratory, Sittingbourne, Kent, England*

C. L. OAKLEY, *Department of Bacteriology, The Medical School, Leeds, England*

A. M. PATON, *Division of Agricultural Bacteriology, University of Aberdeen, Scotland*

L. B. QUESNEL, *Department of Bacteriology and Virology, University of Manchester, Manchester, England*

R. W. SMITH, *Department of Biological Sciences, Purdue University, Lafayette, Indiana, U.S.A.*

HELEN SWAIN, *Milstead Laboratory of Chemical Enzymology, Sittingbourne, Kent, England*

R. O. THOMPSON, *The Wellcome Research Laboratories, Langley Court, Beckenham, Kent, England*

P. D. WALKER, *The Wellcome Research Laboratories, Langley Court, Beckenham, Kent, England*

ELIZABETH WORK, *Department of Biochemistry, Imperial College of Science and Technology, London, S.W.7, England*

ACKNOWLEDGMENTS

For permission to reproduce, in whole or in part, certain figures and diagrams we are grateful to the following—

Edward Arnold Ltd., London; Butterworths, London; Chapman & Hall, London; E. Leitz, Wetzlar; Longmans, Green & Co., London; Nippon Kogaku, Tokyo; Vickers Instruments, York, England; Wild Heebrugg, Switzerland; Karl Zeiss, Oberkochen, W. Germany.

Detailed acknowledgments are given in the legends to figures.

PREFACE

Volume 5 of "Methods in Microbiology" is concerned with the microbial cell—methods of observing it, of studying its properties and behaviour, of analysing it chemically and immunologically, and of purifying and characterizing its various "organelles" and macro-molecular components. Wherever possible, the emphasis has been placed on quantitative methods.

We have tried to cover relatively new techniques such as reflectance spectrophotometry, isoelectric focusing and polyacrylamide gel electrophoresis which appear to us to have considerable future potential in microbiology in addition to more generally used techniques such as those for cell disintegration and hybridization of nucleic acids which are not fully described in a concise form elsewhere.

As with earlier Volumes in the Series we have left the treatment of the different topics largely to the individual authors, restricting our editorial activity to ensuring consistency and avoiding overlaps and gaps between the contributions.

As contributions accumulated it became obvious that there was too much material for a single Volume and the content was divided. Volume 5A contains Chapters concerned with the direct observation or study of whole cells or organelles while Volume 5B is concerned with the disintegration of cells, their chemical analysis and the techniques used to separate and characterize their components.

Our thanks are due to the pleasant way in which our authors have co-operated with us and particularly to those who agreed to update their contributions when delay in the publication process made it necessary.

J. R. NORRIS
D. W. RIBBONS

April, 1971

CONTENTS

LIST OF CONTRIBUTORS v

ACKNOWLEDGMENTS vi

PREFACE vii

CONTENTS OF PUBLISHED VOLUMES xi

Chapter I. Microscopy and Micrometry—L. B. QUESNEL . . 1

Chapter II. Staining Bacteria—J. R. NORRIS AND HELEN SWAIN . 105

Chapter III. Techniques Involving Optical Brightening Agents—
A. M. PATON AND SUSAN M. JONES 135

Chapter IV. Motility—T. IINO AND M. ENOMOTO . . . 145

Chapter V. Production and Isolation of Flagella—R. W. SMITH AND
H. KOFFLER 165

Chapter VI. Antigen-antibody Reactions in Microbiology—
C. L. OAKLEY 173

Chapter VII. The Localization of Bacterial Antigens by the use of
the Fluorescent and Ferritin Labelled Antibody Techniques—
P. D. WALKER, IRENE BATTY AND R. O. THOMSON . . . 219

Chapter VII. Toxin-antitoxin Assay—IRENE BATTY . . . 255

Chapter IX. Techniques for Handling Animals—W. H. KINGHAM 281

Chapter X. The Determination of the Molecular Weight of DNA
Per Bacterial Nucleoid.—J. DE LEY 301

Chapter XI. Hybridization of DNA—J. DE LEY . . . 311

Chapter XII. Hybridization of Microbial RNA and DNA—
J. E. M. MIDGLEY 331

Chapter XII. Cell Walls—ELIZABETH WORK 361

AUTHOR INDEX 419

SUBJECT INDEX 429

CONTENTS OF PUBLISHED VOLUMES

Volume 1

E. C. ELLIOTT AND D. L. GEORGALA. Sources, Handling and Storage of Media and Equipment

R. BROOKES. Properties of Materials Suitable for the Cultivation and Handling of Micro-organisms

G. SYKES. Methods and Equipment for Sterilization of Laboratory Apparatus and Media

R. ELSWORTH. Treatment of Process Air for Deep Culture

J. J. MCDADE, G. B. PHILLIPS, H. D. SIVINSKI AND W. J. WHITFIELD. Principles and Applications of Laminar-flow Devices

H. M. DARLOW. Safety in the Microbiological Laboratory

J. G. MULVANY. Membrane Filter Techniques in Microbiology

C. T. CALAM. The Culture of Micro-organisms in Liquid Medium

CHARLES E. HELMSTETTER. Methods for Studying the Microbial Division Cycle

LOUIS B. QUESNEL. Methods of Microculture

R. C. CODNER. Solid and Solidified Growth Media in Microbiology

K. I. JOHNSTONE. The Isolation and Cultivation of Single Organisms

N. BLAKEBROUGH. Design of Laboratory Fermenters

K. SARGEANT. The Deep Culture of Bacteriophage

M. F. MALLETTE. Evaluation of Growth by Physical and Chemical Means

C. T. CALAM. The Evaluation of Mycelial Growth

H. E. KUBITSCHEK. Counting and Sizing Micro-organisms with the Coulter Counter

J. R. POSTGATE. Viable Counts and Viability

A. H. STOUTHAMER. Determination and Significance of Molar Growth Yields

Volume 2

D. G. MacLENNAN. Principles of Automatic Measurement and Control of Fermentation Growth Parameters

J. W. PATCHING AND A. H. ROSE. The Effects and Control of Temperature

A. L. S. MUNRO. Measurement and Control of pH Values

H.-E. JACOB. Redox Potential

D. E. BROWN. Aeration in the Submerged Culture of Micro-organisms

D. FREEDMAN. The Shaker in Bioengineering

J. BRYANT. Anti-foam Agents

N. G. CARR. Production and Measurement of Photosynthetically Useable Light

R. ELSWORTH. The Measurement of Oxygen Absorption and Carbon Dioxide Evolution in Stirred Deep Cultures

G. A. PLATON. Flow Measurement and Control

RICHARD Y. MORITA. Application of Hydrostatic Pressure to Microbial Cultures

D. W. TEMPEST. The Continuous Cultivation of Micro-organisms: 1. Theory of the Chemostat

C. G. T. EVANS, D. HERBERT AND D. W. TEMPEST. The Continuous Cultivation of Micro-organisms: 2. Construction of a Chemostat

xi

J. Řičica. Multi-stage Systems
R. J. Munson. Turbidostats
R. O. Thomson and W. H. Foster. Harvesting and Clarification of Cultures—
Storage of Harvest

Volume 3A

S. P. Lapage, Jean E. Shelton and T. G. Mitchell. Media for the Maintenance
and Preservation of Bacteria
S. P. Lapage, Jean E. Shelton, T. G. Mitchell and A. R. Mackenzie. Culture
Collections and the Preservation of Bacteria
E. Y. Bridson and A. Brecker. Design and Formulation of Microbial Culture
Media
D. W. Ribbons. Quantitative Relationships Between Growth Media Constituents
and Cellular Yields and Composition
H. Veldkamp. Enrichment Cultures of Prokaryotic Organisms
David A. Hopwood. The Isolation of Mutants
C. T. Calam. Improvement of Micro-organisms by Mutation, Hybridization and
Selection

Volume 3B

Vera G. Collins. Isolation, Cultivation and Maintenance of Autotrophs
N. G. Carr. Growth of Phototrophic Bacteria and Blue-Green Algae
A. T. Willis. Techniques for the Study of Anaerobic, Spore-forming Bacteria
R. E. Hungate. A Roll Tube Method for Cultivation of Strict Anaerobes
P. N. Hobson. Rumen Bacteria
Ella M. Barnes. Methods for the Gram-negative Non-sporing Anaerobes
T. D. Brock and A. H. Rose. Psychrophiles and Thermophiles
N. E. Gibbons. Isolation, Growth and Requirements of Halophilic Bacteria
John E. Peterson. Isolation, Cultivation and Maintenance of the Myxobacteria
R. J. Fallon and P. Whittlestone. Isolation, Cultivation and Maintenance of
Mycoplasmas
M. R. Droop. Algae
Eve Billing. Isolation, Growth and Preservation of Bacteriophages

Volume 4

C. Booth. Introduction to General Methods
C. Booth. Fungal Culture Media
D. M. Dring. Techniques for Microscopic Preparation
Agnes H. S. Onions. Preservation of Fungi
F. W. Beech and R. R. Davenport. Isolation, Purification and Maintenance of
Yeasts
Miss G. M. Waterhouse. Phycomycetes
E. Punithalingham. Basidiomycetes: Heterobasidiomycetidae
Roy Watling. Basidiomycetes : Homobasidiomycetidae
M. J. Carlile. Myxomycetes and other Slime Moulds
D. H. S. Richardson. Lichens

S. T. WILLIAMS AND T. CROSS. Actinomycetes

E. B. GARETH JONES. Aquatic Fungi

R. R. DAVIES. Air Sampling for Fungi, Pollens and Bacteria

GEORGE L. BARRON. Soil Fungi

PHYLLIS M. STOCKDALE. Fungi Pathogenic for Man and Animals: 1. Diseases of the Keratinized Tissues

HELEN R. BUCKLEY. Fungi Pathogenic for Man and Animals: 2. The Subcutaneous and Deep-seated Mycoses

J. L. JINKS AND J. CROFT. Methods Used for Genetical Studies in Mycology

R. L. LUCAS. Autoradiographic Techniques in Mycology

T. F. PREECE. Fluorescent Techniques in Mycology

G. N. GREENHALGH AND L. V. EVANS. Electron Microscopy

ROY WATLING. Chemical Tests in Agaricology

T. F. PREECE. Immunological Techniques in Mycology

CHARLES M. LEACH. A Practical Guide to the Effects of Visible and Ultraviolet Light on Fungi

JULIO R. VILLANUEVA AND ISABEL GARCIA ACHA. Production and Use of Fungal Protoplasts

CHAPTER I

Microscopy and Micrometry

Louis B. Quesnel

Department of Bacteriology and Virology, University of Manchester,
Manchester, England

I.	Introduction	2
II.	Light and its Behaviour	2
	A. Theories of light	5
	B. Light waves	7
	C. Plane waves	8
	D. Interference	10
	E. Phase	10
	F. Diffraction	11
	G. Refraction and reflection	16
	H. Filtration	18
	I. Amplitude specimens and phase specimens . . .	18
III.	The Light Microscope—Bright-field Use	21
	A. Image formation	23
	B. Abbe theory.	24
	C Magnification	26
	D. Resolution and numerical aperture	27
	E. Useful magnification of the microscope	32
	F. Depth of focus	33
	G. Illumination for microscopy	35
	H. Imaging errors	40
	I. Objectives	45
	J. Eyepieces	48
	K. Condensers	53
IV.	Setting Up and Efficient Use of the Microscope . . .	55
	A. Critical illumination	56
	B. Kohler illumination	57
V.	Phase Contrast Microscopy	60
	A. Basic conditions for phase contrast imaging . . .	60
	B. The phase plate and image quality	64
	C. Condensers for phase contrast	68
VI.	Setting-up and Use of a Phase Contrast System . . .	70
	A. Practical procedure	70
	B. Immersion media and phase contrast	73
VII.	Dark-field Illumination	75
VIII.	Setting-up for Dark-field Illumination	78

IX. Filters and Their Use   81
     A. Types of filter   81
X.   Micrometry   88
     A. Measurements on objects   88
XI. Microscopical Journals   101
XII. British Standards   101
XIII. Some Microscope Manufacturers   101
References   102

I. INTRODUCTION

The microscope is an essential tool for the microbiologist; without it he is working in the dark. Modern research microscopes are precision instruments designed with great skill and dedication, yet they are used by many in an indifferent, almost haphazard, way. It is impossible in the space of a single Chapter either to study the microscope historically or to study the scientific theory which lies behind its design, but only to provide the basic knowledge required for the intelligent and accomplished use of the instrument as a tool of microbiological investigation. Basic theory will be discussed only in so far as it is required for the realization of the full design potential of the modern microscope.

The main features of a modern microscope are illustrated in Fig. 1.

II. LIGHT AND ITS BEHAVIOUR

The visible spectrum used in light microscopy forms a very small portion of a much larger spectrum of electromagnetic radiations. This restricted spectrum is of particular use in microscopy because the biological sensitivity of the eye to radiations in this range enables us to use the brain as the "recording" or "display" stage of the process, albeit a temporary record. This does not mean that other sections of the electromagnetic spectrum cannot be used for microscopy—they can—but when invisible radiations are used in the image-forming process a display which "generates" wavelengths in the visible range must be used, e.g. photographic emulsions in U.V. microscopy, fluorescent screens in electron microscopy. The relationship between the visible spectrum and the electromagnetic spectrum (E.M.) is shown in Fig. 2. While the E.M. spectrum includes wavelengths of radiations from about 3000 metres to less than 10^{-12} centimetres visible light fills the very small range from about 7×10^{-5} cm down to 4×10^{-5} cm, or, in the units usually used to define wavelengths, from 7000 Angstrom units (Å) to 4000 which is equivalent to 700 nanometers (nm) to 400 nm ($1\text{Å} = 10^{-8}$ cm; 1 nm $= 10^{-7}$ cm).

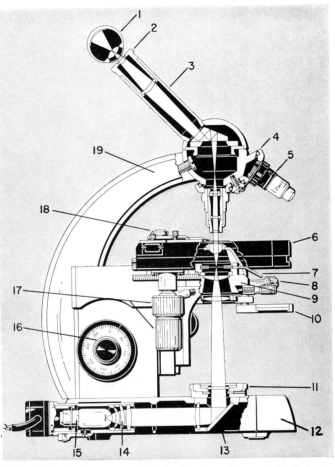

FIG. 1. A modern microscope with inclined oculars and built-in lamp—the Zeiss (Oberkochen) Standard RA Routine and Research Microscope. (Figure by courtesy of Carl Zeiss, Oberkochen, W. Germany.)

1. Eye
2. Eyepiece
3. Eyepiece tube
4. Revolving nosepiece
5. Objective
6. Mechanical stage
7. Condenser diaphragm—control lever
8. Control knob for swing-out top element of condenser
9. Condenser
10. Filter carrier
11. Diaphragm insert
12. Base
13. Mirror
14. Lamp collector lens system
15. Built-in low voltage lamp
16. Concentric coarse- and fine-focus control knobs
17. Concentric x-, y-axis control knobs for slide holder movements
18. Slide holder
19. Limb

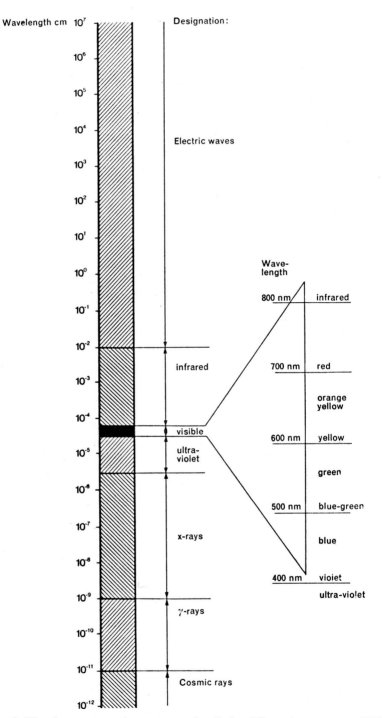

FIG. 2. The electro-magnetic spectrum of radiation (Figure by courtesy of Wild Heerbrugg, Switzerland).

Each wavelength of the electromagnetic spectrum can be associated with a wave number—the reciprocal of the wavelength in cm; and since they are all propagated with the same velocity each will have a different frequency, ν. Each wavelength is also associated with a specific energy E, where $E = h\nu$ is the energy in ergs of a quantum of frequency, ν, where h, is Planck's constant; or the energy value may be expressed as V, the energy in electron volts. In general terms we can say that the shorter the wavelength the higher the energy content and the more penetrating (and dangerous) the radiation; while the longer the wavelength the lower the energy and the less penetrating the radiation. Thus we find a progressive increase of energy from ultraviolet through X- and γ-rays to cosmic rays the most penetrating of all.

A. Theories of light

The earliest theories of light were attempts to explain the phenomenon of vision rather than the nature of light itself. As early as 500 B.C. Greek thinkers had put forward the "tactile" and "emission" theories; the former postulated that the eye sent out invisible sensors or sensitive probes which were able to "feel" objects too distant to be touched physically, while the latter postulated that the object itself emitted something which entered the eye and affected some sensitive part of the eye which was responsible for sight. For reasons obvious to us now, the "emission" theory eventually completely displaced the "tactile" theory, and by giving the emission the term "visible radiation" we can provide a reasonable explanation of the visual process as follows.

Visible radiation emitted (e.g. on heating), reflected or scattered by a body, on entering the eye, is focused by the eye lens onto the retinal surface which contains special sensitive cells connected by nerves to the brain. When light falls on these a chemical and physical reaction takes place involving the transformation of the pigment visual purple and resulting in the emission of electrical impulses to the centre of the brain where the visual image is "recomposed" and results in the sensation of sight. The physiology of vision will not be discussed here.

Quite apart from any physiological consideration it is possible to show experimentally that visible radiation is associated with the transfer of energy and any theory of light must accommodate the energy phenomena associated with it. By analogy with other known methods for the propagation of energy we could propose that light conveys energy in the form of "waves" (as the sea transports energy through its wave motion), or the energy may be conveyed as discrete quantities associated with the movement of particles (as moving billiard balls, for example, possess kinetic energy). It should be noted that the transfer of energy in waves need not involve the physical translocation of the medium through which the wave passes, either of water, or, in the case of light, of air. These considerations led to two basic theories of the nature of light: the "wave" theory and the "corpuscular" theory.

From the 17th century it has been known that the propagation of light could be represented by rays and simple experiment showed that those rays travelled in straight lines translating energy along a path from source to receptor. From the study of the interference phenomena associated with "Newton's Rings", Newton recognized that there was some sort of periodicity associated with light which was evidence for a wave theory; on the other hand, he could not reconcile this with the rectilinear propagation of light and, incorporating the concepts of his laws of motion, preferred to explain light as a procession of corpuscles which either possessed an internal vibration of their own or were in some sense controlled by waves or vibrations of the medium through which the light travelled. This objection to a simple wave theory was removed when it was discovered that light is not propagated in a strictly linear fashion, and that the rays of a beam of light which impinged on the edge of an object were bent away from the direction of propagation—the phenomenon known as *diffraction*.

As a result of many experiments on interference and diffraction a set of determinations of the wavelength of light were made and it was shown conclusively that different wavelengths were always associated with different spectral colours. The wavelengths of the different spectral colours are shown in Fig. 2.

While the simple wave theory could be used to describe the behaviour of light under many circumstances there were many inadequacies of the theory which were resolved when Maxwell formulated the equations of electromagnetism and showed how these could be used to describe the behaviour of light. An essential feature of the theory was the propagation of transverse electromagnetic waves (a vector quantity) as distinct from the simple theory of longitudinal waves in which the direction of vibration is always the same as the direction of propagation, so that longitudinal wave motion could be represented as variations of a scalar quantity.

The work of Maxwell soon led to the enormous expansion of the spectrum and the realization that "light waves" were only a tiny section of a very much larger spectrum of electromagnetic radiation associated with wavelengths from over 3000 m to less than 10^{-13} m.

Even Maxwell's electromagnetic explanation of the nature of light was inadequate to explain certain phenomena such as the energy transferred to the electron of an atom which had been excited by radiation with subsequent ejection of the electron (ionization). The failure resulted from the implication that the energy of electromagnetic radiation was continuously distributed. This difficulty was surmounted by Einstein's explanation that the energy of the radiation was concentrated into separate discrete packets, each packet being called a photon, and that for any particular wavelength all the photons had the same energy value. Coupled with Plank's realization that the energy is emitted in multiples of a single unit, with no fractions of a unit being possible, the unit of energy was called the quantum and the relationships between the energy level of a radiation and the wavelength were given by—

$$E = hc/\lambda$$

Where h is Plank's constant $(6 \cdot 6 \times 10^{-34}$ joule sec), c is the velocity of light and λ the wavelength.

If ν is the frequency then,

$$c = \nu\lambda$$

and

$$E = h\nu$$

The modern theory of light is a composite theory which incorporates both the electromagnetic (wave) theory (which has no place for photons) and the photon (particle or corpuscle) theory which has no place for the waves. The former describes adequately the phenomena of interference, diffraction and polarization while the latter is required for an explanation of the observed interactions of radiation and matter.

Modern quantum mechanics constitutes a single theory incorporating the appropriate parts of the electromagnetic wave theory, the quantum theory and the theory of relativity in a composite explanation of the properties of light and matter.

For an elementary understanding of the behaviour of light in relation to microscopy we can disregard the quantum theory. The quantum aspects are of importance, however, in understanding the phenomenon of fluorescence which lies at the heart of the fluorescence microscopy techniques which are now used to such great effect. Fluorescence microscopy is dealt with elsewhere in this Volume (Walker *et al.*, this Volume, page 219).

B. Light waves

For our present purposes we may depict light as consisting of waves such as that represented by the sine curve shown in Fig. 3.

This form of simple harmonic motion is describable by the trigonometric function—

$$y = A \cos 2\pi \left(\frac{x}{\lambda} - \frac{t}{T}\right)$$

where y is the variable describing the disturbance, A is the amplitude of the motion, λ is the wavelength and t the time variable; while T is the *period*. If t is held constant then the space configuration is repeated every time x increases by a distance λ (wherever the point x is chosen). If x is fixed in space ($x = $ constant) then as time passes, the value of y (the dis-

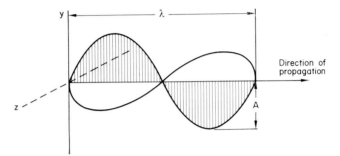

Fig. 3. Light wave. Two sine curves in planes at right angles; the electric field is in the plane of xy, the magnetic field in the plane xz.

turbance in the medium) goes through a repetitive cycle with maximum and minimum displacements of $+A$ and $-A$, a single complete cycle being performed for each increment increase of t from t to $(t + T)$. The number of complete cycles per second is the frequency ν so that $\nu = 1/T$ and the speed of propagation is—

$$c = \lambda/T = \nu\lambda$$

The following two relationships are basic to an understanding of certain aspects of microscopy—

(i) The wavelength determines the colour of the radiation in the visible spectrum.

(ii) The brightness or intensity of the light depends on the amplitude. The intensity is determined by the incident energy per unit area per second and since light is basically a travelling oscillating electrical/magnetic field it can be shown that the *relative energy is proportional to the square of the amplitude.*

C. Plane waves

1. *Inverse square law*

It is easily shown by experiment that a small source emits light in such a way that the flow of energy per unit area is proportional to the inverse square of the distance from the source. Consequently the radiation propagated from such a source is usually represented by a system of spherical waves.

2. Plane waves—Huygen's principle

A continuous surface that is the focus of points where the "particles" are in phase (see next Section) is called a *wavefront*, so that a sphere centred on a point source would represent a wavefront of the radiation from that source. It is also possible to obtain *plane waves* in which the wavefront is always a plane perpendicular to the direction of propagation of the radiation.

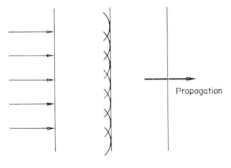

Wave fronts

FIG. 4. Wave fronts of plane waves.

In order to account for wave propagation Huygens suggested that every point on a wavefront itself became the centre of a secondary wave travelling outward from this point source. The forward edges of the secondary wavefronts will all have travelled the same distance after a time interval *t* (distance *ct*) and the boundaries of the wavelets will describe the new advancing envelope of the wavefront. (Fig. 4).

On this model it was possible to explain how a plane wave could yield a spherical wavefront after passage through a small aperture (Fig. 5). By Huygen's Principle the element of wavefront within the aperture acts as a

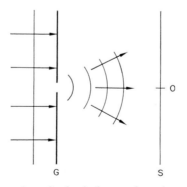

FIG. 5. Formation of spherical wave front by an aperture.

point source so that spherical wavefronts are propagated on the far side of the screen. In practice most of the energy remains within a cone of narrow angle with the hole as apex. Huygen's model is incomplete but forms a useful means of visualizing the behaviour of waves undergoing reflection and refraction.

D. Interference

By means of suitable instruments, such as the Michelson Interferometer, it is possible to produce interference fringes at will. It should be noted that interference fringes can be produced from monochromatic light and it is not necessary to have multicolour fringes (which result when white light is used). For example, if the source of emission is a gas discharge tube yielding the sodium D line alone the fringes will be alternating circles of yellow and "black". It has been shown that the reasons for the differences in illumination which give rise to the appearance of the fringes is due to differences in the path lengths travelled by rays which eventually come together in the plane of the fringes. If two rays are of equal amplitude the relative energy is at a maximum where the path difference is an integral multiple of the wavelength, λ, and the illumination in these regions is also at a maximum, i.e. the yellow bands. On the other hand wherever the path lengths differ by an odd number of half wavelengths then the relative energy in these regions will be zero, i.e. there will be no illumination—the black bands. If the amplitudes of the interfering waves are not equal the visibility is never zero. In those regions where interference occurs between waves which do not super-impose, i.e. they show path differences in the range between λ and $\frac{1}{2}\lambda$, the illumination will be less than the maximum possible but always greater than zero, giving intermediate brightness of illumination.

E. Phase

Returning to our previous consideration of the description of harmonic motion and the sine wave representation of light, we can easily describe the above phenomena in terms of *phase differences*. In mathematical terms, the argument of a trigonometric function is an angle and this angle describes the *phase of the disturbance* at a particular position x and time t (see p. 8). For cyclic phenomena such as we are considering, therefore, the *phase* describes the instantaneous quality of "maximumness", "minimumness", "zeroness", or "intermediateness" of the particular disturbance. We can therefore express the phase relations of the interfering waves considered above by stating that when the waves are in phase (phase angles identical) the relative energies are the combined energies of the waves involved, which therefore "*reinforce*" each other to yield a maximum (illumination). On the other hand the greatest degree of phase angle difference is obtained when the waves are out-or-phase by 180° and under these conditions "*cancellation*" occurs to yield a minimum net relative energy, which is zero when the amplitudes are identical. All the intermediate phase angle differences from 0–180° yield progressively lower illumination as the phase differences approach 180°, i.e. the $\frac{1}{2}\lambda$ out-of-phase situation. It should be noted in passing that these phenomena are only observed when the beams involved are emanating from a small source or are very close together in the light source. Interference fringes are not formed, for example, by interaction between light from two different sources such as two table lamps. In this case it can be shown by photometric measurements that the resultant relative energy obtained at any such doubly illuminated point is the sum of the relative energies produced by each source acting individually— *the law of photometric summation*.

We can now illustrate the interference phenomena which we mentioned by means of the wave trains in Fig. 6. In Fig. 6(a) two sine waves a and b are drawn representing two rays of light. Rays a and b are identical having the same wavelength, and the same amplitude, i.e. the same colour and "brightness". In addition the crests of a and b have the same phase angles, i.e. at any particular time the disturbance in the planes (i), (ii), and (iii) are at maxima of equal relative energy, and the two waves are "in phase". If they were brought together to superimpose at a particular point then the illumination at that point would be the sum of the two waves separately, i.e. "reinforcement" would occur. The situation is represented in Fig. 6(b) (where the rays have a common axis). In Fig. 6(c) wave b is 180° ($\frac{1}{2}\lambda$) out of phase with wave a, crest b_2 following crest a_2, by a $\frac{1}{2}\lambda$ difference. In this situation "cancellation" occurs with a net relative energy of zero and two such interfering rays would not illuminate a "field", i.e. the result would be "darkness". In Fig. 6(d) two waves out of phase by 90° ($\frac{1}{4}\lambda$) are shown. In this case the resultant relative energy (and hence illumination) at a point at which they were focused would be over half that of the maximum obtained when the two waves were in-phase. In other words, this would represent a shade of "grey" relative to the "white" of the reinforced in-phase waves, and the "black" of the "cancelled" fully out-of-phase waves. Similarly, for two waves $\frac{1}{8}\lambda$ out-of-phase we would get a "lighter shade of grey" while if they were $\frac{3}{8}\lambda$ out-of-phase we would have a "darker shade of grey". We shall see later how this is used to yield the graded contrast range in the image of the phase contrast microscope.

F. Diffraction

Light does not always travel in straight lines but will, for example, be bent by the edge of an opaque object; the phenomenon known as *diffraction*. It will occur whenever part of a wave front is so obstructed as to cause a

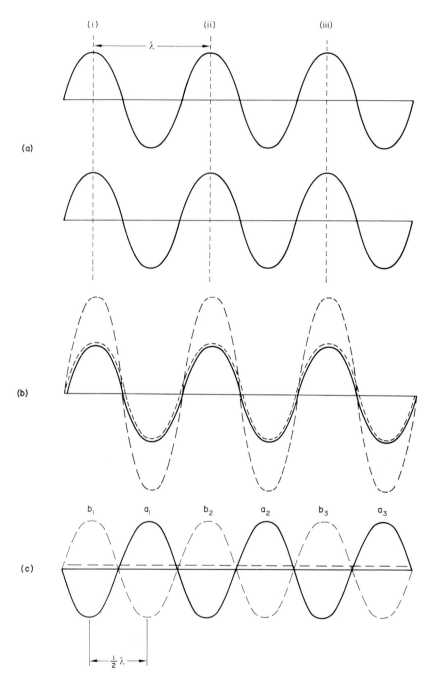

Fig. 6. (a) Two waves parallel and in phase. (b) Two waves superimposed with resultant of double amplitude. (c) Two waves with $\frac{1}{2}\lambda$ phase difference; resultant of zero amplitude.

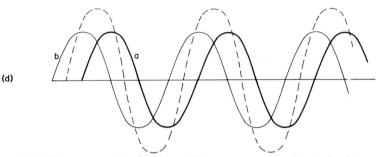

Fig. 6. (d) Two waves with $\frac{1}{4}\lambda$ phase difference; resultant (dashed) of increased but less than doubled) amplitude.

change in amplitude or phase which is not the same over the whole area of the wavefront. Diffraction occurs whenever the width of a beam of light is restricted by an aperture, a mount or frame, a slit, etc., that is, almost all methods of producing light beams will result in some diffraction.

1. *Fresnel Diffraction*

We can illustrate what happens by reference to Fig. 5 which shows a parallel beam of light made to form a narrower beam on passing through a slit in a screen G and which then falls on an imaging screen S. When S is very close to G the illumination appears uniform within the image of the slit while outside this geometric image it appears to be zero, i.e. darkness. However, as the screen S is moved away from G, bands of illumination begin to appear in the dark region outside the geometric image of the slit; these are diffraction fringes and demonstrate the fact that some of the light is bent outside the area of the image. If the size of the slit was 0·05 mm (5×10^{-5} m) and the screen S was 20 metres from G, the illumination across the screen would be that shown in Fig. 7. At this

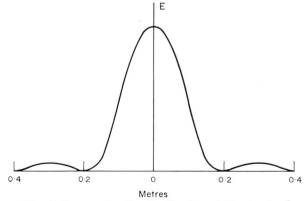

Metres

Fig. 7. Image of a slit: distribution of illumination.[*]

distance the width of the main geometric image is some 8000 times the width of the slit; it is also very unevenly illuminated.

2. *Fraunhofer diffraction*

Consider now the arrangement in Fig. 8, where the screen G lies between two lenses L_1 and L_2, and where instead of a parallel beam and slit, we have

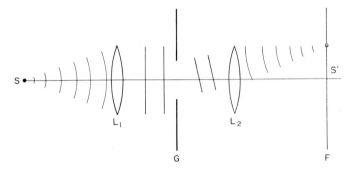

Fig. 8. Formation of Fraunhofer diffraction pattern.

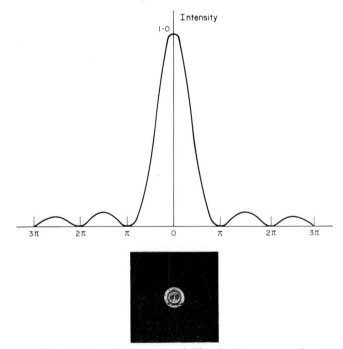

Fig. 9. (a) Distribution of intensity of illumination across the Airy pattern showing maxima and minima. (b) Photo of central disc and the first ring (over-exposed). (Courtesy of Longmans, Green & Co.)

at S a point source in the focal plane of L_1, and a circular aperture in G. In this arrangement the plane waves leaving L_1 will be incident upon G and the image of S will be formed in the focal plane of L_2 centred on S'. However, the aperture in G obstructs the wave front causing diffraction of rays and, in consequence, the image on F is a diffraction pattern of alternating dark and bright bands centred on a bright central disc known as the Airy disc (after Sir G. B. Airy, 1801–1892). In fact the distribution of energy across the area of the image is very uneven and about 84% of the light passed through the aperture falls in the central disc; the outer bright rings representing diffraction maxima get dimmer as one proceeds outwards. The intensity distribution across the Airy pattern is shown in Fig. 9 and has obvious similarities with that in Fig. 7.

We may simplify the important aspects of diffraction connected with microscopy in the following way. Light incident upon an object may be *diffracted* at the edge of the object or *transmitted*, i.e. rays further removed from the edge continue along their rectilinear paths undeviated. Fig. 10(a)

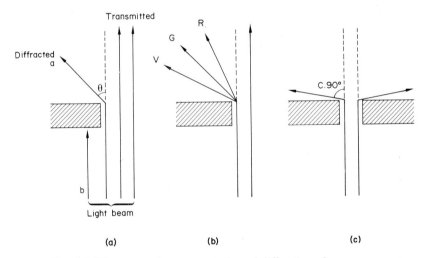

Fig. 10. Diagrammatic representation of diffraction phenomena.

shows the main part of a beam of light being transmitted past the edge of an opaque object; a represents a ray which impinges on the edge of the object and is diffracted at an angle θ from the line of propagation while b represents those rays which strike the object and are absorbed or reflected. The relative extent of the scattering depends on the wavelength of light used; the shorter the wavelength the greater will be the angle of diffraction, represented diagrammatically in Fig. 10(b) where V represents violet, G green, and R red rays. Also the smaller the cross section of the beam (determined by the size of the aperture through which it passes) the greater will be the proportion of rays which are diffracted, i.e. the ratio diffracted/transmitted increases as the diameter of the aperture decreases until eventually there may be no transmitted radiation. In addition, as the aperture becomes smaller the angle of diffraction is increased so that when the edges are very close together the rays are diffracted at an angle of 90° (Fig. 10(c)).

G. Refraction and reflection

The velocity of monochromatic light in a vacuum differs from that in any translucent material. The slower its velocity in a particular medium the more dense the medium is said to be and the ratio of the velocity *in vacuo, c*, to the velocity *in media, V*, is called the *refractive index (n)* of the medium; $c/V = n$. When light travels from one medium into another the frequency of the oscillation remains the same but the wavelength changes and is shorter in the denser medium.

If light falls normally on a surface separating two translucent materials of differing refractive index the light will pass from one medium to the other undeviated (or, in certain situations, a proportion may be reflected normally to the surface). If the incident radiation is inclined at an angle from the normal then *refraction* will take place such that the direction of propagation will be changed on passage from one medium to the next; if one of the media is non-translucent and not fully absorbing then part or all of the radiation will be *reflected*.

When light travelling from a dense to a less dense medium strikes the interface of the two media at an incident angle greater than a certain value—the *critical angle* (dependent on the two media involved)—the light will be

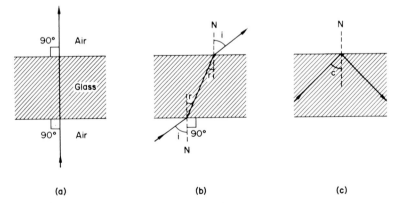

(a)   (b)   (c)

FIG. 11. Transmission (a) of normal incident ray; refraction (b) and internal reflection (c).

reflected back into the dense medium without passing into the lighter medium. In this case the light is said to have undergone *"total internal reflection"* and does so in accordance with the laws of reflection. The laws of refraction are illustrated in Fig. 11.

(a) Light incident upon a plane parallel glass plate at right-angles (normal) to the surface, is transmitted undeviated.

(b) Light incident upon the surface at an incident angle *i* from the normal is *refracted* toward the normal (N) on entering the glass at a refracted angle *r* and on emerging into air again suffers a refraction away from the normal at an angle equal to *i* on re-entering the less dense medium. The emergent ray is parallel to but laterally displaced from the (incident) entering ray; sin *i*/sin *r* = refractive index, *n*.

(c) A ray in glass incident upon the exit interface at an angle greater than the critical angle, *C*, is totally internally reflected; sin *C* = 1/*n*.

It was stated earlier that *c*/*V* = *n*; as the velocity of light *in vacuo* is not appreciably different from that in air the refractive index is usually expressed in terms of the passage of light through air and can be determined more easily from the angles of incidence and refraction according to the simple relationship—

$$n_D = \frac{\sin i}{\sin r} \text{ (constant for any two media)}$$

The sodium D line (5890 Å) is used as standard radiation, hence, n_D.

In complete darkness visibility is zero. Objects can be seen only if they reflect light from a source to the eye. When a perfectly plane reflecting surface is illuminated by a beam of light the beam leaving the surface is similar to the beam incident upon it but reflected at an angle *r* to the normal, N, which is equal to the angle of incidence, *i* (Fig. 12(a)). This is called "specular" reflection and will be exhibited by a good mirror. Most

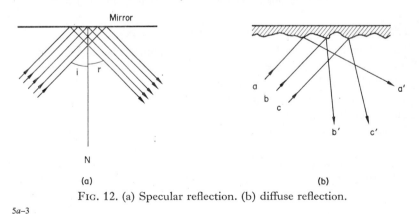

FIG. 12. (a) Specular reflection. (b) diffuse reflection.

everyday objects reflect light beams in a much more imperfect way as a result of irregularities of surface and give rise to "diffuse reflection" in which rays from an incident beam are reflected in various directions depending on the way they fall upon the particular irregularities of the surface and may be illustrated by Fig. 12(b).

H. Filtration

Because of the arrangements of atoms within certain chemical molecules some substances possess the property of absorbing certain wavelengths of radiation while allowing other wavelengths to be transmitted. Thus a substance which passes only wavelengths in the region 5000–6000 Å while absorbing the remainder from a beam of incident white light, will appear green. Similarly, objects which reflect only a restricted range of wavelengths will appear the colour associated with the reflected wavelengths alone. It follows naturally that an object capable of reflecting or transmitting blue wavelengths only, when illuminated with light containing no wavelengths shorter than say, 6500 Å (orange), will appear "black" (Fig. 13). The various filters of use in microscopy are described later (Section IX).

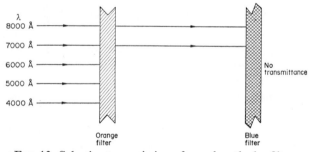

Fig. 13. Selective transmission of wavelengths by filters.

I. Amplitude specimens and phase specimens

The biological material studied by microscopists varies considerably. Many specimens contain light-absorbing pigments, e.g. fungal spores; or other coloured bodies such as chloroplasts. On the other hand the majority of the material interesting to cytologists, microbiologists, etc., appears, when viewed by ordinary bright-field microscopy, comparatively featureless unless steps are taken artificially to enhance the differences in colour or "intensity" of the various constituent parts of the specimen. In other words most biological material examined in the natural state is extremely low in inherent contrast and detail is consequently difficult to determine. In consequence a vast technology of fixation and staining has been developed in order to make substructures visible by increasing contrast and varying colour. This subject is dealt with elsewhere and need not concern us here except to state that tissues which have been subjected to such procedures are almost invariably "dead meat" and frequently exhibit artifacts which may conceal rather than reveal the truths of structure.

The true *amplitude specimen* exhibits images of high contrast as a result of the diminution in amplitude of radiations traversing certain parts of the specimen by comparison with others and with the background. As mentioned earlier, the "brightness" of illumination depends ultimately on the amplitude of the radiation comprising the illumination. We can illustrate this for a hypothetical cell by reference to Fig. 14. In this figure it is

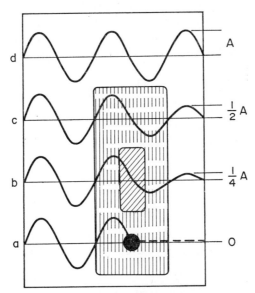

Fig. 14. "Pure" amplitude specimen. Ray a is totally absorbed; b and c are partially absorbed; exit rays have ¼ and ½ the amplitude of the background, unabsorbed ray, d.

assumed that the refractive indices of the cell material and the mountant are the same and that the variations in wave trains are due solely to difference in absorbance of the parts of the specimen. If an included body is totally absorbing (rare) then there is no exit ray, i.e. equivalent to zero amplitude radiation from that particular part of the specimen (ray a). Other parts of the specimen may partially absorb the radiation so that exit rays show different degrees of reduction of amplitude (rays b and c), while rays passing through the fully transparent mountant would emerge with undiminished amplitude (ray d). In this hypothetical specimen the contrast would range from "black" through two shades of "grey" to a "white background".

In practice few materials behave like pure amplitude specimens as refractive index usually differs from one part of the specimen to another and, in addition, selective absorption may occur so that only some wavelengths fully traverse certain parts of the specimen; in which case those parts will appear slightly coloured depending on the wavelengths passed. In a stained specimen, therefore, the emergent radiation may have been changed in three ways, viz, diminished in amplitude, different in wavelength distribution and retarded in phase (see below) from the unmodified rays passing through the mountant background.

In the pure *phase specimen* we assume that no changes in amplitude or wavelength distribution result, but only those changes which depend on the differences in refractive index within the specimen. This state of affairs is represented diagrammatically in Fig. 15(a), where it is assumed

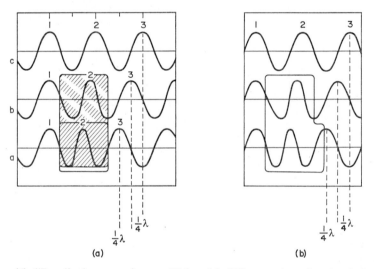

(a)   (b)

FIG. 15. "Pure" phase specimens. Either (a) difference in refractive index or (b) difference in thickness of specimen may retard rays causing phase differences between exit rays (see text).

that the inclusion has the highest refractive index, and the "cytoplasm" has a refractive index less than this but greater than that of the mountant.

Here crest 3 of wave c is in advance of crest 3 of wave b by a quarter wavelength, which is itself in advance of crest 3 of wave a by a quarter wavelength, i.e. waves b and a have been retarded by 90° and 180° respectively relative to wave c. Of course, if the refractive index of the mountant was highest and that of the inclusion lowest, then the succession of crests would be reversed. Since phase differences are the result of the differing velocities of light in dense media there is another way in which emergent rays may be phase-shifted relative to one another and this is illustrated in Fig. 15(b). Here, the hypothetical biological specimen is homogenous with respect to refractive index (and higher than the mountant), but the thickness of the specimen varies. The result is that light "spends a longer time" travelling through the thicker part (ray a) and therefore emerges retarded with respect to the light passing through the thinner part (ray b) which is itself also retarded relative to light passing through the back-ground alone; in fact the same net result as in Fig. 15(a) but for a different reason.

Again, although much biological material falls into the category "phase specimen" at least some radiation will be absorbed with resultant amplitude changes. The way in which the phase properties of the specimen may be used to produce high contrast microscope images was the ingenious discovery of the Dutch physicist Zernicke, and is considered in greater detail in Section V.

III. THE LIGHT MICROSCOPE: BRIGHT-FIELD USE

The living objects studied by microbiologists are often invisible to the naked eye, a staphylococcus for example is only about a 1/1000 mm in diameter and this is about 1/100 the size of particles visible by unaided sight. To enable such objects to be seen the microscope is the instrument of choice. The obvious purpose of the microscope is to magnify, but, in addition, and essentially, it must also *resolve* fine detail in the specimen.

Before looking at the microscope itself it is useful to discuss briefly a few points about human vision. The eye contains a lens whose function it is to project an image of the observed object on the retina. We are all familiar with the fact that objects far away look small and that the nearer we approach the more distinct the individual features become—the better resolved is the detail. There is a limit to the effective distance between the object and the eye and within this limit the detail is once again lost, the image becomes blurred, the eye is physically strained and pain may result. This limit is dictated by the extent to which the curvature of the eye lens can be changed in order to focus on the object; the nearer the object the more curved the eye lens. The shortest distance of clear vision is about 25 cm, although it differs somewhat with age and sex; and the conventional visual distance for clear vision without strain is taken as 250 mm, i.e. the near point is about this distance from the eye and the lens is at maximum curvature. The way in which image size and object distance are related, is shown in Fig. 16.

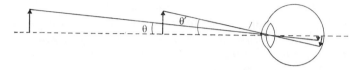

Fɪɢ. 16. Effect of object distance on image size and visual angle.

It can be seen that for any particular object the nearer it is to the eye the bigger is the image formed on the retina and the angle θ increases. (The retina gives good resolution only in the region of the *fovea centralis* which is a small depression in the retina on the visual axis of the eye, and the eye itself must be revolved until the object which is being observed forms its image on the fovea.) In order to distinguish fine details of structure small objects close together must form separate images on the retina and the smallest distance between two points which permits this still to occur is the *limit of resolution* of the eye. Any smaller distance would result in confluent images, i.e. only one image would be formed although two objects were present. For the eye accommodated for vision at the near point of 250 mm the limit of resolution is about 0·1 mm, that is, in order to separate visually the two optical images on the retina of the observer, the two objects must subtend an angular separation of not less than about 1–2 min of arc, angle α in Fig. 17(a) (exaggerated).

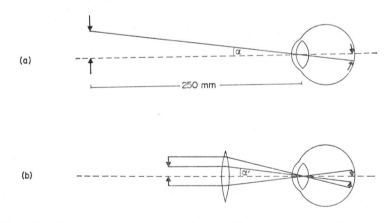

Fɪɢ. 17. (a) The minimum visual angle for resolution of detail at the conventional visual distance; α not less than 1–2 of arc (drawing not to scale). (b) Effect of simple magnifier on conventional visual distance and resolution ($\alpha' = \alpha$).

In order to resolve objects closer together than this, i.e. subtending angular separations of less than 1 min of arc, a magnifier is needed. The result would then be that shown in Fig. 17(b) where $\alpha' = \alpha$ but the detail resolved is finer. The greater the magnifying power of the glass lens the greater will be the inclination of the rays from the lens to the eye. If the power of the lens is $4\times$ then the inclination of these rays with respect to the optical axis will be four times greater than in the absence of the lens, so that angle α is now four times as large and the image separation four times as great at the retina. In other words, if the limit of resolution was previously $0\cdot1$ mm, in the system in Fig. 17(b) it would be $0\cdot1/4$ mm or $0\cdot025$ mm; or simply, we could resolve finer detail with the aid of a magnifier than we could without it.

The magnification of the auxiliary lens is given by the formula—

$$M = \frac{250}{f} = \frac{\text{Least distance of distinct vision}}{\text{Focal length of magnifying glass}}$$

A. Image formation

Figure 18 illustrates the formation of an image by a single bi-convex lens, such that if O is the object there exists a point I which is optically

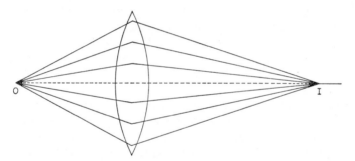

FIG. 18. Image formation by a lens.

equidistant from O no matter by which route the rays traverse the lens. All optical paths from O to I are the same, the time taken for an outer ray to travel from O to I being the same as that taken by one of the paraxial rays (which as drawn appear shorter because it is not possible to indicate the relative amounts of time spent in the thin edge of the lens as compared with the thick middle).

In terms of the waves emanating from O all the disturbances arrive in phase at I. If O is a point source, a spherical wave deriving from the source leaves the lens as a spherical wave converging on I, but since all parts of the spherical wave front from O cannot pass through the lens the image is not perfect.

If, instead of a point source, the object was linear and perpendicular to the optical axis of the lens, then the image would also be linear and perpendicular to the axis, while a plane object would similarly yield a plane image. The relationship between objects and images is said to be *collinear*, the image plane is *conjugate* to the object plane and refracted rays are *conjugate* to incident rays. The image formed in one plane of the system can itself be made the object of a second system which will form a second image in the plane conjugate to it.

B. Abbe theory

Abbe was the first to realize the importance of both the diffracted radiation and the transmitted radiation in the formation of the microscope image. He found that in order to obtain the sharpest and most detailed image possible the lens system must have an aperture wide enough to admit all the light diffracted by the object so that the complete Fraunhofer pattern could be formed behind the objective and all this light must contribute to the image. The image formed by a (perfect) optical system is a Fraunhofer diffraction pattern. While this concept of image formation from diffracted and transmitted radiation may be difficult for the non-physicist to grasp it can be demonstrated experimentally with the aid of an optical bench with lenses corresponding to the various lens systems of the microscope, using a diffraction grating as the specimen. This experiment is beautifully illustrated in a colour film available on loan from Wild Heerbrugg, Switzerland ("The Microscope" Part II, in seven different language versions).

The arrangement can be illustrated diagramatically in Fig. 19 (compare with Fig. 8) where the essential elements of the microscope system are represented by C, the condenser lens system, and O, the objective lens system. The grating G represents the specimen. Here the source S is arranged according to the Kohler system of illumination (see later) so as to emit a single collimated beam of light incident upon the "specimen" parallel to the axis, with the result that light is diffracted in the normal way. If the aperture of the "objective" is sufficiently large, i.e. is able to collect all the diffracted rays, the light will be collected as shown in the figure and will form a "primary image" in the back focal plane of the objective since this is conjugate to the source (see earlier), such that the diffraction maxima are in focus and compose a Fraunhofer pattern of the source in the plane of S'. The rays of light pass on through S' to interfere in the image G' giving an image of G in the plane conjugate to the grating. The image at G' composed from the interfering diffraction maxima is the intermediate image of the compound microscope and forms the object for the eyepiece system. The wave fronts and various image formations are shown diagrammatically in Fig. 19.

From the foregoing discussions it should now be clear how the basic elements of the compound microscope are arranged in order to give the final image. In effect, the microscope consists of the imaging system shown in Fig. 19 used to form an intermediate image which is real and inverted and this image is the object for a magnifier—the eyepiece—which forms a final image on the retina apparently deriving from a virtual image at the conventional visual distance of 250 mm; the arrangement at its simplest is shown in Fig. 20.

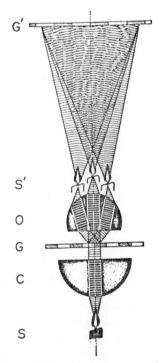

FIG. 19. Image formation and Fraunhofer patterns in the microscope. (Figure by courtesy of Wild Heerbrugg, Switzerland.)

It should be stressed that the eyepiece lens system is essentially a magnifier which does not and cannot increase the resolution of the system. In other words detail in the specimen which is not resolved in the intermediate image cannot be resolved, no matter what the magnifying power of the eyepiece used. However, in order that the eye may appreciate the resolution inherent in the intermediate image a certain minimum final magnification is necessary.

In crude simple terms, then, we may regard the microscope as an arrangement of three lens systems: (1) the objective, the essential purpose of which

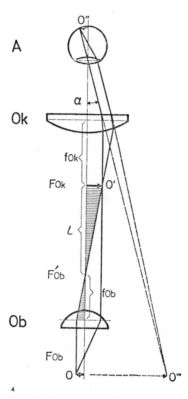

FIG. 20. Image formation in the compound microscope. (Figure by courtesy of Wild Heerbrugg, Switzerland).

is to resolve detail; (2) the eyepiece, to magnify; and (3) the condenser, which is not an essential requisite for compound microscopy, but without which high resolution imaging is not possible. The way in which these components may most effectively serve their respective functions are discussed later.

C. Magnification

The compound microscope provides for high magnifications in two stages (Fig. 20). The object O placed just beyond the front focal point F_{Ob} of the objective lens is imaged at O′, a distance l (the optical tube length) behind the back focal point (F′$_{Ob}$) of the objective. This intermediate image is in the focal plane of the eyepiece lens or ocular, Ok, and gives an image, when viewed by the eye, which appears to be in an image plane at the conventional visual distance and under the angle α. Since the intermediate image is formed in the front focal plane of the eyepiece the rays leaving a given point of the object leave the eyepiece as parallel rays, and would, otherwise, form an image at infinity.

The magnification at the intermediate image is given by,

$$M_{Ob} = \frac{l}{f_{Ob}}$$

i.e. magnification due to objective

$$= \frac{\text{optical tube length}}{\text{objective focal length}}$$

The magnification due to the eyepiece is given by the earlier formula:

$$M_{Ok} = \frac{250}{f_{Ok}} = \frac{\text{least distance of distinct vision}}{\text{focal length of ocular}}$$

The final image magnification will be the product of the individual magnifications viz.

$$M_{Total} = \frac{l}{f_{Ob}} \times \frac{250}{f_{Ok}} = M_{Ob} \times M_{Ok}$$

The magnification factors for objectives and eyepieces are marked on the components, so that a $100 \times$ objective used in conjunction with a $10 \times$ eyepiece gives a final image magnification of $1000 \times$. Since the final image is "virtual" there is no point within the system where a $1000 \times$ image is formed, although one can arrange to form a real image in a plane beyond the eyepiece as indeed one must for photomicrography. The production of the photographic image may require the use of a correction lens in addition to the photo-eyepiece and objective. If the image on the film is to be similarly $1000 \times$ the object then the lens must form a real image in a plane 250 mm from the eyepoint. This would represent a factor 1 camera attachment, and for other factors the optical draw to the image would differ from 250 mm and the final magnification would differ from the product of objective and eyepiece magnifications (see Quesnel, this Series, Volume 7).

D. Resolution and numerical aperture

The relationship between our ability to see detail and the limit of resolution has already been mentioned. It is clear from the foregoing that when one looks down the microscope one is not strictly observing the specimen but the intermediate image which has been formed by the objective. Obviously the eye is unable to resolve detail, that is make separate images, of substructure where separate images do not occur in the intermediate image. The quality of the intermediate image is therefore of primary importance, and we must ensure that all the detail which we require is present in this image. The resolving power of the objective is of crucial importance since its performance cannot be improved by the eyepiece although a minimum total magnification is a necessary factor in realizing the maximum efficiency of the microscope system (see next Section).

The determination of the limit of resolution may be approached in two ways depending upon whether we regard the object as a kind of diffraction grating which is non self-luminous, the Abbe method: or, whether we regard each point in the object upon which light impinges as a point source, i.e. as self-luminous points which give rise to Airy patterns—the method of Rayleigh.

According to Abbe's explanation of image formation the specimen acts like a diffraction grating and so long as the zero order (transmitted) and at least one of the first order diffracted beams enter the lens to give two effective sources in the focal plane (S′ Fig. 19) then the intensity distribution across the (intermediate) image plane will exhibit maxima corresponding to the grating spaces and the rulings of the grating will be resolved. The requirement that the lens must transmit the zero order plus one of the first order beams is known as the Abbe principle.

In Fig. 21(a) a collimated beam of light parallel to the optical axis is incident normally on the grating G. The angle of diffraction for the first order maximum is given by:

$$nd \sin u = \lambda$$

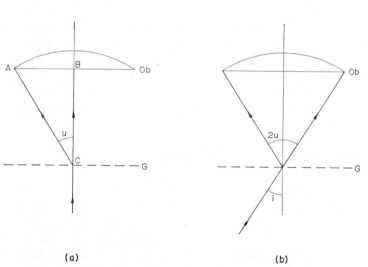

(a)   (b)

FIG. 21. Diffraction angles and angular aperture of a lens.

Where d is the grating interval and n is the refractive index of the medium between the grating and the objective. As we mentioned earlier, the angle of (the first order) diffraction increases as the width of the beam, (which is determined by the aperture through which the beam passes) is decreased; so, according to the Abbe principle the limit of resolution of this objective is the smallest value of d which will yield a first order diffraction not greater than the angle u, in Fig. 21(a) (otherwise it would not enter the lens). The value of u is the semi-angle of the largest cone of rays admitted by the lens, i.e. half the angular aperture of the lens, or the Numerical Aperture (N.A.). The N.A. is defined by $n \sin u$, and the limit of resolution (d) is given by

$$d = \frac{\lambda}{n \sin u} = \frac{\lambda}{N.A.}$$

In Fig. 21(a) $\sin u$ is the ratio AB/AC which is obviously less than unity which is its maximum value, and this ratio increases as BC decreases. Since BC is related to the focal length of the objective and the limit of resolution is inversely proportional to $\sin u$ (above), it is clear that by making a lens of shorter focal length we would increase the resolving power. Similarly, for a given focal length an increase in the diameter of the lens would permit an increase in u and so of resolution. The dilemma is the problem of achieving both short focus and wide diameter. However the effective aperture of the lens can be doubled by changing the direction of the illuminating beam as shown in Fig. 21(b). Here the beam is incident upon the grating G at an angle i such that the zero order rays are just collected by the lens. In this situation the maximum angle available for first order diffraction is $2u$, the full angular aperture and not the semi-angle as before, and the limit of resolution in consequence reduces to:

$$d = \frac{\lambda}{2N.A.}$$

A unilaterally incident beam such as that illustrated in Fig. 21(b) would however yield unilateral illumination of the specimen, and to avoid this and to provide even inclined illumination over the whole area of the specimen a solid cone of light is required. This is the purpose of the condenser, and without it maximum resolution of the objective cannot be achieved except in the case of very low power lenses used in conjunction with spherical mirrors. Equally it follows that for maximum resolution the numerical aperture of the condenser must match the numerical aperture of the objective. (When the apertures differ the equation becomes, $d = \lambda/NA_{ob} + NA_{cd}$ when $NA_{cd} \leq NA_{ob}$.) This implies not only the manufacture of high N.A. condenser systems, but the correct positioning of the condenser relative to the objective when using the microscope.

The Abbe derivation given above is, however, only an approximate theory as it assumes, for example, that light associated with a given maximum has a single direction of diffraction and that all the light of a given order is either accepted or rejected by the objective lens. Indeed, more detailed analysis has revealed that the influence of the condenser is somewhat less important than is represented by the factor 2 in the above equation. This should not, however, be used as an excuse for sloppy microscope technique.

In the approach due to Rayleigh the small items of specimen substructure are considered as incoherent point sources emitting (ideally monochromatic) light. The disturbances from two such point sources arrive in phase at their respective geometrical image points. The image of each point in the image plane is, in effect, an Airy pattern centred on each geometrical image point. The total intensity at any point on the image plane will be the simple sum of the various intensities associated with the Airy patterns overlapping at that point. Fig. 22(a) shows the distribution of

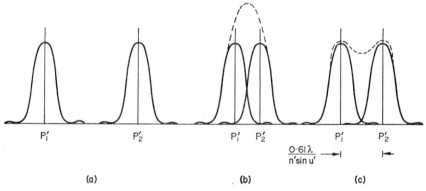

FIG. 22. The Rayleigh criterion for resolution.

intensities of two Airy patterns resulting from two separate points P_1 and P_2. The nearer together these two Airy patterns are formed the greater the degree of overlap of the patterns and hence summation of intensities. Fig. 22(b) shows the situation where the patterns overlap so that the geometric centres are closer together than the diameter of the Airy disc, the total intensity shows no decrease at the mid point, the image appears to be due to a single source and the two object points are not resolved. According to Rayleigh two patterns of equal intensity will be just resolved when the central maximum of one pattern falls over the first minimum of the other (Fig. 22(c)). This situation represents a separation of the centre points by a distance equal to the radius of the first dark ring, when there will be a decrease in intensity at the mid point between the two images.

This distance is the limit of resolution of the particular system and is given by:

$$d' = \frac{0.61\lambda}{n' \sin u'}$$

Where λ is the wavelength of light, n' is the refractive index of the image space and u' is the semi-angle of the emergent cone of rays converging on the geometrical image point. It should be noted here that d' is the distance of separation in the image points not, as in the Abbe theory, of the object points. Also the semi-angle and refractive index terms refer to the image space. However, it can be shown by the sine relation that this corresponds to a limit of resolution (distance between object points) given by

$$d = \frac{0.61\lambda}{N.A.}$$

which is very similar to the expression derived by Abbe.

It has been stated that $N.A. = n \sin u$; in other words, the working aperture is dependent upon the refractive index of the medium in which the lens works. This means that an objective used in air will have an effective $N.A.$ not greater than 1, since the refractive index of air is 1 and $\sin u$ cannot exceed 1. This will be true regardless of what is marked on

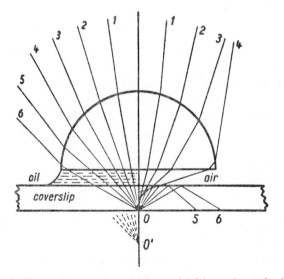

Fig. 23. Optical effects of cover glass thickness (*right*) causing spherical aberration (see p. 59), and homogeneous oil immersion (*left*), where aberration is eliminated and resolution increased. Redrawn after Michel. (Reproduced with permission of Franckh'sche Verlagshandlung, Kosmos—Verlag, Stuttgart.)

the objective. The way in which refractive index affects the working aperture of the objective is clearly demonstrated in Fig. 23. Obviously, if a lens is homogeneously immersed i.e. there is no change in refractive index from specimen to lens face, the light rays will be undeviated on passing through this space. On the other hand, if the space between coverslip and lens is of a lower refractive index (e.g. air) than the coverslip, refraction will take place at the glass/air interface and diffracted rays which might have entered the lens may be "bent" outside the acceptance angle of the lens or may even be totally internally reflected.

E. Useful magnification of the microscope

Earlier on, in discussing the function of a magnifier we referred to its ability to assist in resolving object details which could not be resolved with the unaided eye. It was shown that this meant in effect that the "angular separation" between the two points of the object to be resolved had to be not less than 1–2 min of arc. Now having resolved the two object points so that they form "separate" images in the intermediate image in the way mentioned above, it is the job of the eyepiece, acting as a magnifier, so to angle the rays coming from these two points in the intermediate image that they enter the eye with an angular separation of not less than 1–2 min of arc. If this is not achieved then the eye will be unable to appreciate the resolution achieved by the objective. From Fig. 20 it can be seen that this implies a further magnification stage so that the final image is considerably larger than the intermediate image. How large the final image must be in order to "see" the resolution inherent in the intermediate image will, of course depend on the numerical aperture of the objective used since the N.A. term appears in the formula for limit of resolution of the objective. While some people have a visual acuity good enough to resolve images when the angle is only 1 min of arc, others require a greater angle and for practical purposes those magnifications which give angular separations from 2–4 min of arc are taken as the useful range of magnification. Magnifications below this may mean that some observers are unable to resolve detail while higher magnifications will reveal no further detail and such excessive magnification is often referred to as empty magnification. The upper and lower limits of the useful range are called the maximum and minimum magnifications respectively. Table I lists the maximum and minimum magnifications for objectives of various numerical aperture. From Table I it can be seen that useful magnification = 500× to 1000× N.A. of the objective.

Consequently, when using a 100× objective of N.A. 1·30 a magnification between about 650× and 1300× will give the most suitable result. The eyepiece to be used with such an objective in order to produce a useful

TABLE I

The relationship between numerical aperture, resolving power and useful magnification

Aperture N.A.	Resolving Power d (for λ = 550 nm) in microns	Useful magnification Minimum 2′	Maximum 4′
0·10	2·75	53	106
0·20	1·37	106	212
0·30	0·92	159	317
0·40	0·69	212	423
0·50	0·55	264	529
0·60	0·46	317	635
0·70	0·39	370	741
0·80	0·34	423	846
0·90	0·31	476	952
1·00	0·27	529	1058
1·10	0·25	582	1164
1·20	0·23	635	1270
1·30	0·21	688	1375

magnification is a $10\times$ eyepiece giving total magnification of $1000\times$ while the use of a $20\times$, for example, would give empty magnification and probably loss of image definition.

F. Depth of field

While high magnifications and good resolution are desirable, there is a concomitant rapid decrease in depth of field. The depth of field limits are the planes in front and below the focused plane of the specimen which will still be acceptably sharply imaged. Outside these planes the specimen will be unsharply imaged. Microscopists working with high N.A. objectives are only too well aware of the ease with which the image goes out-of-focus, and of the difficulty in obtaining precise focus for photomicrography. The relationships between depth of field (sometimes referred to as depth-of-focus, but see Quesnel, this Series, Volume 7), resolving power and numerical aperture are given in the graph in Fig. 24.

Precise determination of the depth of field cannot be obtained by strict geometrical optical calculation but working figures may be obtained from the equation:

$$D = \frac{Rn}{M \, N.A.}$$

where D is the depth of field, n is the refractive index of the object space,

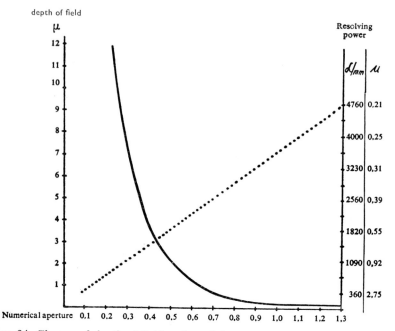

FIG. 24. Change of depth of field and resolving power with N.A. (Courtesy of Wild Heerbrugg, Switzerland.)

M the magnification, N.A. the numerical aperture of the lens, and R the diameter of the "circle of confusion". The circle of confusion is the disc of image formed in the image plane of a focused point by the imaging rays of light which derive from two coaxial points, one in the front limit plane the other in the rear limit plane, of the depth of field of the specimen. It is a disc of finite diameter since the image of a point further than the focused point forms a focused image in a plane closer to the lens than the focused image of the focused point, while a nearer point forms a focused image further from the lens than the image plane of the focused point. It is, in other words, a measure of the amount of image "blurring", and sharpness is defined in terms of the permitted diameter of the circle of confusion. This diameter must be less than 0·2 mm to meet the criterion for acceptable sharpness, and a value of 0·145 mm is used in the calculation of the depth of field distances which are given for three objectives (Carl Zeiss, Jena) in Table II.

The eye can accommodate, to some extent, for depth of field limitation and the values for visual observation are approximately derived from the equation:

$$D_{vis} = D + \frac{250}{M^2}$$

Some values are given in Table II for comparison with the accepted depth of field values for photomicrographic purposes.

TABLE II

Depth of field for three microscope objectives, taking the permitted diameter of the circle of confusion to be 0·145 mm

Magnification	Objective power to N.A.	Eyepiece	Depth of field (μm) Photomicro	Visual
160	16 × (0·32)	10 ×	2·79	12·6
400	40 × (0·65)	10 ×	0·55	2·1
1000	100 × (1·25)	10 ×	0·17	0·42

G. Illumination for microscopy

1. Critical illumination

The Rayleigh theory assumes that the object is self-luminous and that the neighbouring points behave as independent sources emitting non-coherent disturbances which cannot interfere. In an attempt to realize this idealization a diffuse source was imaged in the plane of the object by means of a substage condenser whose numerical aperture was considerably larger than that of the objective. It was argued that the Airy discs of the diffuse source imaged in the object plane were small and below the limit of resolution of the objective, so that, effectively, neighbouring object points were incoherently illuminated. In fact these Airy patterns of points in the source spread to infinity across the object plane and neighbouring patterns overlap so that portions of the light incident upon neighbouring points will be coherent. Because it was thought that the precise imaging of the source in the object plane was critical for effective illumination of the specimen, this type of illumination was called *critical illumination*. The system is shown diagramatically in Fig. 25. Recent analysis has shown, however, that so long as the condenser supplies a wide cone of light, the precise imaging of the source is not particularly critical, the important factor being the numerical aperture of the condenser. The effective N.A. of the condenser can be varied by means of the condenser iris diaphragm I_1. Also since the source is imaged in the specimen plane the area of specimen which is illuminated can be controlled by placing a second iris diaphragm, I_2 (the field diaphragm) in the plane conjugate to the image plane of the source, and this is just in front of the diffuse source (i.e. the opal or ground glass element in front of the lamp filament).

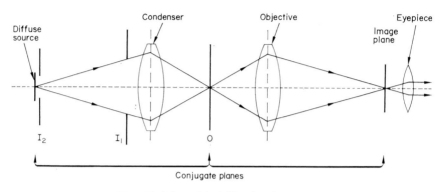

FIG. 25. The critical illumination system.

2. Kohler illumination

Most research microscopes use a different system of illumination due to Kohler, which is based on somewhat different considerations. It is tacitly assumed by microscopists that they are observing the specimen directly, albeit by means of a highly magnified image. When working at the resolution limit of the microscope, however, the similarity between specimen and image is not absolute as no imaging system is perfect and lens aberrations have to be strictly held in check if the "accuracy" of the image is to be of a high order. When there is no specimen in the light path there is nevertheless an image, the primary image—of the light source formed behind the objective due to the zero order rays. The insertion of a specimen (e.g. the grid previously mentioned) results in a duplication of the image as a result of diffraction and interference phenomena as shown in Fig. 19. The first image of the specimen (the intermediate image of the microscope) lies in the focal plane of the eyepiece and is produced by the overlapping and inter-ference of the refracted and zero order rays. In a sense, therefore, we are observing a modified or modulated image of the source and the "quality" of the light source is fundamental to the formation of a definitive image.

A good microscope lamp must produce bright, even and "white" illumination of the field of view. The brightness or light intensity in the specimen plane depends on the condenser aperture and the particular optical components of the condenser, but to be maximally effective the light intensity in the "entrance" pupil must be considered. The entrance pupil is the plane in which the first image of the source is formed in the Kohler system; it is the position of the condenser iris diaphragm since in this system the light source is imaged in the entrance pupil of the condenser, where it functions as a secondary light source and considerably influences the quality of illumination falling on the specimen.

The details of the optical arrangement for Kohler illumination are shown in Fig. 26. It will be noted that the filament of the source is imaged by the collector in the entrance pupil of the condenser, and then again in the back focal plane of the objective and finally in the exit pupil, beyond the eyepiece. Similarly the field diaphragm is imaged by the condenser in the specimen plane, by the objective in the plane of the intermediate image and by the eyepiece in the final image plane. The size of the illuminated field (area of specimen) can be varied by opening and closing the field diaphragm, which is also used to centre the illumination system and to assist in the precise focusing of the condenser.

Kohler illumination has a number of advantages. The specimen is evenly illuminated by parallel radiation since the filament is imaged in the focal plane of the condenser (unlike critical illumination); the centring of the illumination system is very easily performed; and the illumination of the field area can be accurately controlled (by the field diaphragm) to eliminate glare.

3. Types of lamp

A lamp filament emits radiation in all directions and it is the function of the collector lens to direct light on to the condenser. The brightness of the filament image depends on the aperture of the collector lens and is proportional to the square of the aperture for any given filament energy. An efficient collector, therefore, can considerably increase the total light available in the specimen plane.

It is not true that the ideal light source for microscopy is a "point source", since the smaller the area of the source the larger the aperture of the collector lens needed, and this means higher degrees of correction. The surface brightness of the filament is inversely proportional to the area of the source, for any given light intensity. For example, a ribbon filament lamp with a current loading about three times that of a normal voltage lamp produces an average surface brightness which is only about 10% higher, since the total light from the ribbon filament is spread over a much larger incandescent surface. It has the advantage, however, of producing a very even illumination. A comparison of the physical characteristics of a number of different types of microscope lamp can be made from the data in Table III.

The "colour" of the light used is also important for image quality in normal bright field observation. White light is an equal mixture of all the visible colours of the spectrum and is closely approximated by sunlight or the reflections from a clouded sky. If the wavelength distribution of the light used for microscopy differs significantly from this it becomes extremely difficult to determine the true colours of objects or stains. The filament

Projection plane (film plane of camera) with image of specimen and field diaphragm

Exit pupil plane of microscope, with image of aperture diaphragm and filament

Microscope eyepiece

Intermediate image of specimen and field diaphragm

Back focal plane of objective with image of aperture diaphragm and filament (diffraction spectrum plane)

Microscope objective

Specimen plane with image of field diaphragm

Microscope condenser

Entrance pupil plane of microscope – lower focal plane of condenser, with aperture diaphragm and image of filament

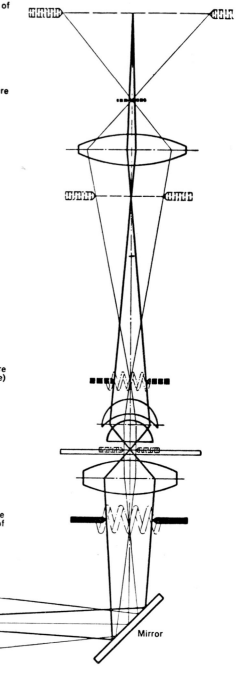

Lamp col-
lector lens

Mirror

Filament

Field diaphragm

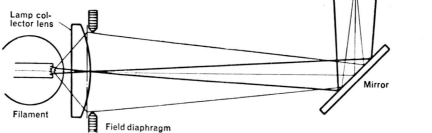

Fig. 26. The Kohler illumination system. (Courtesy of Wild Heerbrugg, Switzerland.)

TABLE III

Physical characteristics of various microscope lamps (Wild, Heerbrugg, Switzerland)

Lamp	Life h	Voltage V	Current A	Loading W	Luminous flux lm	Mean surface brightness sb	Colour temp. °K
Socket lamp P	200	6	0·8	5	5		2500
Built-in lamp S (M20)	200	6	3·3	20	212	1500	2600
Low-voltage lamp	200	6	5	30	480	1800	2800
Ribbon filament lamp	100	6	18	108	2250	2000	3100
Quartz-iodine lamp*	250	12	8·3	100	2700		3300
XBO 162 (Xenon)*	1200	20	8	160	3200	9000	(6050)
HBO 200 (Mercury)†	200	61 (L₁) 53 (L₂)	3·4 3·9	200	9500	25000	—
Electronic flash							c. 6000
Average daylight							c. 5500

h = hours; V = operating voltage; A = amps (operating current); W = watts; lm = lumens; sb = stilbs; °K = degrees Kelvin.

* = d.c. current only; † = a.c. current only.

lamps normally used for microscopy emit a continuous spectrum but the red end of the spectrum preponderates. The relative distributions of spectral energy of a continuous spectrum (emitted by most incandescent solids and liquids) is described by its "colour temperature", the colour temperature of an emitting source being the absolute temperature (°K) to which liquid platinum must be heated in order to emit light with the same energy distribution. For daylight this is about 5600–6500°K while for most filament lamps the value is well below 3300°K. Since a body will emit mainly red at low temperature and will have a blue cast at very high temperature a raising of colour temperature represents a shift towards maximum energy output in the blue region of the spectrum. We can, therefore, simulate the wavelength *distribution* of white light by using an appropriate blue tinted filter to remove excess red. It is not possible to raise the colour temperature of the source to "daylight" values, and there is a great loss of light energy (within the filter) but the colour balance is restored.

For high colour temperature and high surface brightness other forms of lamp must be used such as a carbon arc, mercury vapour or xenon vapour burner. The carbon arc spectrum although better than a filament light still does not produce a "white light" mixture and the mercury vapour burner emits a line spectrum in which only a few specified wavelengths are produced and the others are absent. Colour photography is, therefore impossible with a mercury vapour lamp. However, much of the radiation is in the high energy blue and ultraviolet end of the spectrum and provides a very good spectrum for fluorescence microscopy. On the other hand incandescent xenon vapour provides a spectrum in which, in the visible range, there is almost an even distribution of the wavelengths and, in quality, the xenon burner produces a close approximation of daylight and is excellent for colour photography. Its colour temperature rating at 6050°K indicates that daylight emulsion should be used when photographing with a xenon source. A xenon flash discharge tube is the source in electronic flash equipment for photography and the same comments apply in this case but the surface brightness is even greater from the flash discharge tube.

H. Imaging errors

In this discussion so far it has been more or less assumed that microscope lenses are capable of forming perfect images. This, unfortunately, is not the case and all single lenses have serious inherent imaging errors and the most important of these are outlined below. There is no perfect lens system but the best systems manage to reduce these defects to a negligible amount.

1. *Spherical aberration*

This error results from the simple fact that lens surfaces are curved and, in biconvex lenses for example, the outer edge of the lens is thinner than the middle. Associated with this is the fact that the curvature increases toward the edge while the opposite faces in the axial region are nearly parallel. Consequently the marginal rays are more strongly refracted (and therefore have a shorter focal distance) than the more central rays (Fig. 27).

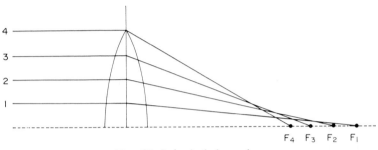

FIG. 27. Spherical aberration.

The effect is to cause "diffusion" of the image plane and therefore a loss of definition. It is corrected (at the simplest level) by combining with the first lens, a second element with opposing properties i.e. a diverging lens, to form a couplet. (See Fig. 34a).

2. *Chromatic aberration*

This error does not depend primarily on the changing curvature of the lens surface but on the fact that for any particular optical material the refractive index is greater for the shorter wavelengths which, therefore, give a greater refraction than the longer wavelengths so that a simple lens has a shorter focal length for blue light than for red. (Fig. 28). As a

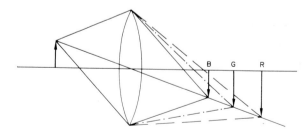

FIG. 28. Chromatic aberration; longitudinal error and chromatic difference of magnification (B, G and R represent the blue, green and red images; see 3 below).

result of the succession of image planes associated with the various colours, when focusing on a single plane the other colour planes of focus appear superimposed on the viewed plane as "circles of diffusion" and give rise to the colour fringes commonly observed with uncorrected lenses. Image quality suffers more seriously from this than from spherical aberration and much effort is devoted to its elimination. For correction, lenses of opposite curvature made from glasses of different dispersions are combined to form achromatic systems. Depending upon the degree of correction such lenses are termed achromats, semi-apochromats or fluorites, and apochromats.

A simplified explanation of the principles of colour correction is illustrated in Fig. 29. Fig. 29(a) shows a couplet in which the positive power of refraction of the converging lens exceeds the negative influences of the diverging lens. The former, of crown glass, possesses a dispersion different from the latter of flint glass such that, although the inclination of the blue rays with respect to the axis is steeper than for the red rays, the flint glass component with its concave surface then has the effect of reversing the inclinations so that the red rays now come to focus at a shorter distance than the blue. By suitable computation it has been selected that for this couplet the red and green rays come to "focus" at the same point leaving the blue "uncorrected". This is a simple achromat.

By suitable choice of materials we can correct for further wavelengths as shown in Fig. 29(b). Here the couplet is composed of a fluorite and a flint glass element. The refractive index of fluorite is almost constant for all wavelengths in the visible spectrum while it rises significantly at shorter wavelengths for flint glass. In other words flint glass is relatively more refractive to blue rays. By the use of such a combination, then, it is possible further to reduce the separation of chromatic focal planes and bring the red, green and blue rays to convergence at nearly the same point. Even in such a fluorite or semi-apochromatic system there is still some residual colour. In a fully corrected or apochromatic system residual colour is eliminated and the focal length has the same value for three wavelengths, i.e. for the middle and ends of the visible spectrum. In its simplest form it may be achieved by the use of a triplet of three different "glasses".

3. Chromatic difference of magnification

This error is inherent in the design of fluorite and apochromatic systems. It is illustrated in Figs. 28 and 30 and arises because the path lengths of rays of different wavelengths within the lens component differ so that although there may be a single image plane, the image sizes for different colours are different, i.e. the "blue image" may be more highly magnified than the red image' This error is eliminated by the use of compensating eyepieces

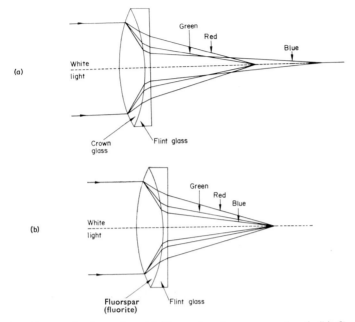

FIG. 29. Chromatic aberration. (a) Correction for green and red. (b) Correction for green, red and blue.

which are so designed as to redress the unbalance in the different colour images to give therefore a single image free of colour fringe.

4. *Astigmatism*

This term describes the defects in imaging objects lying away from the central axis, such that the image of a marginal point object comprises two lines at different focal distances and at right angles to each other. The way this arises is shown clearly in Fig. 31(a) where it can be seen that from a cylindrical entry beam a flattened cone of more or less elliptical cross

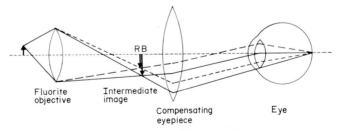

FIG. 30. Chromatic difference of magnification. Correction by compensating eyepieces.

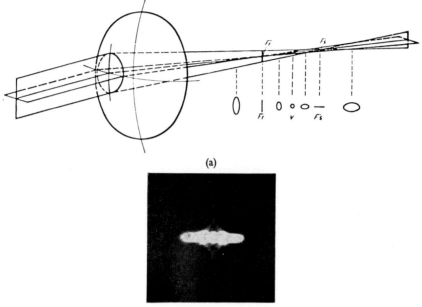

(a)

(b)

FIG. 31. Astigmatism. (a) showing image shape in different planes (Courtesy of Wild Heerbrugg), (b) the image of a point at the focal line (F_t) with about 1 wavelength astigmatism. (Courtesy of Longmans, Green & Co., London.)

section is produced and the flattening is first in the vertical and then in the horizontal sections. In consequence the image of a point source may appear as shown in Fig. 31(b). Since the tangential (vertical) and radial (horizontal) components have focal lengths which change independently as beams increasingly nearer the edge are considered, two separate image shells are formed which give field curvature and increasingly poor edge definition. The astigmatism and field curvature can be corrected to a large extent by the use of suitable concave lenses fixed at appropriate distances from the convex in corrected systems known as *anastigmats*.

Anastigmatic correction (to bring tangential and radial components to focus in the same image shell) may nevertheless leave a residual curvature of field and further correction requires that the components be made from glasses which are very carefully chosen for their refractive index and dispersive properties.

5. *Coma*

This aberration also leads to defects in off-axis imaging of oblique rays. A comatous image of a point source shows a radial "tailing" which increases

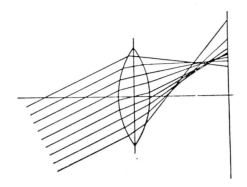

(a)

(b)

FIG. 32. Coma. (a) Image forming rays. (b) Image of a point source with about $\frac{1}{2}\lambda$ of coma (Photo by courtesy of Longmans, Green & Co., London.)

towards the edge of the field. Such an image is shown in Fig. 32(b) and the appropriate ray diagram in Fig. 32(a).

6. Distortion

This defect is always axially symmetrical for normal objects and results from differences in magnification at the marginal and central portions of the image. If the image scale decreases with increasing object size, the image of a square object exhibits "barrel distortion": where the opposite is the case, the image has "concave" sides and the error is known as "cushion" distortion. (Asymmetrical distortion can arise from astigmatic errors). The different types of distortion are shown in Fig. 33.

I. Objectives

Usually objectives are categorized according to their degree of correction for chromatic aberration, in three classes: achromats, semi-apochromats or "fluorites", and apochromats.

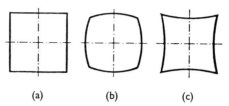

(a)　　　　　　(b)　　　　　　(c)

FIG. 33. Distortion. (a) Undistorted image. (b) Barrel distortion. (c) Cushion distortion.

1. Achromatic objectives

These are corrected for primary chromatic aberration. They are also corrected for coma and for primary and secondary spherical aberration in the yellow-green region, but over-corrected for the blue and under-corrected for the red portion of the spectrum. Colour fringes (the "secondary spectrum") are still evident but may of course be eliminated by use of an appropriate filter and if chosen for the region of greatest correction, i.e. a green filter with transmission peak about 550 nm, the performance of a good quality achromat may approach that of an apochromat. For black and white photography with an achromat, the green filter should be used for maximum image definition. Achromats are not really suitable when the specimen colour is of great importance.

In polarized light microscopy (and interference microscopy) it is essential to use an achromatic objective as the fluorite components of the more fully corrected semi-apochromats and apochromats are birefringent.

2. Semi-apochromatic or fluorite objectives

In construction these lenses are similar to the achromats but, because of the additional corrections resulting from the use of fluorite in some lens components, their quality approaches that of the apochromats. For them the secondary spectrum is almost completely removed and image brightness is greater than that obtained with apochromats as there are fewer air/glass surfaces. There is still residual error due to chromatic difference of magnification and for best performance they must be used with compensating eyepieces which are designed to correct this error. They have a relatively higher N.A. than achromats of equivalent focal length but have a shorter working distance.

3. Apochromatic objectives

The aim in the design of these objectives is to obtain a very high degree of correction for chromatic and spherical aberration. They are corrected for chromatic aberration in the red, green and blue regions and for spherical aberration for two colours. They must, however, be used with compen-

TABLE IV

Data for a selection of Zeiss objectives (Oberkochen)

Primary magnification	Designation	N.A.	Focal length mm	Working Dist. mm
10×	Achromat	0·22	16·7	5·0
	Planachromat	0·22	15·8	4·8
	"Neofluar"	0·30	16·4	4·8
	Planapochromat	0·32	14·6	0·35
40×	Achromat	0·65	4·5	0·47
	Planachromat	0·65	4·4	0·18
	"Neofluar"	0·75	4·5	0·33
100×	Achromat	1·25	1·9	0·09
	Planachromat	1·25	1·66	0·09
	"Neofluar"	1·30	1·92	0·12
	Planapochromat	1·30	1·63	0·09

sating eyepieces because of the residual chromatic difference of magnification. They have high N.A. values and short working distances and when correctly used, provide for microscopy at the highest resolutions and image definition. Because of the larger number of lens components of carefully selected "glasses" of carefully computed shape, they are usually very expensive and the dry apochromats in particular are sensitive to cover glass thickness (see later).

4. Plane objectives

Until the advent of computer programming for lens design the problems of producing lens systems with good field flatness in addition to the other corrections mentioned were almost insurmountable since such systems might contain a dozen or more components of varying chemical composition, refractive index and dispersive power and the number of variables to be considered in the computation were considerable. As a result of the speeding up of design due to the use of computers, manufacturers are now able to offer a wide variety of "plan" objectives in the various categories stated above. It should be mentioned, however, that image sharpness at the edge of the field is seldom as good as at the centre and there are "degrees of flatness of field" to be found among the lenses catalogued by manufacturers.

Some of the points mentioned above are illustrated in the data for a selection of lenses (manufactured by Carl Zeiss, Oberkochen) which are

48   L. B. QUESNEL

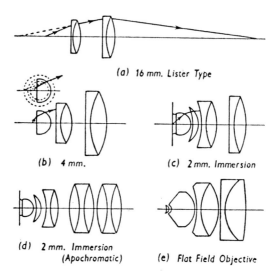

(a) 16 mm. Lister Type

(b) 4 mm.

(c) 2 mm. Immersion

(d) 2 mm. Immersion
(Apochromatic)

(e) Flat Field Objective

FIG. 34. Microscope objectives. (Reproduced by courtesy of Longhams, Green & Co., London.)

given in Table IV. Diagrams of some basic objective systems are given in Fig. 34

J. Eyepieces

A variety of eyepieces are produced by manufacturers and as a general rule the eyepieces used with a particular objective should be those recommended by the manufacturer of the objective. As mentioned earlier, the main purpose of the eyepiece is to produce a magnified image of the intermediate image produced by the objective, and the final image may be virtual for visual observation or it may be a real image as required for photomicrography and projection microscopy. To "form" a virtual image the intermediate image is made to focus at or just inside the lower focal plane of the eye-lens, while a real image is produced when the intermediate image is formed just outside this focal plane (Fig. 35).

Eyepieces are composed of two or more lenses, the upper component or eye-lens is the magnifier, while the lower component is called the field lens. Short descriptions of some of the main types follow:

1. The Huygens (negative) eyepiece

This consists of two plano-convex lenses with the curved surfaces facing the objective and is used mainly with achromatic objectives; the anterior focal plane lies in the plane of the field diaphragm between the two lenses

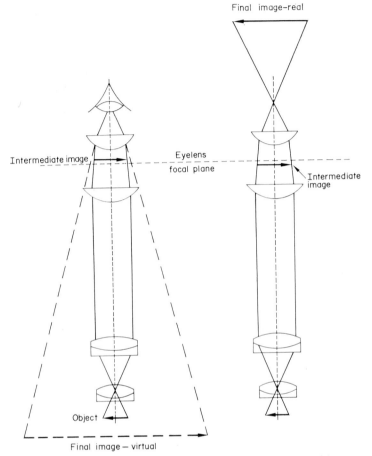

Final image-real

Intermediate image

Eyelens
focal plane

Intermediate
image

Object

Final image — virtual

FIG. 35. Image formation for visual and for photomicrographic use.

(Fig. 36(a)). The Huygens eyepiece is achromatic and fairly free from distortion and spherical aberration. The field lens increases the angular field of view; it forms an intermediate image slightly reduced in size from that which would be produced by the objective alone in its absence. If used for measuring purposes, the graticule is introduced in the focal plane of the eye lens but it should be noted that in this case the magnification of this position is that produced by the objective/field lens combination, and off-axis rulings suffer from some distortion as they are viewed with the eye-lens alone. "Measuring" eyepieces are better effected with a positive type eyepiece such as the Ramsden or Kellner (Fig. 36(c)).

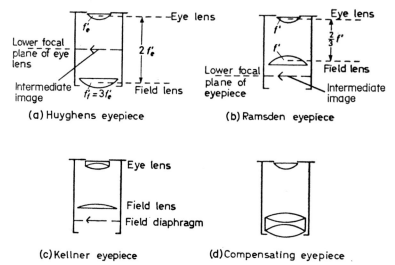

(a) Huyghens eyepiece   (b) Ramsden eyepiece

(c) Kellner eyepiece   (d) Compensating eyepiece

FIG. 36. Microscope eyepieces. (Reproduced from G. W. White, "Introduction to Microscopy", Butterworths, London.)

2. The Ramsden (positive) eyepiece

This eyepiece is made up of two plano-convex lenses of equal focal length and with the convex surfaces opposed. They are so arranged that the lower focal plane lies just below the field lens, so there is no reduction in the size of the intermediate image. An eyepiece scale may be placed in the plane of the field diaphragm which is the site of the intermediate image (Fig. 36(b)). In this way the distortion of the eyepiece scale which occurs with the Huygenian system is avoided and any aberrations due to the eyepiece will apply to both specimen and graticule images simultaneously.

3. Compensating eyepieces

These eyepieces are designed to correct for the chromatic difference of magnification common to all high power objectives. Some manufacturers introduce chromatic difference of magnification into their medium and low power achromatic objectives so that a single set of compensating eyepieces may be used with the complete range of objectives (e.g. Zeiss, Oberkochen). They are usually constructed from two separated components, one or both of which may be compound (Fig. 36(d)). In addition to correcting for chromatic difference of magnification so as to produce an image free of colour aberrations to the edge of the field, they can be designed to correct some of the image curvature as well.

They can be distinguished from ordinary eyepieces by the colour fringe around the inside edge of the diaphragm or stop when daylight is viewed through them; they give a yellow, orange or red fringe, while the Huygens or Ramsden type, for example, show a blue fringe.

4. *Wide field eyepieces*

Wide field eyepieces are now supplied by most manufacturers. Because of the great increase in field-of-view number (or field-of-view index) achieved in these combinations an increase of up to 50% or more in the diameter of the field of view may be obtained. They also provide for good flatness of field and are recommended for use with flat field objectives when critical photomicrography is contemplated.

The field-of-view number (or index) is the diameter of the intermediate

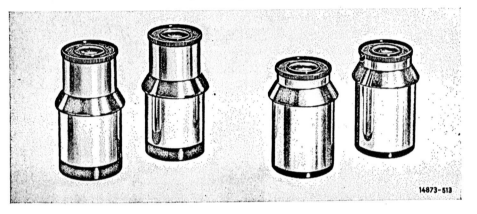

Top — Paired eyepieces, 30mm diameter. On the left GW 6.3x (field-of-view index 28), on the right GW 10x (field-of-view index 24).

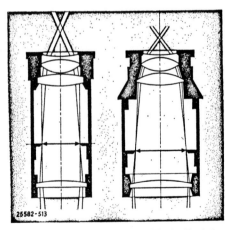

Cross section and beam path, on the left in the GF eyepiece, on the right in the GW eyepiece. It is readily seen that the GW type accommodates a considerably larger intermediate image (indicated in the diagram by an arrow).

Fig. 37. Comparison of Leitz Periplan wide-field and normal eyepieces. (Courtesy of E. Leitz, Wetzlar.)

image, measured in mm, which is visible with the eyepiece; this in turn is limited by the diameter of the eyepiece tube and eyepiece diaphragm. Obviously, the "flatness" of the intermediate image is the factor deciding whether or not it is profitable to increase the area of specimen included in a single field since it would be pointless to provide more image if all the extra image was so blurred as to make detail indiscernible. However, modern plan-objectives provide images free from field curvature over a much larger field of view than was previously possible. In order to make full use of this improved image it is necessary to alter the basic design of the microscope so that wide-tube eyepieces may be used. The modification necessary is clearly indicated in Fig. 37 for Leitz periplan GW eyepieces, and in Table V, which demonstrates the greatly increased field diameter obtained by increasing the eyepiece diameter from 23 to 30 mm.

TABLE V

Eyepiece	Eyepiece diam.	Field-of-view Index	Field-of-view (Area)
Periplan GF 10 ×	23·2 mm	18	100%
Periplan GW 10 ×	30 mm	24	178%
Periplan GW 6·3 ×	30 mm	28	242%

Note: A field-of-view index of 18 is considered to be wide field by some manufacturers since until recently eyepiece indices were often much lower for equivalent magnifications.

5. *High eye-point eyepieces*

A number of manufacturers supply high eyepoint eyepieces in which the eyepoint may be as high as 17 mm from the eyelens face and which are especially convenient for microscopists who use spectacles for the correction of vision defects due to myopia, hypermetropia, astigmatism and anisometropy. People suffering from these defects should retain their spectacles for microscopy as they may obtain less than the best results otherwise. In particular hypermetropic persons performing microscopic observations without spectacles tend to tire quickly. On the other hand, spectacles worn only for close work to correct visual defects due to ageing, e.g. presbyopia, should be removed for microscopy. High eyepoint eyepieces of compensating and wide-field type are both readily available and many are supplied with retractable sleeves and plastic protecting discs to avoid damage to the spectacles.

6. Negative projection eyepiece

By using a concave lens eyepiece instead of the usual eyepiece for visual microscopy an enlarged real image suitable for display on a "frosted" screen can be produced. The negative characteristics of the lens can be used to assist in the correction of field curvature and for chromatic difference of magnification and when used with apochromatic objectives provides a final image which is both flat and free from colour aberrations. Especially, when used in conjunction with a quartz-iodine lamp source, very bright yet enlarged image displays are possible.

K. Condensers

Far too little attention is paid by the casual microscopist to the use of the condenser system of the microscope, but, since the illuminating beam must pass through it before the objective, its quality and method of use are decisive for really good microscopy and photomicrography. Of course, a condenser, is not strictly necessary when very low power objectives are used (less than $10 \times$) when the cone of light provided by a concave mirror is adequate. It is also not used in observation of solely epi-illuminated specimens. However, it must be stressed that if a condenser forms part of the optical system the plane mirror must be used (unless the condenser is on the same axis as the source e.g. in some inverted microscopes, when no mirror is necessary).

As pointed out in the earlier discussions the primary purpose of the condenser is to supply a cone of light capable of filling the objective aperture so that the maximum resolving power of the objective may be realized. To do so it must also be in the correct position relative to specimen and objective as, e.g. if it is too low the maximum diffraction angles and hence N.A. for the system will not be obtained, and there will be a loss of field brightness as some of the transmitted rays will not enter the objective even when the N.A. ratings for condenser and objective are equivalent.

In addition, by concentrating more light on the specimen greater magnifications are possible for a given lamp intensity, i.e. image brightness is improved. It is also necessary to supply special types of illuminating beams, for example, contrast may be heightened by the use of an oblique beam, or, with the aid of appropriate annular diaphragms hollow illuminating cones may be provided for use in dark field illumination or phase-contrast illumination.

1. The Abbe condenser

This is a simple two lens unit with numerical apertures usually 1·25 or 1·30 when used oil-immersed; obviously, when used dry the N.A. is less than 1·0. The system suffers from spherical and chromatic aberration and it is not possible to provide really uniform illumination of the objective and should be avoided for visual microscopy with objectives of N.A. greater than 0·4.

2. *Aplanatic condensers*

Aplanatic condensers are corrected for spherical aberration and for curvature of field and give a good image of the light source when used with monochromatic light, i.e. with say a yellow-green filter which avoids the errors due to the uncorrected chromatic aberration. They are supplied with numerical apertures up to 1·4 when used with oil. They may have swing-out top elements to allow for better matching of N.A., between condenser and objective, when 40 × or less are used.

3. *Achromatic condensers*

Further correction of the condenser system is required for the most critical work and the achromatic condenser is corrected for spherical aberration (usually at one wavelength about 555 nm for use with yellow-green filter) and for chromatic aberration at two wavelengths in the red and blue regions. Dry and oil-immersion designs are available and by changing the front element a variety of different N.A.'s may be catered for. The Leitz (Wetzlar) achromatic variable focus condenser (code 602) of NA 0·9, for example, may be converted to an "Achromatic-aplanatic variable focus condenser" (code 603) by replacement of the top element with a "4-lens condenser top" to bring the N.A. to 1·25 when used under oil. Other replacement top elements provide condensers of NA 0·45, 0·60, and 0·70,

Fig. 38. The Wild achromatic aplanatic condenser. (Courtesy of Wild Heerbrugg, Switzerland.)

respectively with working distances of 20, 11 and 4 mm. The 0·45 and 0·60 N.A. systems are non-achromatic.

The arrangement of lenses in the Wild Achromatic-aplanatic condenser with N.A. range 1·30, 0·95, and 0·70 is shown in Fig. 38. Achromatic aplanatic condensers are indispensible for colour photomicrography at high N.A. and should be used for all critical microscopy and photomicrography at maximum resolution (semi-apochromat, apochromat or plan-high power objectives). Dark-field and phase-contrast condensers will be considered later.

Condenser assemblies usually incorporate an iris diaphragm and a filter holder. The iris diaphragm when open fully allows the maximum N.A. of the condenser to be realized, and narrows the angle of the illuminating cone of light as it is progressively closed. Its use in the proper illumination of the specimen is very important yet often disregarded and will be considered later. Obviously, as it is closed the N.A. of the condenser is artificially restricted and this will determine the effective N.A. of the objective in use.

IV. SETTING UP AND EFFICIENT USE OF THE MICROSCOPE

Before outlining the practical steps to be taken in order to set up the instrument for use it is as well to recall in broad outline the purpose of the exercise. The research microscope is a carefully designed instrument with a large number of optical units in four main groupings: (a) the lamp unit with associated field or collector lenses, (b) the substage condenser (which may have a "flip-out" top element, or turret mount for phase contrast and darkfield annular diaphragms), (c) the objective lens system and (d) the eyepiece lens system. For optimum performance it is essential that all the optical components have a common axis. (c) and (d) are fixed relative to each other by the manufacturer and are not "centrable" once the objective is properly screwed into the nosepiece and the nosepiece registered in its "click-stop". However, the lamp filament must be centred with respect to the collector lens system and the best condenser systems are supplied in centrable mounts, although many are in fixed mounts pre-centred by the manufacturers (not always accurately). In addition, in many systems a mirror or prism is inserted in the train between (a) and (b) and this must be appropriately adjusted to maintain the proper optical axis of the system.

If we regard off-axis positions of components as lateral movements there are then the vertical movements to be considered. The lamp assembly must be the correct distance away from the condenser which must be correctly placed with respect to the specimen, which is then focused upon by the objective, and finally, the tube length may be adjustable, so that the correct distance between eyepiece and objective may have to be set by the observer. In broad summary the setting up of the microscope consists of the arrangement of the optical components on a single axis and the adjustment of the distances between these components.

A. Critical illumination

1. Arrange the lamp so that the light path from source to object is about 30 cm.

2. If necessary set the tube length to 160 mm (or other figure which may be specified by the manufacturer for that microscope).

3. Remove the microscope optical components and adjust the mirror so that light travels up the tube.

4. Replace all the lenses.

5. Place a specimen on the stage and bring into focus (using a dry objective).

6. If the lamp unit is fitted with an iris diaphragm close this and rack the condenser up or down until the iris inner edges are imaged in the specimen field.

It was pointed out earlier that in the critical illumination system this iris is placed in the plane conjugate to the specimen and this is the plane of the diffuse source, i.e. the frosted glass screen or frosted surface of the collector lens present.

If a simple opal bulb is used as source then the source may be imaged by holding a wire loop against the surface of the bulb and focusing on this. Centre the iris aperture (or centrally fixed loop) in the field of view by adjusting the mirror.

7. Open the field (lamphouse) iris until the field of view is just fully illuminated. This must be readjusted for each objective so as to eliminate light from non-observed parts of the specimen.

8. Remove the eyepiece and observe the back of the objective. Close the substage (condenser) iris until its edges are just visible and centre the condenser by means of its screw adjustments until the iris aperture is concentric with the objective lens aperture. Then adjust the iris aperture so as to fill 2/3 of the back aperture of the objective with light.

9. Replace eyepiece and make final adjustments to focus.

10. To obtain the best results from the microscope it may still be necessary to insert an appropriate filter to eliminate aberration, reduce the intensity of illumination, help increase contrast etc., but as far as the "lateral" and "vertical" positions of the components are concerned the microscope has been arranged for its best performance.

Many microscopes nowadays may be fitted with simple socket lamps of one form or another, working either from a mains supply or on D.C. via a transformer. In these cases the source is brought near the condenser and the lamp has a heavily frosted field lens which acts as a collector and diffuser. The correct position of the condenser may be set by focusing the condenser to give an image of the frosted surface superimposed on the focused specimen image; then rack the condenser down just sufficiently to blurr the image of the "frosting" and so produce an evently illuminated field.

In centring the condenser (8 above) and for checking the concentricity of iris apertures (both lamp and substage) with the objective aperture, the best method is to insert an auxiliary telescope of the type used in the setting of the phase system (see later) and focus this onto their respective images. If this is done at low light intensities a good glare-free view of the various apertures is obtained and the adjustments can be made quite precisely.

To change to other objectives. Objective sets are usually par-focal and dry objectives may be brought into use simply by rotating the nosepiece in the appropriate direction and finally adjusting the focus. Since the different power objectives will have different field areas the iris diaphragm must then be readjusted.

To set up an oil immersion objective, first rack up the tube then place a small droplet of the immersion oil of the correct refractive index (1·515) onto the specimen and while observing from the side of the instrument rack the oil-immersion objective (marked HI for homogeneous immersion) down to make contact with the oil and continue racking carefully until the lens face is almost touching the specimen. It is essential that no air bubbles, however small, are trapped in the oil layer. Look through the instrument and rack slowly upward until the specimen is brought into focus.

B. Kohler illumination

All research microscopes now employ the Kohler system of illumination for bright field microscopy and photomicrography. The light source is usually a high power lamp run on low voltage D.C. current via a regulating transformer by means of which the voltage, and so light intensity, may be controlled over a continuous range. Of course if a gas burner (xenon or mercury) is used, this will not be possible.

The lamp may be housed in a separate unit with its collector lens system and field diaphragm or it may be accommodated in the microscope stand. The principles used in setting up are the same although the mechanics involved in realizing them may differ. The procedure for an independent lamp unit will be described.

1. The essential first step is to ensure that the filament of the lamp is centred with respect to the collector lenses. Adjust the lamp to shine on a

white card (or wall) fixed at a distance of 50–100 cm. Close the field diaphragm by about $\frac{3}{4}$.

2. Move the lamp/socket assembly towards or away from the collector until a clear image of the filament is obtained.

3. Test whether the filament is on the optical axis of the collector by carefully rotating the socket and lamp. If it is centred the image will rotate about its central point; if not, the whole image will move round in a circle. Adjust until central by means of the three adjusting screws (these may be large and extremely accessible or may be small screws between socket and bulb-base adjustable by means of a screwdriver. Also, this procedure will not be possible if the long axis of the lamp is at right angles to the optical axis of the collector as is the case with some quartz-iodine bulbs and the centring process may be somewhat more tedious). Once the lamp has been accurately centred it should not need attention for long periods of time.

4. Arrange lamp unit so that the field diaphragm is about 23 cm from the microscope mirror and the beam falls on the centre of the plane mirror; adjust mirror so as to reflect the beam up the tube (see A3).

5. Place specimen on the stage and bring into sharp focus.

6. Close field diaphragm and adjust condenser up and down until a sharp image of the field diaphragm aperture is obtained in the specimen field. This is more readily achieved if the substage diaphragm is also stopped down to eliminate glare.

7. Centre the image of the field diaphragm in the field of view by adjusting the mirror (or, as for example in the case of the Wild M20 with built-in lamp, by moving the centralizing inset lens).

8. Focus the image of the lamp filament on the lower surface of the substage diaphragm as in 2 above. (For less critical adjustment the filament image may be focused in the plane of the mirror by placing a frosted glass or piece of lens tissue against the surface of the mirror).

9. Check again that the image of the field diaphragm aperture is in the centre of the field and sharply defined.

10. Open the field diaphragm until its image just disappears from view.

11. Remove eyepiece and adjust substage iris aperture so that 2/3 of the objective aperture is filled with light.

12. Replace eyepiece and make final critical adjustments to focus.

It should be remembered that the resolution of the system is dependent upon the NA of the condenser and that this in turn is controlled by the substage iris. With some specimens, however, the contrast is very poor unless the condenser iris is closed to give less than the full aperture of the objective, otherwise the iris should be opened just enough to give the maximum resolution without glare.

Steps 6–12 must be repeated for each change of objective.

Before switching on the lamp the voltage regulating knob should be positioned at zero. After switching on the loading on the filament is gradually increased until the desired intensity is obtained. Similarly, when the lamp is being operated near its rated maximum voltage the voltage should be reduced gradually to zero before switching. In this way the life of the lamp will be extended and its performance maintained.

It is well established that the best results are obtained with a system in which the condenser numerical aperture is matched to that of the objective. If the best resolution of the oil immersion objective is to be obtained a condenser of equivalent N.A. must be used and the space between the top lens of the condenser and the base of the slide must be oil-filled. To do this remove the specimen, then lower the condenser slightly and apply a large drop of immersion oil, replace the specimen and rack up the condenser so that the oil forms a bubble-free layer between the two. The Kohler illumination principles will still apply and the various adjustments previously listed must be carried out.

In certain circumstances the slide and coverglass thickness can affect the performance of the microscope. If a double immersion system is used in which the oil and glass have the same refractive index—no deterioration in image quality will result. However, in darkground illumination with high N.A. condensers the thickness of the slide must not exceed certain limits specified by the manufacturer. (For values see under Dark-field illumination.)

When dry objectives of high N.A. are used the thickness of cover-glass may be important. Most objectives are designed for use with cover-glasses 0·17–0·18 mm thick (at a mechanical tube length of 160 mm). Thicker coverglasses may introduce spherical aberration into the image (Fig. 23). Dry objectives in the range $20 \times$ to $60 \times$ are the most sensitive, but the error can often be corrected by careful adjustment of the tube length of the instrument, or, objectives with a *correction collar* should be used. The collar is a rotatable ring which, when turned, alters the relative distances between the objective components. The thickness of the layer of mountant between specimen and coverglass may also contribute to this error since the refractive index of the material between specimen and objective is important in correcting for the aberration. This layer of mountant should, in general, be as thin as possible.

Since the dry higher power objectives are corrected for minimum spherical aberration with a coverglass thickness of 0·17, uncovered specimens will not yield good images. Some suppliers (e.g. Wild, Heerbrugg) provide a coverglass mount which can be attached to the objective so that the glass can be introduced into the light path without the need for covering the specimen.

For critical work with dry objectives of high N.A. the precise thickness of the coverslips used should be determined with a special micrometer (coverslip guage) and only those measuring 0·17 mm. retained, as the fairly poor manufacturing tolerances for coverslips usually result in a variety of thicknesses, even in a specified grade; e.g. Chance No. 1½ range from 0·16–0·19 mm. Moreover, in use the instrument may have to be readjusted to achieve minimum spherical error under working conditions since the effective coverslip thickness is a composite of the thickness of the mountant layer between specimen and coverslip, and the coverslip itself. Where necessary a mechanical tube length adjustment may be made on the basis of a "star" test (Slater, 1957).

The optical phenomena of the "star" test are best observed using a silvered slide with minute pinholes to act as "point sources" of light. The test slide with mountant and coverslip is placed on the stage and the microscope focused on one of these point sources. The image is then carefully observed as the microscope is de-focused upward and downward. If spherical aberration is absent the upward and downward out-of-focus images will have an identical appearance. If, on de-focusing upward a hazy ring with bright central spot is seen, while de-focusing downward produces a bright ring (see Airy Disc, section D), then the coverglass/mountant combination is too thin and the under-correction of the system must be eliminated by increasing the tube length. If the converse is observed, i.e. the bright ring appears on de-focusing upward, then the combination is too thick, and the over-correction must be eliminated by decreasing the tube length.

The biological specimen is unlikely to have pinhole apertures of this sort, but the same test can be applied on small dark "specks" in the preparation, either of foreign matter or specimen substructure. Bring such a "dot" to the centre of the field and de-focus upward and downward as before; on one side of focus the dot will disappear in a hazy spot, on the other side of focus the dot will expand into a more or less well defined ring. If the bright ring is formed on de-focusing downwards—increase the tube length; if on de-focusing upward—decrease the tube length. (A correction collar would be turned clockwise in the former, and anticlockwise in the latter, case.)

It should be noted that binocular microscopes are not adjustable for tube length (except slightly by change of interpupillary distance) and these corrections may not be possible. The problem is totally avoided by using, say, a 50× oil immersion objective.

V. PHASE CONTRAST MICROSCOPY

A. Basic conditions for phase contrast imaging

Earlier in this Chapter the meaning of "phase" was discussed, and it was shown that the illumination at a point was dependent upon the "phasing" of the rays falling on that point. From the outline given we can see that light rays in-phase produce the maximum brightness by "reinforcing" each other, while rays with a phase angle of 180° or a $\frac{1}{2}\lambda$ difference produce maximum interference (or cancellation) and yield minimum brightness for those rays. Any other phase difference from 0°–180° (0–$\frac{1}{2}\lambda$) results in varying degrees of interference, the net brightness decreasing for differences from 0–180°. Clearly the phenomenon of interference provides us with the opportunity of using "displaced" rays to produce a complete range of intensity from maximum ("white") to minimum ("black") with a continuous spectrum of greys in between; in other words the opportunity of producing contrasts in an image derived from an object which does not itself introduce amplitude differences in the radiation. (But, of course, on interference between two rays the resultant energy flux may be represented as a single ray of reduced amplitude).

Contrast differences must exist in an achromatic (un-coloured) image if image detail is to be made visible. Most biological material is inherently low in contrast as the cellular constituents usually do not produce significant amplitude differences in the image-forming rays. Contrast and colour may be introduced by staining, with the consequent injurious or lethal effects upon living matter, and only with specific staining techniques under specific illuminating conditions is it possible to study living "stained" material. Phase contrast microscopy, however, enables unstained, apparently featureless material to be observed with good contrast over indefinitely long periods with no injurious effect provided precautions are taken to prevent overheating of the specimen and to maintain the other physiological conditions for life. It is, therefore, the system of choice for long term observation of living material.

Many students despair of understanding the phase microscope and in consequence this system is used far less than it should be. Yet so long as the points outlined under the "Phase" and "Phase specimens" sections above, are grasped, it becomes an extremely simple matter to set up and use any of the instruments made for this purpose.

In broad concept the phase microscope is an ordinary microscope set up for bright field use but with two additional "plates" which enable some of the image forming rays to be phase-shifted with respect to the others. These additional "plates" are (i) the substage annulus fixed in the lower focal plane of the condenser and (ii) the phase plate (or phase ring in some manuals) fixed in the back focal plane of the objective. In order to produce the interference phenomena described, two "sets" of rays are required, one "set" being out of phase with the other; equally, it was discussed earlier how, in order to obtain defined images of an object, two types of radiation were necessary—the transmitted (or background) rays and at least the first order of diffracted rays. The purpose of the two additional items required for phase microscopy is to "separate" these two types of radiation within the microscope and then deliberately to introduce a phase difference between them. When they are recombined in the formation of the image the resulting interference phenomena bring about an enhanced contrast equivalent to that obtained as a result of amplitude differences. It is achieved in the following way.

Imagine a microscope correctly set up for Kohler illumination but without any specimen on the stage. A solid cone of light would be produced with its "apex" in the specimen plane and (if the N.A.s were matched) the cone would be reversed above the specimen and fill the aperture of the objective; light would travel up the tube and, on observation, a totally filled bright field would be seen.

Imagine now that the substage diaphragm is replaced by an annular diaphragm, that is, a metal or glass plate, which stops the passage of light to the condenser except for a ring-shaped (or annular) zone concentric with the axis of the lens system. In this case the illumination on the condenser would be a hollow cylinder, then focused by it, to give a hollow cone below and above the specimen plane (Fig. 39(a)). Assuming that the annular aperture was less than the aperture of the objective an image of this aperture would be formed in the plane conjugate to the diaphragm plane, viz. the back focal plane of the objective. Looking down the microscope now, we would still see that the field was fully and flatly illuminated (but the intensity of illumination was reduced). By removing the eyepiece and replacing it by an *auxiliary telescope* (also variously referred to as an "auxiliary microscope", "auxiliary focusing magnifier", or "centring telescope") it is possible to focus upon the image of the annulus so that a sharply defined circle of light in a black field is seen. In other words at this position within the system all the illumination is collected into a hollow ring-shaped zone.

Suppose now that we made a plate of glass of suitable diameter that could be mounted within the objective sleeve in this plane, and we etched a circular groove in this of such a size that the entire illuminating "annular" beam at this point passed through the etched ring, we would have made a phase contrast microscope out of our ordinary one. The etched glass plate is the "phase plate". Of course the depth of the etched ring is important (see below), but the point to be made here is that a system can be devised in which we know where the illuminating beam is and we can arrange that all the light has been made to traverse a "ring" of known thickness of glass before proceeding on to form the subsequent image (Fig. 39(a)).

If a specimen is now inserted at the appropriate place all the radiation which is undeviated by it travels, as described in the previous paragraph,

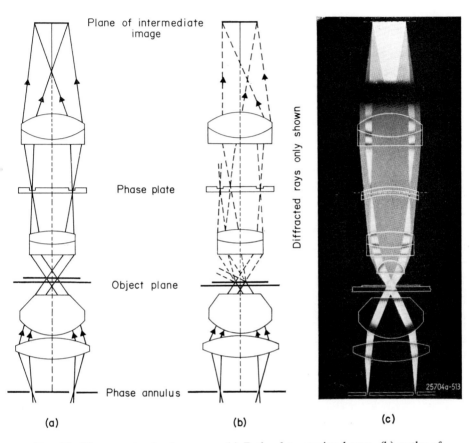

FIG. 39. Phase contrast microscopy. (a) Path of transmitted rays, (b) paths of diffracted rays, (c) transmitted and diffracted rays. (Photo by courtesy of E. Leitz, Wetzlar.)

through the etched ring and this is the transmitted or background (zero order of diffraction) radiation. But a proportion of the radiation incident upon the specimen is inevitably deviated to give the various orders of diffracted light mentioned in Section II F. The angles of diffraction will be determined by the factors there mentioned and the result will be that diffractions from various parts of the specimen field will be scattered over the face of the objective (and, of course, some outside the objective also). Those diffracted rays which are accepted by the objective then travel up the microscope and since they were incident upon the front element at all angles they will pass through all parts of the phase plate including the etched ring. Since this "ring" is a comparatively small area of the total aperture area most diffracted rays will pass through the un-etched, thicker parts of the phase plate (Fig. 39(b)).

We, therefore, now have undeviated transmitted radiations, all of which pass through the phase ring, and diffracted radiations passing through all parts of the plate on their way to the plane of the intermediate image where they are "recombined" to form the overlapping interference patterns of which the image is made. Rays emanating from the same (strictly, very nearly the same) point of the specimen return to the "same point" in the image and if one set is phase displaced with respect to the other they interfere to produce variations in intensity of illumination, and hence contrast, in the way described earlier.

B. The phase plate and image quality

Let us look a little more closely at the phase plate and the amount of displacement it produces between the transmitted and (the bulk of) the diffracted radiation. In Fig. 15 it was diagramatically shown how changes in refractive index or of thickness within the specimen can lead to the same net result, namely the retardation of one wave with respect to another. The principle demonstrated in Fig. 15(b) can be used to form a phase plate of the type described above which retards the diffracted waves with respect to the transmitted waves. It is possible to cut the groove to a variety of depths, so producing phase plates with a whole range of different phase retarding values; we could, in fact, retard rays by any chosen fraction of a wavelength. In practice it is found that the most useful value is $\frac{1}{4}\lambda$ retardation; the so-called quarterwave plate, or 90° plate. If the diffracted radiation is retarded with respect to the transmitted it is (or should be) called a positive phase plate; thus, a 90° + ve phase plate.

On the face of it this would seem to be an insufficient phase displacement to cause maximum interference (darkness) in the image plane. It should be remembered, however, that the phase plate is only the secondary stage of phase displacement, as the specimen itself will cause retardations in diffracted radiations wherever it contains structures of higher refractive index or thickness than the surrounding material. We may easily envisage the situation in which a portion of the specimen yields diffracted rays a quarter of a wave-length retarded with respect to the adjacent background rays. If such diffracted rays then pass through the thick part of the phase plate they will be further retarded by a $\frac{1}{4}\lambda$ to give a total retardation of $\frac{1}{2}\lambda$ and, therefore, the possibility of maximum interference and darkest area of the image. Other specimen-produced retardations will similarly be augmented by the phase plate and produce the brightness range found in the final image.

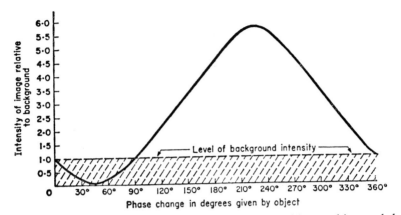

FIG. 40. Relationship between the phase change produced by an object and the intensity of its image when viewed under a phase contrast microscope with 90°+ve phase plate with zero absorption. (Reproduced with permission from K. F. A. Ross, "Phase contrast and Interference Microscopy for Cell Biologists", Edward Arnold (Publishers) Ltd, London.)

The above explanation is an over-simplification since, in fact, there is a non-linear relationship between phase change and image brightness relative to background. For example, with a non-absorbing 90°+ve plate the relationship is of the form shown in Fig. 40. Here, object-produced retardations of not more than 90° will produce darker-than-background image areas, while retardations in excess of 90° lead to "reversal" and give bright image areas, which are maximally bright for a "retardation" of 225° when the intensity may be several times that of the background. Hence the phrase well known to bacteriologists—"highly refractile bodies", e.g. spores, lipid droplets.

Now since only a small proportion of the total illumination is diffracted, there is a great preponderance of transmitted light in the image plane and a relatively small fraction of it is "matched" by diffracted rays with which it interferes. Unless the intensity of this background illumination is reduced image contrast will be poor. In order to increase contrast the phase ring area is coated with a light absorbing material to a greater or lesser extent. The resulting increase in contrast is not, however, linearly dependent upon the percentage absorption. A separate relationship between image bright-ness (relative to background) and the phase change produced by the speci-men, exists for the different absorption values which may be given to any quarter-wave plate. From the graphs plotted for such relationships (see Ross, 1967) it can be seen that as the absorption of the phase ring is increased three things happen: (i) the object appears darker than the background over a decreasingly small range of phase change (retardation relative to background) produced by the object: (ii) the minimum brightness (maximum darkness) of specimen image decreases further; and (iii) this minimum brightness occurs for a lesser degree of phase change produced by the specimen. In other words, at higher absorption values of the phase ring, small changes of (say) refractive index in a specimen will yield image areas of greater darkness (and so contrast) than for lower absorptions. This is why the most commonly used phase plate in microbiology is of the positive type high absorption value, usually in the range 70–90%.† High absorption values have the drawback that they accentuate the "halo" which is commonly observed around structures at the change of refractive index. Data for some phase objectives are given in Table VI.

So far we have spoken only of the positive phase plate, the main characteristic of which is that diffracted rays are retarded with respect to background rays. The contrast as we saw is dependent upon both the degree of retardation and the absorption. There are other combinations which are possible and which give different image characteristics. For example the image contrast may be reversed by making the plate with a thicker annular ring so that the background (transmitted) illumination is retarded with respect to the diffracted radiation. This will cause those parts of the object formerly dark to appear bright and vice versa, i.e. *negative phase contrast*. In a third type of objective the diffracted radiation is retarded and absorbed. (Unfortunately there is no agreed nomenclature for these three different types of phase objective and the manufacturers' literature presents a somewhat confusing picture. An attempt to distinguish these types is given in Table VI.) Positive phase plates are now made by depositing an annular layer of magnesium fluoride on a glass disc, coating this with absorbing material (e.g. vaporized stainless steel), and covering the whole area with a cement of higher refractive index than magnesium fluoride.

The variations in image obtained with the different types of phase objective for different phase retardations of specimen are clearly brought out in Fig. 41, while the various types of objective produced by Leitz (Wetzlar) and their working conditions are given in Table VII.

The positive phase contrast objectives are the most useful type for microbiology. Those with 70–75% absorption are recommended for general purposes and the high absorption plates are recommended for use in displaying structures which have a small difference in refractive index with their surroundings, e.g. some granular inclusions in bacteria, fine fibres etc. Negative phase contrast objectives produce an effect similar to dark

† From what has been said it will be very evident that the vast majority of the light leaving the collector system plays no part whatever in image formation, hence the essential need for a high intensity source for phase-contrast microscopy.

TABLE VI

Properties of phase plates supplied by different manufacturers

Structure of phase plate (black = absorbing; shaded = retarding)

Maker
(a) = retardation, (b) = absorption, (c) = catalogue designation

Leitz

(a)	$\frac{1}{4}\lambda$	$\frac{1}{4}\lambda$	$\frac{1}{4}\lambda$
(b)	$75\% \pm 5\%$	$88\% \pm 2\%$	$88\% \pm 2\%$
(c)	positive phase type n	intensified positive phase type h	negative phase type $-$h

Nikon

(a)	$-\frac{1}{4}\lambda$	$-\frac{1}{4}\lambda$	$+\frac{1}{4}\lambda$
(b)	50–60%	86%	86%
(c)	dark contrast DLL	dark medium contrast DM	bright medium contrast BM

VEB Carl Zeiss*

(a)	$-\frac{1}{4}\lambda$
(b)	70%
(c)	Ph v.

Vickers

(a)	$\frac{1}{6}\lambda$	$\frac{1}{6}\lambda$	$\frac{1}{6}\lambda$	$\frac{1}{6}\lambda$
(b)†	70%	86%	70%	86%
(c)	positive micro-plan phase (10 × only)	positive micro-plan phase and positive phase fluorite	negative micro-plan phase (10 × only)	negative phase microplan and negative phase fluorite

Wild*

(a)		$\frac{1}{4}\lambda$	$\frac{1}{4}\lambda$
(b)		$75\% \pm 3\%$	$80 \begin{array}{l} -0\% \\ +5\% \end{array}$
(c)		wild fluotar phase contrast	achromatic phase contrast

* The figures are entered as supplied by the manufacturers on a standard form; the quality of image in both these cases is similar to the positive phase or dark contrast image of other manufacturers.

† Derived from transmission values (30%, 14%) supplied.

field illumination and are suggested for use in the examination of the flagella of bacteria or protozoa, minute granules, etc.

The difference between positive and negative phase contrast is excellently summarized by Ross (1967) in the following paragraph. "Most commercially marketed phase contrast objectives have $90° + \text{ve}$ phase plates (which

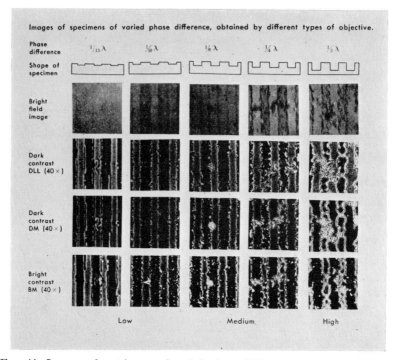

FIG. 41. Images of specimens of varied phase difference obtained by different types of objective, c.f. Table VI. (By courtesy of Nippon Kogaku, Tokyo.)

means that they are constructed so that the diffracted light is retarded one quarter of a wavelength behind the directly transmitted light) and, if these are used, a homogeneous object will appear darker than the background if its refractive index is slightly greater than the mounting medium, and brighter than the background, or "reversed" if its refractive index is slightly less than the background. Negative phase plates, however, in which the diffracted light is advanced relative to the direct light, are also sometimes used and with those the opposite is true. Thus if one knows the characteristics of the phase plate in the objective one is using, one can usually tell at a glance whether the refractive index of the mounting medium is higher or lower than the object being measured".

C. Condensers for phase contrast

The optical components of these condensers may be the same as those used for normal bright field microscopy but the substage iris diaphragm is replaced by an annulus or annular diaphragm. The diameter of the annular opening required for the different powers of phase objectives varies so that

TABLE VII

Phase contrast objectives manufactured by E. Leitz (Wetzlar)

Type	Designation[1]	Free working distance	Micrometer value measured with eyepiece H 6×	Cover glass correction[2]	Type of eyepiece[3]	Designed with absorption of		
						75 ± 5%	88 ± 2%	*88 ± 2%
Dry system	Pv 10/0.25	5.8	15	DO	P	n	h	—
	Immersion attachment for Pv 10/0.25	0.3						
Dry system	Pv 20/0.45	2.8	7.6	D	H(P)			-h
Dry system with very long working distance	Pv Apo L 40/0.70 in correction mount with automatic focusing compensation	0.38	3.8	D!	P	n	h	-h
Dry system with very long working distance	Pv Apo L 63/0.70 in correction mount with automatic focusing compensation	0.35	2.4	D!	P	n	.h	-h
Water immersion objective	Pv WE 22/0.60	0.05	6.5	O	P	n	h	—
Water immersion objective	Pv WE 50/0.70	0.05	2.8	O	P	n	h	—
Water immersion objective	Pv WE 80/1.00	0.06	1.9	O	P	n	h	—
Oil immersion	Pv Fl Oil 70/1.15	0.20	2.0	DO	P	n	h	-h
Oil immersion	Pv Apo Oil 90/1.15	0.12	1.6	DO	P	n	h	—

[1] The number before the oblique stroke gives the initial magnification, while the figure after the stroke gives the numerical aperture.

[2] D: with cover glass = 0.17 mm (cover glass thickness should be observed accurately to within ±0.05 mm). O: without cover glass. DO: can be used with or without cover glass. D!: Cover glass thickness should be observed accurately to within ±0.01 mm., or should be accurately set with the correction mount.

[3] H = use Huygens eyepieces; P = use Periplanatic eyepieces.
*Available on special request.

a phase contrast outfit made to cover a range of objectives from $10\times$ to $100\times$ will contain several of these annuli, although in some outfits the same annulus will serve for $50\times$ and $100\times$ objective, and another for both $20\times$ and $40\times$. For convenience the various annuli are usually mounted in a rotating turret below the condenser so that any required annulus may be quickly swung into position. Almost all turret condensers possess an iris diaphragm position for normal bright-field as well, and some (e.g. Wild, Heerbrugg) have in addition a position with the annulus ("field stop") required for dark ground. With such "universal" condensers one may view a single specimen under bright-field, phase contrast, or dark-ground conditions in very rapid succession. As will be seen in a moment a necessary feature in a phase contrast condenser is the ability to move the annulus in any desired direction so that its aperture may be centred with respect to the phase ring in the objective. Centring screws are usually provided but in some systems (e.g. the Nikon) the annuli are set in friction-glide mounts and are centred simply by pushing the mount in any desired direction with the finger-tips.

Many manufacturers now supply long-working-distance condensers for phase use. Obviously these will have a comparatively low maximum numerical aperture but their great usefulness in studying the behaviour of cells and tissues in bottles, Petri plates or micro-culture chambers (see Quesnel, this Series, Volume 1) far outweighs the disadvantages of slightly reduced resolution. Typical specifications are: working distance 20 mm, N.A. 0·52 (Wild, Heerbrugg) and working distance 12 mm, N.A. 0·7 (Nikon).

A unique type of Universal condenser is supplied by Leitz, called the Heine condenser, which by a simple vertical movement of one component of the condenser permits the observation of objects by brightfield, phase contrast or darkfield illumination. The extremely interesting feature of this instrument is the fact that one may pass continuously from one type of illumination to the next with "intermediate types" of illumination quality which can be highly informative, the principles underlying the performance of the Heine condenser are illustrated in Fig. 42.

VI. SETTING-UP AND USE OF A PHASE CONTRAST SYSTEM

A. Practical procedure

The description given relates to the use of a condenser with rotating turret for the various annuli, and iris diaphragm for normal bright field use as this is probably the most commonly used research accessory.

1. Set up the microscope for bright-field Kohler illumination in the

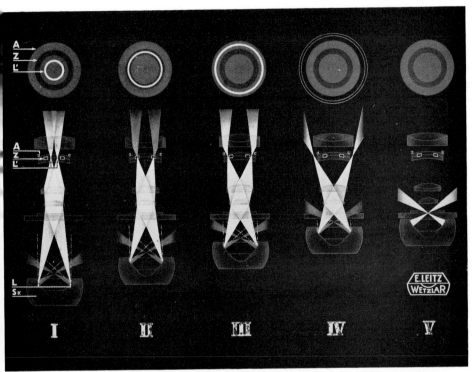

Fig. 42. The various types of illumination achieved with the Leitz Heine Condenser on raising the mirror component SK. I, Bright field illumination. II, Phase contrast. III, high contrast bright field. IV, Special dark field. V, Normal dark field. (By courtesy of E. Leitz, Wetzlar.)

normal way (Section IVB) but with phase objective (and appropriate eyepiece), and phase condenser in brightfield position. Focus on a lightly stained specimen (to make final adjustments easier while learning to set up the system).

2. Remove the eyepiece and replace it by the auxiliary telescope supplied with the phase contrast outfit (or insert the Bertrand lens if provided). Focus the telescope on the phase plate in the back focal plane of the objective. A clearly defined "grey" ring (the phase ring) will be observed against a bright field background. Notice in particular the outer circumference of the ring and the annular width between inner and outer borders.

3. Revolve the turret mount so that the appropriate annulus for the objective in use is brought into register below the condenser.

4. Look down the telescope and note that the field is now dark except for a brightly illuminated ring which is the image of the annular aperture.

5. Most likely this bright ring will not be in the centre of the field and its "track" will probably intersect the "track" of the phase ring. (When the annulus is swung into place it is often quite difficult to discern the phase ring except in the region where the two tracks intersect. However, by careful adjustment of the telescope one can usually discern a faint very thin ring of light emerging from the outer or inner circumferences of the phase ring). Manipulate the annulus (by turning the centring screws or by pushing the annulus mount as the case may be) until the brightly illuminated image of the annulus is exactly concentric with the phase ring. If the correct annulus has been used the illuminated ring will fall completely within the outer and inner boundaries of the phase ring so that no illumination falls outside of this area. This is the essential requirement for good phase contrast; if any part of the annular light field falls outside the phase ring area, image quality will be seriously affected or even totally useless.

6. Now remove the telescope and replace the eyepiece.

7. Adjust fine focus and intensity of field illumination (transformer).

8. Check that the conditions of Kohler illumination are still met in the way described earlier: adjustment to condenser height and position may be necessary. These in summary are: with image of centred lamp filament focused on substage (annular) diaphragm, closing the field iris diaphragm almost fully should give a sharp image of this iris in the centre of the focused specimen field of view. If it is not central then centre it by adjusting the (axis of the) whole condenser component (if a centrable condenser mount is present, or by means of the centring lens insert or mirror depending on the particular instrument), and re-set field iris to illuminate just the field of view (see IV B above).

9. Re-check the phase-plate and annulus alignment by means of the telescope, replace eyepiece and observe specimen.

A few special points are worth noting in regard to the use of phase contrast illumination. (a) All glass surfaces in the optical path must be scrupulously clean; dirt particles, finger-prints, etc. especially on the surfaces of slide or coverslip will diffract light and seriously affect image quality. (b) Only thin specimens are really suitable for the phase system— up to about 10 microns thick. (c) It is a considerable help in setting up and aligning the system if a "trinocular head" is used. In this case the auxiliary telescope may be left permanently in the third eyepiece sleeve and kept in focus on the phase plate, while the binocular portion is reserved solely for viewing the specimen field. (d) For those inexperienced in microscopy it is a considerable help in setting up the system, for phase specimens which show almost no image by brightfield, if some more obvious marker is introduced in the specimen plane for the purpose of initial focusing. This may be a small scratch on the surface of the slide or underside of the coverslip, placed well away from the fields of subsequent observation, or a thin wax pencil mark in a similar position. In the procedure above a lightly stained preparation was suggested but, more often than not, one uses phase contrast because the specimen is unstained. A very useful practice specimen may be prepared by swabbing the inside of the cheek and making a wet-mounted preparation of the swabbings. Buccal mucosa cells are invariably present (as well as many other interesting creatures) and are large enough and possess sufficient inherent contrast to be easily focused upon by bright field. A switch over to phase-contrast then shows the different quality of image to great effect.

B. Immersion media and phase contrast

Since differences in refractive index between specimen and mountant give rise to differences in contrast in the final image it follows that under appropriate conditions mounting the specimen in a medium of equivalent refractive index will eliminate the contrast. If the refractive index of the mountant is known, therefore, the refractive index of the specimen material can be found. This is a very useful measure, since refractive index varies linearly with the concentration of dissolved cell solids and therefore provides the means by which the degree of hydration of tissues may be measured and estimates for dry weights of cell sub-structures obtained.

It also follows that the degree of contrast in a specimen image may be modified by modifications of the refractive index of the mountant. The whole field of immersion refractometry and its application to the quantitative "analysis" of single cells has been reviewed and enlarged by the recent publication of Ross (1967) to which the reader is enthusiastically referred for details. In this Chapter a short note only on the qualitative applications of immersion media changes will be given.

Liquid media for immersion refractometry of living cells and micro-organisms must be non-toxic and non-penetrating and must not affect cell volume. Bacteria, with their rigid cell walls, are less susceptible to volume changes but membrane constrained cells must be mounted in media of equivalent isotonicity. The pH of the mountant should also be adjusted to the requirements of the cells' metabolism if continuing observations are to be made on the same cells. Often, however, one may use such mountants simply to demonstrate more clearly the changes in cell structure which have resulted after a particular period of treatment. Such changes are well illustrated in the example which was given in Chapter X, Volume 1 of this series, which showed the effect of Mitomycin C on *Escherichia coli* cells.

To match the full range of refractive indices which may be encountered in living cells a series of mountant media covering the refractive index range from 1·333 (water at room temperature) to c. 1·540, which is the

74   L. B. QUESNEL

value for the dehydrated protein of some bacterial spores, is needed. No single mixture of suitable substances covers the whole range but for general purposes various aqueous solutions of bovine plasma albumin will be found most suitable and have been extensively used. For these solutions bovine plasma albumin, fraction V (Armour Laboratories, Kanakee, Illinois, U.S.A., or Eastbourne, England) is dissolved in distilled water or saline in concentrations of up to 50% w/v (refractive index 1·424). The solutions should be membrane filtered and stored under sterile conditions at low temperature (c. 5°C) if they are to be used over an extended period since contaminating micro-organisms would readily degrade these preparations.

The very wide range of contrast variations which may be obtained by

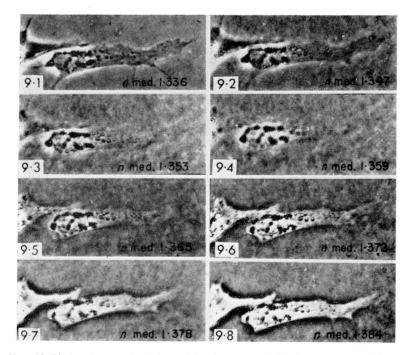

Fig. 43. Photomicrographs, taken with a 4 mm +ve 90° phase contrast objective, of living chick myoblast growing in one-day-old culture, mounted in the culture medium, 9·1, which had a refractive index of 1·336, and in a succession of isotonic saline/protein media, 9·2–9·8, of increasing refractive index. In 9·1, 9·2 and 9·3 the cytoplasm of this cell appears darker than the background indicating that these mounting media had lower refractive indices than the cytoplasm. In 9·5–9·8 the cytoplasm conversely appears bright or 'reversed', indicating that these mounting media had higher refractive indices than the cytoplasm. In 9·4 the cytoplasm is almost invisible, indicating that it must have a refractive index very close to 1·359.

adjusting the refractive index of the immersion medium within quite small limits is clearly demonstrated by the photomicrographs in Fig. 43. It will be seen from these that the cytoplasm of the chick myoblast cell had a refractive index of 1·359 since the contrast between cytoplasm and mountant was zero at this value of mountant refractive index. Yet a very small change in refractive index of mountant to 1·384 caused the formation of a "reversed" contrast image of considerable brightness. A similarly small refractive index change in the opposite direction shows the cytoplasm as a dark area relative to background. Other substances which have been used for immersion media include human plasma albumin, dialysed commercial egg albumin, carboxy-haemoglobin, and bovine plasma globulin, fraction II (Armour Laboratories). Solutions of acacia gum (gum arabic) (Barer and Joseph, 1955) and polyglucose (Allen, 1958) have also been used. Further information on all aspects of phase contrast and interference microscopy will be found in the work by Ross (1967) already cited.

VII. DARK-FIELD ILLUMINATION

Dark-field illumination may be defined as a system set up so that the minimum "effective" numerical aperture of the condenser exceeds the maximum effective numerical aperture of the objective. If the foregoing explanation of the operation of the phase microscope has been understood the arrangement for dark-field illumination will present even fewer problems and no laboured details will be necessary. The illuminating system resembles that used in phase since an annular stop or diaphragm is necessary in order to produce a hollow cone of light incident upon the specimen but in this case the objective works within the area of the dark cone. However, the diameter of the annular aperture will be greater since the maximum numerical aperture of the condenser lens system is utilized. The smallest angle of divergence of the rays bordering the enclosed dark cone may be conveniently termed the "minimum effective aperture". The condenser and objective are arranged relative to each other in such a way that none of the zero order rays are allowed to enter the objective and the image is formed entirely from the diffracted orders of radiation produced by the specimen. Normal objectives are used and the ray paths for "dry" dark-field illumination are shown in Fig. 44. For dark-field use at the highest magnifications double-immersion is essential and special condensers of very high N.A. are available for this purpose. Even at these high N.A. values (1·4) and the wide angle of the dark cone, some light does enter the objective and steps must be taken to absorb this background illumination or a dark-field will not be obtained. To exclude these radiations from the image either a funnel stop or an objective iris diaphragm must be used.

L. B. QUESNEL

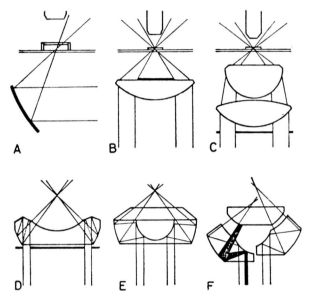

FIG. 44. Different types of dark field condenser B–F and concave mirror used for dark field, A. (Reproduced with permission from A. L. E. Barron, "Using the Microscope", Chapman and Hall, London.)

The funnel stop is a tube which screws into the back of the objective, having at its lower end an aperture whose diameter reduces the effective aperture of the objective to a value below that of the minimum value of the condenser aperture, thus blocking those direct rays accepted by the objective and permitting only the diffracted rays from the central area of the field in the back focal plane to form the intermediate (and final) image.

The working conditions for oil-immersion dark-field are far more critical than for dry-objective use. Two sets of centring controls are needed; the position of the field stop must be centrable with respect to the condenser lenses, and the whole condenser/diaphragm substage should be centrable with respect to the objective system axis. Also, the focusing movement for adjusting the condenser vertically must be smooth and precise as the position of the condenser is critical. In spite of the fact that the condenser is oiled to the slide the thickness of the latter is also critical if the cone apex is to fall exactly in the specimen plane. Some values for slides to be used with various dark-field condensers are given in Table VIII.

The refractive index of the mounting medium is also important since the great obliquity of the rays make total internal reflection a real possibility if there is a mountant of low refractive index between slide and cover-slip; the result is clearly illustrated in Fig. 45. If the viability of the specimen is an important consideration then the mountant fluid should not

TABLE VIII

Conditions of use for various types of dark-field condenser

Type of condenser	Maker	N.A. of light cone	Maximum N.A. of objective	Thickness of slide
Paraboloid		1·2–1·4	1·0	critical (usually 1 mm)
Bispheric	Vickers	1·25–1·33	1·1	1·2–1·4 mm
	Leitz	1·20–1·33	1·0	
Cardioid	Zeiss	1·10–1·40	1·0	1·1 –1·3 mm
Cassegrain		1·40–1·50	1·35	critical
Focusing	Vickers	1·20–1·44	0·95	0·75–1·5 mm

injure the cell in any way, a fact which restricts the choice of mountant severely. However, it has already been mentioned under phase-contrast illumination that solutions of bovine plasma albumin, up to 50% w/v (refractive index 1·424) may be used. A wide range of other (and often injurious) mountant media have been used. Some of these are listed in Table IX.

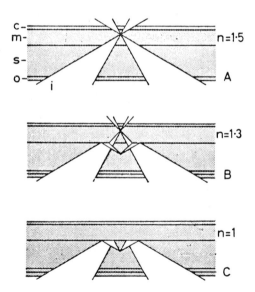

FIG. 45. Effect of the refractive index of the mountant on the dark field illumination cone. c, coverslip; m, mountant; s, slide; o, oil layer. (Reproduced with permission from A. L. E. Barron, "Using the Microscope", Chapman and Hall, London.)

TABLE IX

The refractive indices of some mounting media

Medium	n	Medium	n
Distilled water	1·33	Tensol ("Perspex"	
Ethanol	1·36	cement, ICI)*	1·49
Hexane	1·38	Xylene	1·50
n-Butyl alcohol	1·40	Sandalwood oil	1·51
Amyl alcohol	1·41	Cedarwood (immersion)	
Chloroform	1·45	oil	1·51
Kerosene	1·45	Clove oil	1·53
Eucalyptus oil	1·46	Oil of Wintergreen	1·54
Carbon tetrachloride	1·46	Canada balsam*	1·54
Olive oil	1·47	Aniseed oil	1·55
Glycerol	1·47	"Styrax"	1·58
"Euparal"*	1·48	Aniline	1·59
Castor oil	1·48	Polystyrene*	1·59

* Permanent mounting media.

VIII. SETTING-UP FOR DARK-FIELD ILLUMINATION

To obtain really good dark-field illumination absolutely strict measures must be taken to ensure cleanliness of all the optical components and the slide and coverslip must be washed and polished free of grease and dust particles. In addition steps must be taken to ensure that no bubbles, even of microscopic size, are trapped either in the immersion oil layers or in the mountant as stray diffractions have a considerable effect on the quality of the final image. The procedure, when a dry objective is used, is as follows.

1. Align the light source as previously described, ensuring that the light falls centrally on the mirror, if present, and that it travels up the tube so as to illuminate a ground glass screen placed on top of the eye-piece tube, intensely and uniformly when the optics are removed.

2. Install the dark-field immersion condenser so that the centring screws are conveniently placed.

3. Adjust condenser so that the top surface is about a ½-cm below the level of the stage and very carefully place a large drop of immersion oil onto the top surface, observing carefully to ensure that no bubbles are present.

4. Place the object slide in position and "quickly", but carefully and smoothly, rack the condenser upward until the oil drop just contacts the under-surface of the slide and allow to spread over this surface by "cohesion" before proceeding to rack the condenser any higher. (By

"quickly" here it is meant that one should not delay long enough for the oil to run over the edge of the top surface of the condenser lens; nor should the drop be so large that when closely applied to the slide the majority is forced to run down the sloping side of the condenser lens). No bubbles should be present in this immersion layer.

5. Observe specimen plane with low power objective (10×) and eyepiece in place. A central solid circle of light *should* be observed when the field iris diaphragm is reduced, but it is more likely that the condenser will be in the wrong position and a dark central spot and diffuse ring of light will be seen. Adjust the condenser vertically so as to produce a solid circle of light; this may even be observed by carefully examining the preparation from the side. Open and close the field diaphragm to check the position of its image in the specimen plane (Kohler illumination).

7. Open the field diaphragm so that the whole specimen field is only just illuminated when the required dry objective has been selected and make final adjustments to focus.

The centring and adjusting procedures described above assume that there is a sufficiently widespread specimen which will make the field visible by diffracting light from various parts of its area. In bacteriological examinations, however, the objects to be viewed may be extremely small and rather widely spaced, which means that there will be a few "points" of light in an otherwise non-diffracting field and it will not be possible to see an image of the field diaphragm. Low-power (20×) dark-field examination of bacteria is often a futile pursuit in any case and a method for correctly setting up the system for use with an oil immersion objective is necessary. For those who are familiar with bright-field use of oil-immersion objectives I would recommend the following procedure when an objective with built-in iris is used.

1. Perform stages 1–4 already described.

2. Viewing the specimen plane from above and to the side, carefully observe (most conveniently with a hand lens) the illumination of this plane while raising the condenser very slowly. Keep raising the condenser until an intensely illuminated bright spot can be seen—almost a "pin-point" of light. This will be the approximately correct position of the condenser.

3. Place a small drop of oil on the surface of the coverglass and lower the objective (still viewing from the side) until it nearly touches the coverglass.

4. Open the objective iris fully so that light *does* enter the objective. (Field iris fully open.)

5. Replace the eyepiece by an auxiliary telescope (see under phase-contrast) and focus this on the back of the objective so that the ring of light passing the substage annular stop is in focus, and close down the objective iris until only a very fine ring of light is visible. (More conveni-

ently, as mentioned before, insert the telescope in the third arm of a "trinocular head".)

6. Viewing through the eyepieces adjust the fine focus until diffracted light from the specimen is observed in the "half-dark" field which is produced by allowing (in 5 above) for some of the direct illumination to travel up the tube. Focus as nearly as possible on any object items so located.

7. Now find the correct position of the condenser (Kohler illumination) by fully opening the objective iris (to give ordinary bright field) and stopping down the field (lamp) diaphragm until its image is visible in the field. Make this image sharp, as for the usual Kohler method of illumination, by carefully adjusting the position of the condenser up or down as required.

8. Centre the image of this diaphragm by adjusting the annular substage diaphragm of the dark-field condenser by means of the screws provided.

9. Now open the field diaphragm until its edges are only just visible in the field of view.

10. Close down the objective iris diaphragm noting as you do so that a dark spot appears in the centre of the field and spreads outwards to the edge. Close down this diaphragm until the whole field just becomes black.

11. Adjust fine focus on the now brightly illuminated specimen objects.

These centring procedures can be instructively observed through the auxiliary telescope, especially step 8 above. It is also a good idea to re-check the condenser focus after the final focusing of the objective and this is simply done by opening the objective iris, when the image of the field iris immediately becomes visible if set as stated in 9.

From the brief description of dark-field microscopy given above it will be realized that there are a number of special requirements. Since only the diffracted radiation is involved in the image a very intense light source is required; indeed it was the custom to use the sun for this purpose and it remains a good idea to do so in countries where it is dependable. The condenser must be fitted with a "stop" to provide hollow-cone illumination and where highly inclined rays are required for high N.A. work the design of the condenser is different from the usual type and is usually a reflecting condenser. Oil immersion is imperative when objective apertures greater than 1·0 N.A. are to be used. The numerical aperture of the objective relative to that of the condenser must be carefully controlled, and high N.A. objectives must therefore be fitted with either a funnel stop or an iris diaphragm. The great benefits of the dark-field method are the high image contrast and the fact that the light scattered by fine structures just below the resolving power of bright field or phase-contrast systems make these organelles visible, e.g. the flagella of bacteria, spirochaetes, etc. The great drawback is the loss of definition which results from the elimination of the zero-order rays—so that objects tend to be slightly blurred; and no increase in resolution over other forms of microscopy is obtained.

IX. FILTERS AND THEIR USE

The conscientious microscopist cannot avoid the use of filters and comes to appreciate their necessity for a variety of purposes quite apart from their use in fluorescence microscopy. Their variety and purpose is often overlooked by the less enthusiastic.

A. Types of filter

1. *Neutral density filters*

These are used to reduce the light intensity of the source without changing the colour temperature (see III G, 3). If a regulating transformer is not available for controlling the output of the lamp, or if the source cannot be controlled by changes in voltage, e.g. xenon vapour burners, then neutral density filters must be used. Even if the lamp is transformer-controlled reduction of intensity by this means should be restricted to situations where colour is unimportant since the spectral quality of the light emitted will change markedly (towards red) and voltage regulation should not be used to modify the light for colour photomicrography. For this purpose neutral density filters must be used and their transmission curves must be as near horizontal as possible. The best modern types are made by deposition of thin metal films on glass. The combinations of % transmission which are obtainable with the 3%, 12% and 50% transmitting filters supplied by Zeiss, Oberkochen are given in Table X. Note especially the factors by

TABLE X

Combinations of neutral density filters (Zeiss, Oberkochen); three filters of transmission—3%, 12% and 50%

Filter combination	% light transmission	Photographic exposure factor increase
50	50	2
50 + 50	25	4
12	12	8
12 + 50	6	16
3	3	32
3 + 50	1·5	64
3 + 50 + 50	0·75	128
3 + 12	0·375	256
3 + 12 + 50	0·185	512
3 + 12 + 50 + 50	0·095	1024

which exposure must be increased when used for photography. It should again be stressed that it is never beneficial to alter illumination intensities by re-adjusting the position of the condenser from the correct Kohler situation, nor should the diaphragm be varied from its optimum position.

Opal and ground glass screens should not be used for the above purpose. They are frequently used, however, to break up the filament image—especially in critical illumination systems where it lies in the plane of the specimen. In this sense they act as diffuse emitters and must be placed immediately in front of the bulb—not in the substage filter holder for example, where they will frustrate the use of the field iris in focusing the condenser.

2. Polarizing filters

The characteristic feature of these is that because of their lattice-like structure they permit the transmission of rays oscillating in a single plane only. Thus two polarizers placed in the same light path and rotated with respect to each other will alter the transmitted intensity from a maximum to zero for a rotation of 90° relative to their planes of polarization. In the latter position they are said to be "crossed". Such a system placed in the light path between lamp and specimen will give continuous control of the intensity of illumination reaching the specimen without altering the colour balance of the radiation. In this way they serve as neutral density filters. In addition when one of the pair is used to polarize the light falling on the specimen and the second is rotated relative to this at the eyepiece end of the microscope system, the wide range of colours associated with polarization microscopy is obtained. (Polarizers may also be used to great effect in the elimination of glare from epi-illuminated specimens, e.g. in metallurgy.)

3. Light balancing and contrast filters

When stained material is viewed in the microscope the eye easily distinguishes between the different colours over a very wide range of the visible spectrum. However, a photographic reproduction on black-and-white film may not distinguish between different colours, rendering them in the same tones of grey, i.e. with no difference in contrast. The sensitivity of the eye to different wavelengths (colours) varies and is different from the sensitivity of film emulsions, which themselves may vary enormously (Fig. 46). By means of coloured filters the contrast qualities of a specimen image may be varied considerably both visually and for photography. So long as the colour of the filter (i.e. the wave-length range of the light transmitted; the defining colour corresponds to the centre of the transmitted waveband, Fig. 47) falls within the sensitivity range of the emulsion it may be used as a contrast filter.

TABLE XI

The wavebands of the spectral colours and their complementaries

Waveband (nm)	Hue	Complementary	Waveband (nm)
400–440	Violet	Yellow-green	550–575
440–490	Blue	Yellow	575–590
490–510	Blue-green	Orange	590–650
510–550	Green	Red	650–730
550–575	Yellow-green	Violet	400–440
575–590	Yellow	Blue	440–490
590–650	Orange	Blue-green	490–510
650–730	Red	Green	510–550

The principles governing the choice of a contrast filter are: specimen colour corresponding to the preferred transmitted waveband of the filter will be rendered light; specimen colour complementary to the filter colour will be rendered dark. To introduce contrast into a photograph of a specimen in which two adjacent colours are reproduced in equal greyness, the filter selected should be complementary to one of the colours. However, if the filter colour is the complementary of one area and the same as the other the effect will be to make one "black" and the other "white" and the contrast would be over-severe and detail lost. In practice the best results are obtained by choosing a filter colour in the range near the complementary of one of the colours so that contrast is enhanced without the loss of image detail. The transmission wavebands of various coloured filters and their complementary colours are given in Table XI.

There is no easy way of judging in advance the exact qualities that will be obtained from a particular filter but a good idea may be gained by studying the emission spectrum of the lamp, the transmission spectrum of the filter, the nature of the specimen and the spectral sensitivity of the emulsion on which the record is to be made. The sensitivity curves of the eye and of panchromatic and orthochromatic film emulsions, are given in Fig. 46. Obviously film is more sensitive to the blue/violet end of the spectrum than the eye and therefore records a greater "density of colour" in this region for a given intensity. Whether or not the eye or film will respond in a given region, and to what extent, therefore, depends upon the intensity and wavelength of the light reaching these "recorders", and this in turn will depend on "the combined light filtering" effects of any filter used and of the specimen itself. The transmission curves of three "green" filters of widely different wavebands are shown in Fig. 47.

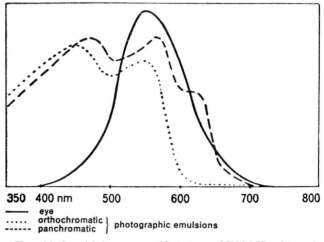

350 400 nm   500   600   700   800

—— eye
····· orthochromatic ⎫
------ panchromatic ⎬ photographic emulsions

FIG. 46. Sensitivity curves. (Courtesy of Wild Heerbrugg.)

Light balancing (compensation or conversion) filters are necessary in order to give the emission spectrum of the source, a wavelength distribution which "corresponds" with the colour sensitivity of the film material. Thus, the distribution of wavelength in the emission from a low-voltage tungsten lamp is weighted toward the red (low colour temperature) and a blue balancing filter is required if it is to be used for colour photography on daylight emulsion. The alternative is to change the sensitivity spectrum of the emulsion—hence tungsten or artificial light emulsions (see Quesnel, this Series, Volume 7).

It is also possible to increase the colour contrast rendering of colour emulsion by the use of certain interference filters ("didymium" filters) which transmit a series of wavelength bands while suppressing the transmission of intermediate wavelengths. In this way the gradual blending of

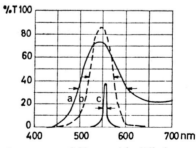

FIG. 47. Transmission curves of filters with differing wave-bands. (Courtesy of Wild Heerbrugg.)

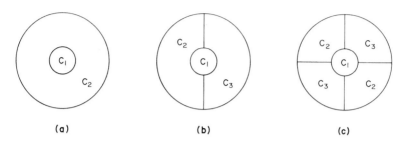

(a)   (b)   (c)

FIG. 48. Rheinberg-type filters.

adjacent colours is prevented and a series of "pure" colours of increased colour contrast results.

Optical staining of colourless specimens may be achieved by the use of Rheinberg filters. In this technique the filter itself is made up of two complementary colours or, in modified Rheinberg, a combination of colours. (Fig. 48).

The central disc (Fig. 48(a)), colour C_1, has a diameter $1 \cdot 1 \times 2f_c(N.A.)_o$, where f_c is the focal length of the condenser and $(N.A.)_o$ is the N.A. of the objective. The outer area, colour C_2 ,gives the colour to the object and the C_1 colour is the colour given to the background. These filters may be home-made by cutting appropriate gelatin filters, or may be obtained from Eastman Kodak (Rochester, N.Y.) as Wratten Rheinberg Differential Colour Filters.

With all coloured filters there is a reduction in the total amount of light, and with some filters this may be considerable. In photography, therefore, compensation must be made by increasing the exposure time. Since in some photographic applications the exposure time required may be many minutes, further allowance may have to be made for reciprocity failure, i.e. as exposure time is increased there is no longer a linear relationship with light intensity decrease for a given "activation" of film emulsion, so that even longer exposure times are required (see Quesnel, this Series, Volume 7).

4. "Correction filters"

The peak sensitivity of the eye is in the region around 5500 Å (Fig. 44), i.e. the yellow-green region of the spectrum. Achromatic objectives are designed to give minimum spherical aberration at this wavelength, and are corrected for chromatic aberration at only two wavelengths so that at focus there is still some residual colour. Panchromatic emulsions are also highly sensitive in this region. From this it is clear that the insertion of the appropriate green filter in the path of the incident light will considerably

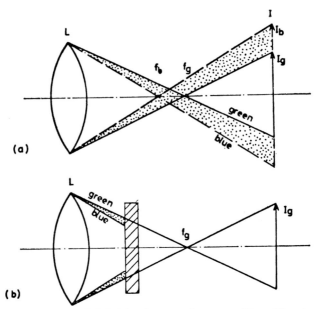

Fig. 49. Improvement of definition by use of a green filter. (Courtesy of Wild Heerbrugg.)

improve the performance of these objectives. Colour fringes will be eliminated, less eye-strain will be involved, and better definition will be achieved in photography as well as contrast.

In phase contrast the $\frac{1}{4}\lambda$ retardation due to the phase ring is calculated for a wavelength of about 5500 Å, so again the green filter will increase contrast because those wavelengths for which contrasts are lower (red and blue ends of spectrum) are eliminated.

The way in which definition is increased by use of a green filter is shown in Fig. 49. An achromat used under these conditions performs nearly as well as the much more expensive fluorite and apochromat objectives.

Improvement in resolution may also be obtained by use of filters since the resolving power is a function of the wavelength of light employed:

$$d = \frac{\lambda}{NA_0 + NA_c}$$

where d is the limit of resolution, λ the wavelength of light, NA_0 the NA of the objective and NA_c the NA of the condenser. Clearly the use of filters transmitting only blue or UV would greatly improve resolution of the fine detail, (but would lead also to great intensity losses, and some increases in spherical error depending on the objective used).

5. Interference filters

It is sometimes desirable to use a very narrowly selected band of wavelengths, i.e. virtually monochromatic light, for such purposes as determination of refractive index. For this a lamp emitting a line spectrum is used in conjunction with an interference filter of the appropriate narrow band transmission. They are also used in UV and infra-red photomicrography. Some interference filters, as a result of multiple reflection and interference of the light, transmit a spectrum consisting of several very sharp peaks with virtually no transmission of the wavelengths in between.

6. Heat filters

With high intensity light sources the intense heat generated may be injurious to the specimen, the optics or the filters, and these should be protected by a suitable heat filter. Several types are possible. Some act by absorbing the long wavelength end of the spectrum (red-infrared) while transmitting the remainder of the visible spectrum with almost no loss (Chance O.N. glasses), but in doing so they become very heated themselves and it may be necessary to install these in a small glass cuvette of boiled water. Likewise, liquid solutions may themselves be used as heat absorbing filters, a common one being 1–3% aqueous copper sulphate but this may be safely replaced by Schott filter BG 12. The Calflex interference heat filters produce by Balzers of Leichtenstein may be particularly recommended. They reflect heat and ultraviolet rays rather than absorb them yet have a very high transmission for all wavelengths between 4000 and 7000 Å.

7. Barrier filters

The heat filters mentioned are really types of barrier filter but this word is more usually understood to refer to the eliminating filters used in fluorescence microscopy. UV barrier filters should also be used to protect the eyes when using a lamp source with high UV output for normal observation, e.g. the HBO 200 mercury vapour lamp. A typical example would be the Schott GG-13 which absorbs all wavelengths below 3500 Å with high transmission throughout the visible. The Kodak Wratten 2B gelatin filter is also much used for the purpose. Barrier filters for eliminating background blue light are also necessary in blue light fluorescence techniques.

It is as well here to mention the exciter filters used in fluorescence work, which are barriers in the reverse sense, i.e. their purpose is to prevent radiation other than the short wavelength fluorescence-exciting rays from reaching the specimen. Examples are the Schott UG 1, UG 2 and BG 3; Corning 5860, 5850; Kodak Wratten 18A, 18B, among others. The different filters are selected for use according to their particular transmission curves and the nature of the fluorochrome used for staining the preparation.

The Schott GG 17 uranium glass filter falls into a special class; and is a fluorescent glass which is used to facilitate the centring of the light source in fluorescence microscopy. (The techniques involved in fluorescence microscopy are described by Walker *et al.*, this Volume, page 219.)

X. MICROMETRY

Two types of measurement can be made with the microscope. Those relating to the microscope itself, such as focal lengths of the lens systems, working distance of the objective, field diameter and depth of field, magnification, numerical aperture and resolving power, as well as the height and diameter of the eyepoint, are not of primary experimental interest to the microbiologist (although they are, of course, related to all other types of measurement) and will not be considered in this Section. On the other hand, measurements of various parameters of objects viewed under the microscope usually are, and this Section will be devoted to the more commonly used methods of measuring objects.

A. Measurements on objects

With the best optical microscopes, limits of resolution below 0·2 microns can be achieved. In general the smaller the particle being measured, the greater is the error likely to be, and even measurements of objects as long as 3 or 4 microns, may involve appreciable errors. Particles of millimetre rather than micron dimensions can also be conveniently measured with the aid of a microscope.

1. *Linear Measurements*

(a) *Macroscopic measurements by stage vernier scales.* The measurement of small macroscopic objects can be carried out with the aid of the stage verniers. The method is simple and requires only a cross-hair type eyepiece graticule in addition to the usual stage movements and verniers. Firstly, focus the eyepiece graticule, then focus on the specimen with a convenient low power lens system. Align the object to be measured so that its long axis (say) is parallel to the X axis movement of the stage, and position it so that the "end" of the object just impinges against the eyepiece cross-hair. Now take the reading on the X axis vernier of the stage. Turn the stage control knob so that the object moves across the cross-hair and stop exactly when the object's following end just impinges on the cross-hair. Take the vernier reading again. Assuming that the true length of the vernier intervals is known (they are usually in 1 mm divisions) then the length of the object is derived by subtraction of the two readings. It should be noted that vernier readings must always be taken for a single approach direction only. In other words, if the first reading is taken after moving the stage from left to right on the X axis, then the second reading must also be taken after movement in this direction. If one "overshoots" the end of the object the readings must be started again, since reversing the direction of travel in order to align object and cross-hair may lead to appreciable error due to the backlash in the rack and pinion or worm and nut mechanism by which the stage is driven. The same precaution applies in the recording of locating co-ordinates of specimens required for future reference. (The precision of stage control is a point worth checking when buying a microscope as great variations are to be found among different instruments).

The usual stage verniers read to 0·1 mm and a second decimal place may be estimated. Micrometer stages are made by some companies and are provided with micrometer screw gauge controls. These enable readings down to 0·01 mm and estimates approaching 0·002 mm (2 μm) to be made. Even the micrometer stage is too gross a device for linear measurements on microscopic objects. Vickers Instruments (York, England) claim their micrometer stage enables precise length measurements to be made to an accuracy of 5 μm.

(b) *Measurement by eyepiece graticule.* All measurements made by eyepiece graticule require that the eyepiece scale be first calibrated against a standard scale of known dimension. The standard scale is usually inscribed or "photographed" on a microscope slide and is called the stage micrometer (not to be confused with the micrometer stages mentioned above).

The procedure in all these cases is, (1) using the standard scale as object calibrate the eyepiece scale; (2) replace the stage micrometer by the specimen and measure the required distances in units of eyepiece scale; (3) convert eyepiece scale units to standard scale units to obtain the real distance.

The standard scale of the stage micrometer varies in pattern of subdivision but it is usually a real length of 1 mm divided into 100 equal parts or it may be divided into 10 equal parts and one of these parts further subdivided into 10 equal parts. Whichever way it is ruled the smallest unit is usually 10 μm. Similarly, the eyepiece graticule scale may vary widely from one manufacturer to the other, but the total length of scale is frequently 10 mm divided into 100 subdivisions each of 100 μm. The real length of the scale or of its subdivisions are of no consequence so long as the units are accurately calibrated.

To calibrate the eyepiece care must be taken to avoid errors due to parallax. For best results, therefore, a micrometer eyepiece should be used as these are provided with movable lens elements to enable the graticule scale to be accurately focused. Micrometer eyepieces are usually of the Kellner type (achromatized Ramsden) in which the single eye lens of the Ramsden is replaced by a doublet giving better correction (Fig. 36(c)). The graticule is placed in the lower focal plane which is below the field lens so that any aberrations present affect both primary image and eyepiece scale to an equal extent. When calibrating this scale against the stage micrometer the graticule must be brought into sharp focus by adjustment of the eyelens before the stage micrometer is focused. Using the objective that will subsequently be used in the measurement of the specimen the stage micrometer is then brought into focus so that the long axes of the two scales are aligned in parallel and the vertical bars of the calibration lie over each other. With both scales precisely in focus the stage is then moved so that one of the graduations of the stage micrometer exactly coincides with a graduation of the eyepiece scale. Examine the scales and find where such an exact coincidence is repeated and determine how many divisions of the eyepiece scale are equivalent to how many on the stage micrometer; then from this derive the real distance equivalent to one eyepiece scale smallest unit. It can be seen in Fig. 50 that these two coincidence points occur, for example, at 2·0 and 7·2 on the eyepiece scale (marked from 1 to 10). These 52 small units correspond with nine small units of the stage micrometer scale. If each small division of the stage micrometer is a real distance

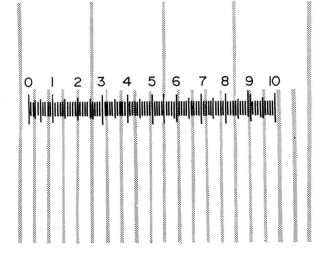

FIG. 50. Stage micrometer scale in the field of view of the measuring eyepiece 10 × .

of 10 microns then the value of a single eyepiece unit (E) is, for those optical conditions

$$E = \frac{9}{52} \times 10 = 1 \cdot 73 \; \mu\mathrm{m} \; \text{(to 2 decimal places)}$$

To perform the measurement on the specimen this is inserted in the place of the stage micrometer and the object brought sharply into focus and underlying the eyepiece scale. The required dimension of the specimen is then read in small units of eyepiece scale and this number multiplied by the calibration value E, for those optical conditions, to give the dimension of the object in microns.

Any particular eyepiece scale must be calibrated for each objective to be used. It is not sufficient to calibrate against a single objective and then use the other inscribed objective powers as a correction factor since true magnifications for a particular objective may differ from the "rounded-off" value inscribed on the objective casing. Also, as mentioned earlier, magnification is affected by tube length and it is sometimes convenient when calibrating the eyepiece, to alter the tube length in order to get exact coincidence between the two scales. The tube length, obviously, must be the same for the measurement as for the calibration. It is also an aid to accuracy if the calibration is made over a wide range of the scale as was done in the example quoted above; it would obviously be much more inaccurate to try to determine directly from the two scales (Fig. 50) the number of eyepiece units equal to one micrometer unit or, say, the number of micrometer units equal to 5 eyepiece units. Again, in general, calibration and measurement should be performed in the same field position if possible, remembering that optical performance is best in the centre of the field and that aberrations toward the edge of the field may be considerable.

(c) *The Filar micrometer.* The Filar or screw-micrometer eyepiece is used in essentially the same way just described for fixed eyepiece micrometers but differs in its much greater accuracy and ease of use. In this instrument the eyepiece graticule or cross-hair is movable across the field of view under the control of a screw micrometer. Different types of ruling are found but basically a single etched line would suffice; in practice a finely etched double line (for precision of eyepiece focusing) is the most convenient, with or without side scales. No graduations are required on the eyepiece graticule since the units to be calibrated in this case are those on the screw micrometer attached to the eyepiece. The micrometer has two scales, a linear one on the spindle, each small unit corresponding to one revolution of the drum; and a "circular" scale around the circumference of the drum, divided into 100 smallest units. Each small unit on the spindle

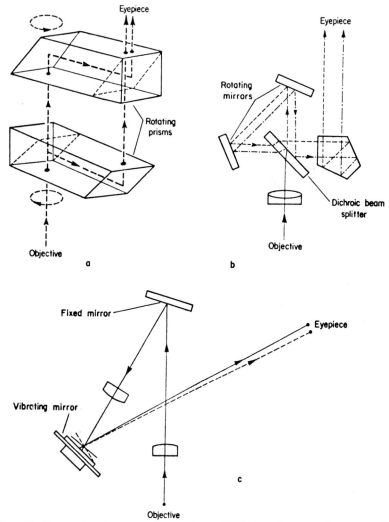

FIG. 51. Image shearing eye-piece systems. (a) Vickers image-splitting eyepiece; (b) Watson image-shearing eyepiece; (c) Fleming particle eye micrometer.

is thereby subdivided into 100 parts. In some of the latest designs the micrometer scale is read internally, i.e. whilst observing through the eyelens.

To calibrate the Filar micrometer focus the graticule and stage micrometer as before, then select a length of stage micrometer scale in the central area of the field, say 10 stage micrometer units in length (i.e. real distance overall = 100 μm). Now screw the micrometer in the required direction so that the cross hair (or any appropriate calibration mark) is exactly coincidental with the stage calibration mark at the beginning of the selected scale length. Record the micrometer reading (preferably in terms of the smallest units), say, for example, 342 units. Now screw the micrometer across the field until it has traversed exactly the 100 μm distance selected. Take the micrometer reading again, say, 688. The distance travelled, in "drum" units, is therefore $688 - 342 = 346$ units. The true distance travelled by the cross hair for a single drum unit of movement is, therefore,

$$\frac{100}{346} = 0 \cdot 29 \ \mu\text{m}$$

Now replace the stage micrometer by the specimen, as before, and traverse the specimen with the cross-hair taking the drum readings at the beginning and end, and take the difference between them to give the total length in drum units. Multiply this by the value for a single drum unit to obtain the real length of specimen measured.

By means of this instrument distances may easily be measured to tenths of a micron and estimated to a few hundredths. The precautions to be taken to ensure maximum accuracy are, as before, to calibrate using the objective to be used for specimen measurement, the avoidance of parallax, use of fixed tube length, readings taken after a single approach direction to avoid backlash, calibration and measurement to be performed in the central area of the field.

(d) *Image-shearing eyepieces.* In these devices the optical design enables two separate images of a single object to be formed in the same field of view plane, the degree of separation or shear being controlled by a calibrated revolving knob. Three different methods of obtaining image-shear are shown in Fig. 51.

The Vickers image-splitting eyepiece may be used with any microscope of standard eyepiece tube diameter and 114 mm unobstructed tube length. The setting accuracy increases with objective magnification from 2·03 μm with 5 × objective to 0·127 μm with 100 × objective.

Light entering the instrument is split into two at the partially reflecting surface of the lower prism block, the undeviated part of the beam travelling to the top prism where it is internally reflected and directed to the eyelens, whilst the reflected portion of the beam is internally reflected in the lower prism and then traverses the upper prism before entering the eyelens (Fig. 51). The two prisms are connected via a linkage system to a micrometer screw and are rotatable about a vertical axis by equal amounts in opposite directions. The two images can be made to cross each other by rotating the screw, the amount of shear being directly proportional to the screw movement. Red and green filters can be introduced in the light

Fig. 52. Stages in image shearing with the Vickers eyepiece. (Courtesy of Vickers Instruments, York, England.)

paths to enable clear differentiation between the two images. Figure 52 shows the appearance of the split image of a thin wire filament.

The graduations on the micrometer drum (or cyclometer scale in the latest model) are calibrated against a stage micrometer in the usual way.

In the Watson image-shearing eyepiece (W. Watson & Sons Ltd., Barnet, England), separation of the images is obtained by rotation of two mirrors about a horizontal axis (Fig. 51). In place of the micrometer drum there is a revolving knob which drives a "cyclometer" counter from which the readings are made. It must be calibrated as before. Each unit on the cyclometer scale is subdivided into ten by a vernier on the control knob.

TABLE XII

Limits of measurements—Watson Image Shearing Eyepiece

Objective magnification	Approx. largest measurable dimension (μm)	Approx. value of one division of drum vernier (μm)
4 ×	560	1·10
10 ×	225	0·45
20 ×	110	0·22
40 ×	56	0·11
100 ×	25	0·05

The range of measurement which is possible and the value of a vernier unit for different objective powers is given in Table XII. Once the eyepiece has been calibrated for a particular set of optical conditions it is a very simple matter to discover from a sample the proportion of objects of a particular size. Simply set the cyclometer and vernier to the required size as found by calibration against a stage micrometer, then bring the speci-

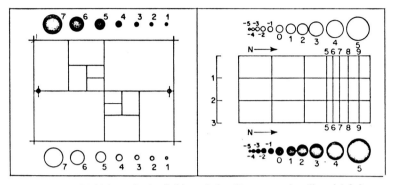

FIG. 53. BS3625 graticule (left) and the Guruswami ruling (right).

men field into focus. Those objects which are of the critical dimension will appear as fully sheared images, one green one red, with the images just not overlapping. All objects below this dimension will show partial shear only and the overlap area of the red and green images will be black, whilst larger objects will each give a red and green image, separated by "free space" i.e. by a "white" gap. Size distributions are easily obtained in this way by setting the upper limit of each size group in succession and recording the number of unsheared or just sheared at each stage until all particles are classified.

In the Fleming Particle Size Micrometer and Analyser Type 526 (Fleming Instruments Ltd., Stevenage, England) image shearing is achieved by the vibration of a mirror attached to a coil which is activated by a variable current of square waveform. When the current amplitude is increased the vibrator rotation increases until a point is found when there are two just-separated images of the object. Although the mirror produces only a single image at any instant the periodicity is such that the eye records two images as a result of the persistence of vision. The degree of shear is proportional to the current and this can be calibrated against a scale of known length as before.

(e) *Eyepiece graticules for sizing by comparison.* When the objects to be measured vary widely in size it is often more convenient to classify the particles into size groups by comparison of the particle with a "globe" or "circle" of known diameter and area. Various designs of etched graticules are available for this purpose which differ in the size ratios chosen and the positioning of the circles and graticule. Two examples are shown in Fig. 53. In the British Standard Specification 3625 the graticule is ruled in a subdivided rectangular format which provides seven different sizes of rectangle, and a sequence of seven globes and circles whose sizes decrease in a $\sqrt[2]{}$

progression. In the design by Guruswamy (Fig. 53) the ratio between the diameters of the circles is 1·2589. The method of marking the circles, and the fact that log 1·2589 = 0·1 facilitate the calculation of results (Guruswamy, 1967).

In use the sample to be sized is brought to focus within the area of the ruled grid. Each object is then compared "by eye" with the circles or globes outside the grid and allocated to a particular size class. Obviously this is a much more indirect method than the previously described measurements by eyepiece scale, filar micrometer or image shearing device, but with experience will give consistent results and is of great benefit when the prime interest is in the distribution of particle size. The greatest errors in allocation of size class are obtained with the smallest sizes of particle so that images should be magnified to as great an extent as is permitted by the character of the sample.

(f) *Measurements from drawn images.* There are now available a number of devices which enable more or less exact drawings of microscopic images to be made. For example, a "micro-projection drawing mirror" may be attached to the eyepiece tube and its position so adjusted that it projects the microscopic image on to a drawing surface. In a darkened room the ordinary microscope lamp may give enough light for the purpose. Microscopes with quartz-iodine illuminators are more effective but the effect of heat on the specimen should be given due consideration. Reflecting mirrors of this type must be metallized to reflect from their front surface only. Alternatively, a drawing prism may be clamped to the eyepiece so that on observation the microscopic image appears superimposed on the drawing surface. Two types of camera lucida are also available. In the case of the "large camera lucida" (Zeiss, Oberkochen) the accessory is fitted over the eyepiece tube and the microscopic image reflected into the Abbe cube housed in the arm of the camera lucida via a prism. Looking through the Abbe cube, the observer sees an image of the specimen over the drawing surface. Two polarizers are incorporated so that the brightness of the microscopic image can be balanced relatively to the brightness of the image of the drawning surface.

The camera lucida designed for use with binocular microscopes is usually fitted into the body of the microscope below the eyepieces. An image of the separately illuminated drawing surface is produced within the microscope by means of the beam combiner and the investigator observes through the binocular eyepieces in the normal way whilst tracing the outlines of the image on the drawing board. Neutral density filters are used for varying the brightness of the specimen image.

To obtain the real dimensions of the specimen from a drawn image it is necessary to calibrate the drawing paper; if squared paper is used, by projecting the image of a stage micrometer under identical optical conditions and making a calculation of the number of paper squares equivalent to a known real length of the micrometer scale. It is also necessary to check that distortion is absent from the drawn image, and this is done by drawing out the micrometer scale for two positions of the scale axis at right angles. The intervals between graduations of the drawn scale (measured by rule) should be equal. If distortion exists the mirror of the drawing aide should be adjusted or the drawing surface appropriately inclined.

(g) *Measurements on projected images.* Most manufacturers now supply viewing screens which replace the conventional eyepiece. This accessory incorporates a projection eyepiece and mirror or prism system which projects the image on to a ground glass screen. Such viewing screens can be equipped with scales etched on plastic and the "screen scale" when calibrated against a stage micrometer will enable direct measurements to be made on the projected image. Similarly, the use of a projection prism and strong light source to cast images on a distant screen, may in some cases be appropriate for measurement purposes. In this latter case enormous final magnifications may be obtained (but with loss of definition).

(h) *Measurements from photomicrographs.* When the microscope is used for the production of drawn, projected or photographed images, it is usual to refer to the size relationship between final image and the specimen as the "image scale" or "picture scale" rather than the magnification. To calculate the image scale in photomicrography we need to know the objective magnification (M_{ob}), the eyepiece magnification (M_{ok}), the camera factor which is the distance in mm between the photo eyepiece and the film plane divided by 250 mm, and the magnification factor of the tube (if any) F. The image scale of the photomicrograph negative is then given by:

$$M_{ob} \times M_{ok} \times \left(\frac{b}{250}\right) \times F$$

where b is the distance in mm from eyepiece to film plane.

If, for example, a $40 \times$ objective was used with a $10 \times$ photoeyepiece, which was inserted in a photo tube with a magnification factor $1 \cdot 5$ (F), and the camera factor was $0 \cdot 5$ (i.e. the distance from eyelens to film $= 125$ mm), then the image scale would be:

$$40 \times 10 \times 0 \cdot 5 \times 1 \cdot 5 = 300 : 1$$

This means that the size of the image on the negative is 300 times that of the specimen. If, in printing, the image is further enlarged, the first

5a–6

FIG. 54. Scale projected into image plane and photographed simultaneously with image. Each division of scale ≡ 2 μm.

value must be multiplied by the enlargement factor as well. It should also be pointed out that the objective magnification may not have the precise value inscribed on the casing and it would be a good idea to photograph a stage micrometer scale in order to check the accuracy of the theoretical formula for any particular combination of elements.

In some of the more elaborate microscopes now available (e.g. the Nikon inverted research microscope) it is possible to project an illuminated image of a scale into the microscope image plane. This then replaces the conventional eyepiece graticule and can be calibrated against a stage micrometer. Taking photographs with the illuminated scale focused in the field will yield a negative with a scale as an integral part of the image. It does not then matter how manyfold the subsequent enlargement of the print, since the known real value of a unit of the photographed scale will still remain the same and can be used directly for determining specimen measurements on the photograph. An example is shown in Fig. 54.

2. Measurement of area

(a) *Estimation using a square-ruled grid.* An eyepiece micrometer rule with a grid of small equal-sided squares can be used for rough measurement of the area of a specimen. The size of a grid square is obtained by calibration against a stage micrometer and with the specimen focused the number of squares enclosed by the outline of the specimen is counted. Partly enclosed squares must be estimated.

Similarly, a grid ruled on transparent plastic sheet can be used in the same way by superimposing it on the projected, drawn or photographed image. So long as the image scale is known the real area can be calculated.

(b) *Areas of definable geometry*. If the specimen area required is of a regular geometrical shape, rectangular, circular, etc., then obviously the areas can be calculated from measurements of linear dimensions using the usual formulae. This has been done for example, assuming staphylococci to be spheres, when only the diameter need be measured; and rod-shaped bacteria to be cylinders or cylinders with hemispherical ends when lengths and widths are required. Obviously in these cases both cross-sectional area and surface area are calculable (as well as volume).

(c) *Measurement by polar planimeter*. The polar planimeter is a mechanical device for measuring the area under a given curve. The movable point is traced over the contour of the curve and this operates a wheel which registers the number of revolutions in a complete circuit of the curve. The area is a function of the revolutions of the wheel and the dimensions of the instrument. In practice it is best to calibrate the planimeter using a piece of graph paper of known area. The planimeter is used on enlarged drawings or photomicrographs and the area in square mm is read from the measuring drum. To obtain the real area of the specimen this value must be divided by the square of the image scale ratio.

(d) *The cut-and-weigh method*. From a sheet of good quality millimeter-ruled graph paper cut a piece exactly (say) 10 cm × 10 cm (area $= 10^4$ mm^2) and weigh precisely. By division derive the weight of 1 mm^2. On the reverse (unruled) side of the same sheet make an enlarged drawing of the object whose area is to be measured, using a camera lucida or other drawing aid Cut out the drawn area and weigh. From the weight derived for 1 mm^2 of paper calculate the area of the paper image. Again, from the value of the image scale derive the real area of the object (see previous paragraph). The same method may be applied to a photographed image but the sample weighed to calibrate the paper must be ruled by hand before cutting.

The image scale of a drawn image may easily be obtained by drawing the image of a stage micrometer scale of known length (and measuring the drawn scale with an accurate millimeter rule) under the same optical conditions subsequently used to draw the specimen image.

It is claimed that if metal foil is used instead of paper considerable accuracy can be achieved.

3. *Measurement of volume or thickness*

Volumes may be calculated from linear dimensions if the object has the shape of a regular geometric solid; it is a very much more difficult matter if the object has an irregular shape. In the latter case volumes are better obtained by means of an electronic particle sizer such as the Coulter counter which, however, will give the distribution of particle volumes and not the volume of any specific individual particle.

(a) *Depth measurement by fine focus adjustment.* In order to estimate volumes the thickness of the specimen may be required. There is no easy or accurate way of doing this but reasonable approximations may be obtained with the fine focus control if precautions are taken. In outline the method is as follows. Using an objective with high numerical aperture and smallest possible depth of field, bring the underside of the object just into focus (approaching from a focus below the specimen). Take the reading on the drum of the fine focus control. Continue moving the fine focus control in the same direction until the top of specimen is just in focus. The drum is read again, and the difference between the readings will give the apparent thickness of the specimen, if the true vertical movement of the specimen for a single revolution of the drum is known. The approximate true thickness is obtained by multiplying this value by the refractive index of the mounting medium.

Unfortunately, the value for a single revolution of the fine focus control is not necessarily constant over the whole range of movement and the fine control must therefore be calibrated in advance (White, 1966). Because of the accommodation of the eye and the depth of field of the objective, determinations of the top and bottom focus positions are very subjective and these measurements are therefore not easily reproducible or very accurate.

If greater accuracy of thickness measurement is required this may be done by phase or interference microscopy, and for details of these methods the reader is recommended to consult the book by Ross (1967).

4. *Measurement of angles*

Angles are conveniently measured with a goniometer eyepiece. This instrument has a circularly ruled 360° graticule in 1° gradations. By means of an external knurled ring a cross-line pointer can be rotated about the central axis enabling the scale to be read. To perform a measurement the apex of the angle to be determined is brought to the optical axis (by manipulation of the stage controls) and the pointer "arm" then aligned first along one side of the angle then along the other. The difference in the scale reading directly read is the value of the angle. The eyepiece scale can be read directly to 1° and estimated to about 20 min of arc.

Methods of size analysis, especially those used for the measurement and classification of irregularly shaped grains, have been recently reviewed in detail by Humphries (1969) and need not again be considered here.

NOTE: The list of References and Bibliography has been deliberately restricted. It should be stated, however, that all the major manufacturers of microscopes and microscope accessories publish comprehensive descriptions of the instruments available, as well as instruction manuals for their use, and numerous other publications concerning the principles and practice of the microscopic arts. These extremely useful documents are readily available free of charge, and I feel sure that I speak for microscopists in general in expressing my thanks to a co-operative and efficient industry.

XI. MICROSCOPICAL JOURNALS

Applied Optics
Bulletin de la Société Française de Microscopie
Bulletin de Microscopie Appliquée
Journal de Microscopie
Journal of Photographic Science
Journal of The Optical Society of America
Journal of The Quekett Microscopical Club
Journal of The Royal Microscopical Society
Microscopie
Mikroskopie
Quarterly Journal of Microscopical Science
Stain Technology
The Microscope and Crystal Front
Transactions of The American Microscopical Society
Zeitschrift für Wissenschaftliche Mikroskopie und für Mikroskopische Technik

XII. BRITISH STANDARDS

B.S. 233 (1953)   Glossary of Terms used in Illumination and Photometry.
B.S. 3406 (1963) Methods for the Determination of Particle Size of Powders. Part 1. Sub-division of gross sample drawn to 0·2 ml. Part 4. Optical microscope method.
B.S. 3836 (1964) Components of Microscopes. Part 1. Microscope cover slips and slides.
B.S. 3836 (1965). Components of Microscopes. Part 2. Dimensions and marking of microscopes.

XIII. SOME MICROSCOPE MANUFACTURERS

American Optical Corp., Eggart and Sugar Rds., Buffalo, New York 14215.
Bausch & Lomb Inc., 635 St. Paul St., Rochester, N.Y. 14602.
R. & J. Beck Ltd., Greycaine Rd., Watford WD2 4PW, England.
Ernst Leitz GmbH, Wetzlar, W. Germany.

Gillett & Sibert Ltd., Battersea, London.
Meopta, Kovo, Praha, Czechoslovakia.
Microscopes Nachet, 17 Rue St-Severin, Paris V.
Nippon Kogaku K. K. (Nikon), Tokyo, Japan.
Olympus Optical Co. Ltd., Kanda-Ogawamachi, Chiyoda-ku, Tokyo, Japan.
W. R. Prior & Co. Ltd., Bishop's Stortford, England.
C. Reichert Optische Werke A.G., 1171 Wein, Austria.
Strübin Optical Corp., 4000 Basel 11, Schweiz.
Swift Instruments Inc., San Jose, California, U.S.A.
Union Optical Co. Ltd., Tokyo, Japan.
Vickers Ltd., Vickers Instruments, Haxby Rd., York YO3 7SD, England.
Wild Heerbrugg Ltd., Heerbrugg, Switzerland.
Carl Zeiss, Oberkochen, Wuertt., W. Germany.
VEB Carl Zeiss, Jena, DDR.

Note added in proof: Since this Chapter was written the firm of W. Watson & Sons has ceased the manufacture of microscopes and accessories.

REFERENCES

Allen, R. D. (1958). Cited by Ross, K. F. A., *In* "Phase Contrast and Interference Microscopy for Cell Biologists" (1967). Edward Arnold, London.
Barer, R., and Joseph, S. (1955). *Q. Jl. microsc. Sci.*, **96**, 1.
British Standard Specification 3625 (1963). "Eyepiece and Screen Graticules for the determination of Particle Size of Powders". B.S.I., London.
Guruswamy, S. (1967). "Particle Size Analysis". Proceedings of a Conference. Soc. for Analytical Chemistry, Loughborough, 1966, pp. 29–31.
Humphries, D. W. (1969), *In* "Advances in Optical and Electron Microscopy" (Ed. Barer, R., and Coslett, V. E.), Vol. 3, pp. 33–98. Academic Press, London.
Ross, K. F. A. (1967). "Phase Contrast and Interference Microscopy for Cell Biologists". Edward Arnold, London.
Slater, P. N. (1957). *J. Quekett microsc. Club*, **4**, 415.
White, G. W. (1966). "Introduction to Microscopy". Butterworth, London.

BIBLIOGRAPHY

Allen, R. M. (1962). "Practical Refractometry by means of the Microscope". Cargille, New York.
Barer, R. (1953). "Lecture Notes on the Use of the Microscope". Blackwell, Oxford.
Barron, A. L. E. (1965). "Using the Microscope". Chapman and Hall, London.
Bennett, A. H., Jupnik, H., Osterberg, H., and Richards, O. W. (1951). "Phase Microscopy". Chapman and Hall, London.
Clark, G. L. (1961). "Encyclopaedia of Microscopy". Chapman and Hall, London.
Coslett, V. E. (1966). "Modern Microscopy". G. Bell and Sons, London.

Deflandre, G. (1947). "Microscopie Pratique". Paul Lechevalier, Paris.

Eastman Kodak Co. (1962). "Photography through the Microscope". Rochester, New York.

Ehringhaus, A. (1953). "Das Mikroskope". Verlag Teubner, Berlin.

Françon, M. (1967). "Progress in Microscopy". Pergamon Press, London.

Françon, M. (1967). "Introduction to Recent Methods of Light Microscopy". G. Braun, Karlsruhe.

Gause, H. (1966). "Enige neue Mikroskopische Untersuchungsverfahren fur angewandte Naturwissenschaften". Kernian Teolisius, Helsinki.

Lamore, Lewis. (1965). "Introduction to Photographic Principles". Dover Publications, New York.

Langeron, M. (1949). "Précis de Microscopie", Tomes I et II. Masson et Cie, Paris.

Lawson, D. F. (1960). "The Technique of Photomicrography". Newnes, London.

Longhurst, R. S. (1967). "Geometrical and Physical Optics", 2nd Edition. Longmans, London.

Marigault, P. "Microscope (Techniques d'emploi)". Ed. de la Tourelle, St-Mande (Seine).

Martin, L. C. (1966). "The Theory of the Microscope". Blackie, London.

Martin, L. C. and Johnson, B. K. (1958). "Practical Microscopy". Blackie, London.

Needham, G. (1958). "The Practical Use of the Microscope including Photomicrography". Blackwell, London.

Payne, B. O. (1957). "Microscope Design and Construction". Cooke, Troughton and Simms, York. (Now, Vickers Instruments Ltd.)

Peris, J. Aguilar. "Teoria y Practica del Microscopia". Editorial Saber, Valencia.

Romeis, B. (1943). "Taschenbuch der Mikroskopischen Technik". Oldenbourg, Munchen.

Ruthmann, A. (1970). "Methods in Cell Research". G. Bell and Sons, London.

Seguy, E. (1951). "Le Microscope, emploi et Application", Tomes I et II. Paul Lechevalier, Paris.

Selwyn, E. W. H. (1959). "Theory of Lenses". Chapman and Hall, London.

CHAPTER II

Staining Bacteria

J. R. NORRIS AND HELEN SWAIN

Milstead Laboratory of Chemical Enzymology, Sittingbourne, Kent, England

I.   Introduction   105
II. Microscope Slides and Cover Glasses and Their Use   .   .   .   106
     A.   Slides and cover glasses   106
     B.   Wet mounts   108
III. Staining Procedures   111
     A.   Preparation of smears   111
     B.   Fixation   112
     C.   Staining solutions   112
     D.   Application of staining reagents   113
     E.   General stains   113
     F.   Cytological stains   120

I. INTRODUCTION

The routine staining of bacteria for microscopic examination is much less important today than it was during the period up to the Second World War. The advent of the phase contrast microscope (see Quesnel, this Volume, page 1) and the increasing use of the electron microscope for routine examination of micro-organisms have rendered many of the staining techniques redundant. It is not unusual today to find a microbiological laboratory functioning entirely without the use of stains, making all routine examinations by phase contrast microscopy. Nevertheless, some stains remain important for diagnostic work and others still find application in specific research projects concerned with cell structure and morphology.

There are many stains described in the literature and many variations of technique. In this Chapter we present a selection of the methods which we find to be of most value today and which are, in our experience, most reliable and reproducible in practice. We include only stains applied to preparations of bacteria themselves; the specialized stains and handling techniques used for micro-organisms in tissue sections are beyond the scope of the present Volume and the reader is referred for these to a standard work such as "Handbook of Histopathological Technique" by C. F. A. Culling, Butterworth, London (1969).

Bacterial stains are of two kinds; those which are no more than histo-chemical reactions, like the Gram stain and the acid-fast stain which are of diagnostic significance, telling the observer how the bacterial cell reacts to certain chemical manipulations but giving no real information concerning the structure of the cell, and those which are designed to throw some light on the nature of cell component structures. With the former type of stain fixation technique is not of great importance; the cells are subjected to vigorous treatment by powerful stains and reagents and fixation implies mainly the firm attachment of the cells to the microscope slide under conditions in which the stains can react. With the "cytological" stains fixation may be a more important part of the procedure since it is necessary to retain the cell structures in as natural a state as possible. With these stains the fixation procedure must be carefully designed. It becomes an intimate part of the staining operation and the details will be described in each case.

For the sake of completeness the preparation of a wet mount of bacterial cells will be described. This is not, of course, a staining procedure but it is used extensively both for the observation of bacterial motility and to prepare micro-organisms for phase contrast observation. It is essentially a simple technique but there are several variations and we considered it worthwhile describing the methods which we find to be most effective.

II. MICROSCOPE SLIDES AND COVER GLASSES AND THEIR USE

A. Slides and cover glasses

Modern microscopes are computed to work with glass slides and cover glasses of carefully specified thicknesses. For microbiological work we recommend 3 in. × 1 in. slides of thickness 0·8–1·0 mm and 18 mm square cover glasses of thickness 0·13–0·16 mm. Several manufacturers produce glassware meeting these specifications but there are still inferior products on the market. The use of sub-standard (usually too thick) glassware can materially affect the quality of the image seen under the higher powered lenses of the microscope (see Quesnel, this Volume, page 1). The makers specifications should always be checked carefully before purchasing slides and cover glasses. Most general laboratory suppliers stock a comprehensive range of glassware for microscopy.

1. Cleaning of glassware

Most bacterial preparations are made by spreading a drop of aqueous solution containing cells over the surface of a slide and either covering it directly with a cover glass or allowing it to dry out at room temperature over a period of a few min. Difficulty is sometimes experienced in persuading the drop to spread evenly over the glass surface and remain as a thin film during drying. The problem is caused by trace amounts of grease on the surface of the slide; it is surprising how quickly a new microscope slide will become unusable on exposure to the normal environment of a laboratory. In dealing with this situation, there is no doubt that prevention is better than cure. No difficulty will usually be experienced if slides are stored in their original box which is kept closed, or in a sealed container, and a fresh slide is used for each microscopic preparation. Open the container, remove a slide taking care to handle it only by the edges, breathe sharply onto one surface and give this surface a quick wipe with a fresh paper tissue, being careful not to touch the surface with the fingers. A surface prepared in this way will almost always be satisfactory for general bacterial smear preparation.

For more rigorous cleaning a slide may be held in forceps and passed slowly several times through a bunsen flame until the flame begins to show slightly yellow around the edges. The slide is then placed on several layers of blotting paper on the bench to cool. If the slide has been sufficiently heated the paper will scorch immediately below the glass. A little experience is necessary to achieve sufficient heat without cracking the slide on cooling. Slides cleaned by this method are very clean indeed and are usually suitable for more critical stains such as the flagella stains (see below). They become contaminated rapidly on exposure to the laboratory atmosphere and it is important to use them as soon as they are sufficiently cool.

If a microscope slide has been previously used and cleaned by immersion in chromic acid or some other cleaning solution, its surface has usually been damaged and cleaning for subsequent use is more difficult. Some laboratories store slides in cleaning solutions or solvents often labelling the containers "grease free slides" with more hope than accuracy. Probably the best of these solutions is 50% alcohol in water and the slides are removed using forceps and dried either by rubbing with a paper tissue or by burning off the alcohol in a bunsen flame. If a method of this kind is employed, it should be remembered that the solvent will rapidly accumulate grease from the slides placed in it and in a short time it will be a serious source of contamination for fresh slides. A covered jar containing 100 ml or 50% alcohol may have capacity for 20 microscope slides but certainly no more. The solution must be replaced completely when these slides have been used.

By comparison it is much less important to clean cover glasses since it is rarely necessary in routine microbiological work to prepare a wet mount from which all air bubbles are rigorously excluded (see, however, the chapter by Quesnel on "Micro-culture techniques", this Series, Volume 1).

The practice of storing cover glasses immersed in 50% alcohol, or some other solvent solution, in shallow covered containers can scarcely ever be justified and is not recommended. The sheer physical difficulty of removing a delicate glass from the container increases the chance of contamination and leads to considerable breakage. Again, the most effective way of working is to use a fresh cover slip for each preparation, removing it from its original box when required and removing surface dust by a quick, gentle rub with a fresh paper tissue.

2. To clean or throw away?

The practice of cleaning and recirculating used slides and cover glasses was common until relatively recently and is still operated by some laboratories. If purely economic criteria are applied, it is probably never possible to justify cleaning microscopic glassware as opposed to throwing it away after one usage. However, the decision is often not one of simple economics. Where particular kinds of slides, such as well slides (see below) or slides specially marked for particular staining procedures, are needed these are more expensive and should certainly be used more than once. A cleaning routine in general use is to remove the cover slip, place both cover slip and slide in a chemical sterilizing fluid (Chloros or 4% formaldehyde solution) and leave for at least 12 h to sterilize. Wash and place in a cleaning agent such as Rapidex (Tool Importers Limited, Pharmaceutical Division, 41/43 Praed Street, London W.2) or hot chromic acid. Wash thoroughly, rinse in distilled water and dry in a hot oven or wipe dry.

Warning: It has been amply demonstrated that some bacteria are not killed by routine staining procedures. This is particularly the case with spore-forming bacteria, such as the anthrax bacillus and acid-fast organisms such as the tubercle bacillus. When organisms of this kind are being handled, it is absolutely essential that any glassware used in their examination should be decontaminated and destroyed immediately after use. This is best done by autoclaving or incineration.

B. Wet mounts

The wet mount has two main functions in microbiology; to determine whether or not a micro-organism is motile and to present unstained material for phase contrast microscopy. The method of preparing the mount is similar for both purposes and, indeed, the two are often combined.

1. Determination of motility

Preparations for motility determination may be made using ordinary plain glass slides and cover slips or by using well slides specifically made for the purpose. These latter have a circular depression in the upper surface. We recommend the use of slides 1·25 mm thick with a 1 mm deep depression, 15 mm in diameter with circular cover glasses 0·8–1·0 mm in thickness and 22 mm in diameter. In practice there is little to choose between the two methods. Well slides are rather expensive and we prefer to use plain slides.

(a) *Use of plain slides.* A drop of bacterial suspension is placed in the centre of a clean slide and a cover glass lowered onto it. It is not necessary to do this carefully since the inclusion of a few air bubbles in the aqueous film will assist both in locating micro-organisms under the microscope and, more significantly, in stimulating and retaining active motility in some aerobic organisms. If too much liquid is present so that the cover glass floats freely above the surface of the slide, a piece of blotting paper may be placed over the cover slip and gently pressed to absorb liquid exuding from the edges of the cover glass. *N.B. Remember that this blotting paper is heavily contaminated and it should be discarded accordingly with appropriate precautions.*

The resulting thin film of bacterial suspension will be suitable for observation but it will tend to dry quickly and unless the preparation is to be examined and discarded immediately the preparation should be sealed. This may be accomplished conveniently by placing a few dabs of vaseline at the corners and along the edges of the cover slip using, for instance, the end of a match or a swab stick and then running a hot wire gently along the edges of the cover slip to melt the vaseline and enable it to seal the preparation. A simpler, quicker and equally effective way is to use nail varnish which can be applied along the edges of the cover slip using the brush provided by the manufacturers. Pigmented nail varnish is more suitable for the purpose than the clear type. The preparations will subsequently be discarded into a decontaminating solution and it is important to remember that the anti-bacterial agent will not penetrate to the cells unless the vaseline or nail varnish seal is broken before disposal. This is effectively done by using a second microscope slide to push the cover glass away from its original position and into the disposal solution, subsequently dropping both slides into the solution.

The preparation is then examined under the phase contrast or ordinary light microscope. Using a phase contrast instrument, the bacteria will be easily located but there may be initial difficulty using simple optics. The easiest way is to reduce the light intensity by closing the sub-stage diaphragm, or racking down the sub-stage condenser, and then locating the edge of an air bubble using a low power objective. Bacterial cells are usually immediately visible on switching to a high power dry ($\times$ 60) objective and it is not usually necessary to go to an oil immersion objective.

Cells often congregate at the edge of an air bubble and may be motile immediately next to an air bubble but non-motile in other parts of the preparation. A little experience is necessary to distinguish motility from random streaming movements and Brownian movement in the preparation.

Irradiation with white light can have detrimental effects on bacterial motility and heat from the light source can also have adverse effects. The use of a green filter and a heat filter is recommended for motility determination. There are also general background considerations to be borne in mind when examining a culture for motility. These include the establishment of a suitable growth environment for the organism. One factor which is sometimes forgotten is the effect of temperature shock on motility. A culture removed from an incubator at 37°C for instance will suffer considerable temperature shock on being brought to room temperature. Motility may cease for a period but might return later. It is important to handle bacteria as gently as possible both before and during observation.

(b) *Use of well slides.* Many bacteriologists prefer to use a "hanging drop" preparation, claiming that there is less of a problem from streaming movements in the liquid and that aeration is more efficient. A clean cover slip is placed on the bench and a small blob of vaseline placed in each corner. A small drop of bacterial suspension is carefully deposited in the centre of the cover slip. It is advisable to hold the cover slip down firmly by use of the end of a match or a swab stick whilst applying the drop since there is a possibility of picking up the cover slip by capillary attraction during the process. A well slide is then inverted and lowered over the cover slip so that the vaseline forms an attachment between the cover slip and the plain area of the slide outside the well. The preparation is then rapidly inverted so that the drop hangs down from the cover slip into the well. It is important that the drop should not be too large or it will tend to run during the inversion process. Drying is much less rapid with a hanging drop preparation than with a plain slide preparation and it is not normally necessary to seal the cover slip to the glass slide.

The preparation may be examined using phase contrast or conventional illumination but the shape of the well interferes with phase optics and phase contrast is usually less effective with a preparation of this kind than with a preparation on a plain slide. Again, it is necessary to reduce the intensity of illumination for normal light microscopy in order to see the bacterial cells clearly. The edge of the drop is located under low power and the cells are visible on switching to a high power dry objective. Cells are most readily examined in the vicinity of the boundary between the edge of the drop and the cover glass.

2. Phase contrast observation

Specimens for phase contrast observation are prepared by using the method described in 1(a) above. Phase optics work best when the film of bacterial suspension is as thin as possible and deteriorate sharply as film thickness increases. Firm pressure applied to a piece of blotting paper placed over the cover slip before sealing will usually succeed in drawing sufficient liquid from the edge of the preparation to provide a film of suitable thinness. This has the added advantage that the cover slip will tend to adhere firmly to the slide and the preparation can be examined under an oil immersion objective without the problem of the cover glass moving in relation to the slide during manipulation of the microscope stage.

If the object is simply to examine and note the form of bacterial cells under phase contrast, a little movement of the cells, either motility or by Brownian movement or streaming in the preparation is not particularly disconcerting but it may be important to hold cells stationary if they are to be examined in detail or photographed. The simplest way of achieving this is to allow a bacterial smear to dry at room temperature onto the surface of the slide and then place a small drop of water or other suitable fluid such as sterile culture medium onto the surface of the dried smear and place the cover slip on top of the liquid drop. Sufficient cells will probably remain adhering to the glass slide to enable them to be examined and photographed as required. A slightly better approach is to prepare the smear on the under-surface of the cover glass and allow it to dry before mounting in a drop of fluid. If the cells are still difficult to immobilize, the application of a thin film of protein, such as a 5% solution of albumin or a drop of serum, to the glass surface before the preparation is made will usually give a satisfactory result.

III. STAINING PROCEDURES

A. Preparation of smears

For most routine staining procedures smear preparation is a simple matter. A small drop of bacterial suspension is deposited onto the centre of a clean (see II A above) slide and spread as thinly as possible by means of a sterile loop. It is, of course, important to avoid too high or too low a cell concentration on the slide. Only experience will enable the dilution of the initial suspension and the area of slide covered to be judged effectively. If it is important to observe the arrangement of cells to one another—as it often is in diagnostic work—the preparation should be made with the utmost care from a culture grown in liquid medium in order to preserve the cells in their natural arrangements.

A smear will dry in air at room temperature in a few minutes. For most stains there is no harm in speeding the drying process by warming the slide over a bunsen flame but excessive heat will cause cell damage and distortion when applied to cells in aqueous suspension.

In some cases (see, for example, flagella staining below) simple smear preparation is inadequate and specialized techniques will be described under the appropriate staining procedures.

B. Fixation

An effective fixation procedure will attach the cells firmly to the surface of the slide and retain the structure of the cells in as near natural a state as possible. Fixation procedures are essentially methods of controlled protein denaturation and a wide range of chemical fixatives is described for histological work with plant and animal tissue. The reader is referred to Hopwood (1969) for details of these and for a discussion of the nature of fixation. Chemical fixation is of relatively little importance in bacteriology since most bacteriological stains are crude histochemical reactions and simple fixation by heat is adequate to attach cells to the microscope slide under conditions in which the stain can react. Where more elaborate fixation is required, the details are given under the particular stain below. The reader is also referred to articles in Volumes 3B and 4 of this Series for descriptions of stains applied to specific groups of micro-organisms, other than bacteria, where fixation procedures are detailed.

Heat Fixation. An air-dried smear is heat fixed by passing the slide, smear side upwards, through a bunsen flame. The usual method is to hold the slide by the edges at one end and lower it about one-third of the way into the flame, withdrawing it from the side in one smooth movement and repeating three times, each pass taking about 1 sec. Overheating can lead to charring of the cells and a useful rule of thumb method of determining the correct exposure to the flame is to place the back of the slide against the fleshy part of the back of the hand near the base of the thumb. If the slide has been adequately heated it can just be held in this position without discomfort.

Heating this way will normally fix cells firmly to the slide but problems are sometimes encountered when the original suspension is particularly alkaline (for example, some urine specimens). Here it may be necessary to adjust the pH by adding a little dilute hydrochloric acid before preparing the smear.

C. Staining solutions

Stains are generally expensive and their properties can change dramatically if they are stored in unfavourable conditions. Storage cupboards should be dry, cool and away from direct sunlight. Stains in solution will often tend to develop moulds; this can be prevented by adding a little chloroform, sufficient to leave a few globules in the bottom of the bottle. Some stains tend to oxidize (or ripen) on storage. This is not always a bad thing (cf. the Romanowsky stains and Methylene Blue below). Particular care should be taken to store stains well away from ammonia or acids.

Distilled water is normally used to make up staining solutions and in some staining procedures. It is a variable reagent tending to absorb carbon dioxide on standing and stale distilled water can easily spoil stains. The pH can be checked by adding a little Bromo Thymol Blue and very dilute alkali (1% sodium carbonate) until the colour is greenish-yellow. Unless specified the stains referred to below are made up in good quantity *distilled water* and water used during staining is *tap water*.

Osmic acid requires special attention on storage. Solutions of osmic acid are particularly prone to spoilage by contamination. Traces of dirt or dust on the lip or stopper of a bottle will ultimately ruin the solution. Solutions should be stored in scrupulously clean containers the tops of which are protected by covering with, for example, inverted beakers. Better is to buy the reagent in small sealed ampoules in which it will keep indefinitely.

D. Application of staining reagents

Staining reagents are conveniently stored in amounts of 25–30 ml in dropping bottles. Completely closed bottles with a teat-controlled dropper passing through a ground glass stopper are preferred to the type which has a rotating glass top. The latter can become rapidly encrusted with dry stain deposit.

E. General stains

The stains described in this Section are mainly of diagnostic significance. They give simple information concerning the size and shape of bacterial cells and, in some cases, additional information about the gross histochemical reactions of the cells. The literature contains many variations of these widely used staining procedures and most workers have their own preferences. There are particularly large variations in the times quoted for exposure to the various reagents but these are seldom based on any proper study of the interaction between cells and reagents and we make no attempt to list all the variants of each of the methods. Rather we describe procedures which have proved most satisfactory in our own hands and which we have taught successfully to other people for many years.

Slides to be stained are usually placed on staining racks consisting of two glass (or better metal) rods 2 in. apart over a small sink which should be used only for staining. Stains are then applied directly to the smear and removed by washing using a wash bottle or tap water applied via a rubber tube and glass nozzle. If heat is required a bunsen flame may be applied gently, and briefly, from below or a methylated spirit flame from a cotton wool swab on a metal holder may be used.

Where immersion in a reagent is required, or where prolonged exposure is necessary, conventional staining jars should be used.

Many staining solutions produce deposits on standing and it is important to inspect dropping bottles and stock solutions regularly. Stain deposit will often interfere seriously with a preparation and suspect solutions should be filtered before use.

At the end of the staining procedure, it is often sufficient simply to press the slide, smear down, onto a sheet of coarse blotting paper in order to remove excess liquid. Immersion oil is applied directly to the surface of the smear for examination (cf. Quesnel, this Volume, page 1).

It is rarely necessary to preserve microbiological slides but this can readily be done if immersion oil is first removed with xylene and mounting medium is applied direct to the dry smear. Most commercially available mountants are suitable and the maker's instructions should be followed. Most preparations will survive well if the dry stained smear (from which all traces of immersion oil have been removed) is stored in the dark. Many microbiological stains tend to fade rather quickly and a good photomicrograph is often the best way of recording a particular preparation.

1. Loeffler's Methylene Blue

This is a simple stain which is mainly used to demonstrate the presence of bacteria and to determine their gross morphology. It is particularly useful where animal or plant cells are present since it stains these well, often showing considerable detail and enabling the types of cells present to be identified—which is often useful in diagnostic work. For this reason Loeffler's Methylene Blue is sometimes used as a counterstain to differentiate micro-organisms from background material. It is also useful for observations on protozoa.

The stain has a special affinity for metachromatic granules which stain dark blue or violet in contrast to the lighter blue of the cell.

Reagents

Methylene Blue	0·5 g
1% potassium hydroxide solution	1 ml
Ethanol	30 ml
Water	100 ml

Warm the water to about 50°C, stir in the Methylene Blue and add the other ingredients. Filter. The staining solution improves with keep. Store at room temperature.

Method

1. Prepare and heat fix smear in the usual way
2. Stain for 5–60 sec
3. Rinse with water
4. Blot dry and examine

Bacteria and cells are dark blue.

2. *Dilute carbol fuchsin*

This is a simple stain which is rather popular for simply demonstrating the presence and morphology of micro-organisms. It is useful as a counter-stain—particularly in the Gram stain but will show little detail of cellular background material.

Reagents

Ziehl–Neelsen's Carbol Fuchsin
(see 4 below) is diluted 1 in 10 with water.

Method

1. Prepare and heat fix the slide in the usual way
2. Stain for 5–10 sec
3. Wash with water
4. Blot dry and examine

Bacteria are uniformly pale red against a similar background.

3. *Gram stain*

Although the Gram reaction is a character of fundamental importance in bacterial classification, and the stain has been studied extensively, the basic mechanism is still only incompletely understood. The basic dye, Crystal Violet, penetrates the bacterial cell and reacts with acidic components of the protoplasm. When iodine is added a complex is formed. This complex remains in Gram-positive bacteria when a decolourizer is applied, the cell wall apparently interfering with extraction of the dye. Gram-negative bacteria are readily decolourized—probably because of their higher cell wall lipid content which is soluble in the decolourizer—and these are counterstained to give a contrasting red colour.

There are many variations of the Gram stain and experience with a particular method is more important than the choice of minor variants of technique. The decolourizing step is particularly important for good differentiation and the nature and time of exposure of the decolourizing solution must be varied when different materials are examined. We recommend the use of acetone/ethanol mixtures for decolourization. For staining smears prepared from cultures a 50/50 mixture is advocated. Increasing the acetone content gives a more powerful decolourizer and 100% acetone is used for Gram staining such material as sputum or pus.

Smears must be prepared in such a way that cells lie separately since overcrowding prevents proper decolourization and makes it difficult to observe cell shapes. Decolourization time depends on several factors. We recommend that the slide should not be allowed to dry at all during the staining procedure and that the decolourizer is allowed to drip onto one end of the inclined slide so that fresh drops flow evenly over the smear. With a little experience, the extent of decolourization can be judged from the amount of violet stain washing out of the smear. Observation is facilitated by holding the slide over a white background.

Important. When determining the Gram reaction of an unknown organism it is advisable to smear it next to a control smear consisting of a mixture of a Gram-positive (e.g. *Staphylococcus aureus*) and a Gram-negative (e.g. *Escherichia coli*) organism.

Reagents

A.   1% aqueous Crystal Violet
B.   Iodine solution: Iodine   1 g
                         Potassium iodide   2 g
                         Distilled water   100 ml
    (Dissolve in about 25 ml of water and add remainder when solution is complete.)
C.   Decolourizer: Acetone   50%
                         Ethanol   50%
(or other combination as indicated above.)
D.   Counterstain: Aqueous Safranin O   2%
                         or Dilute Carbol Fuchsin (see 2 above)

Method

1. Prepare a thin smear and heat fix in the usual way
2. Stain with Crystal Violet for 1 min
3. Wash stain off with iodine solution and stain with fresh solution for 1 min
4. Wash in tap water.
5. Decolourize judging time from appearance of washings (see above). About 5 sec for a normal smear

6. Wash briefly in water.
7. Counterstain with Safranin or Dilute Carbol Fuchsin for 10 sec
8. Blot dry and examine.

Gram-positive cells are purple, Gram-negative cells red. In interpreting the results of a Gram stain you should remember that some organisms which are Gram-positive in young culture readily become Gram-negative as the culture ages. The reverse change also occurs. Also that some species contain prominent granules in the cells which resist decolourization for relatively long times.

Some workers advocate drying the slide before decolourizing but in our experience this very much prolongs decolourizing time to the detriment of differentiation. Dye bound to protoplasm is not entirely responsible for Gram staining and it is important not to remove unbound dye from the cells before iodine treatment. For this reason we do not advocate washing with water immediately after the Crystal Violet stage.

Some workers omit the wash before decolourization. This appears to have no detrimental effect and we wash simply because the removal of excess iodine solution facilitates observation and timing of the decolourization step.

Rapid Gram method

It should be recognized that the times quoted above have little basis in critical examination of the Gram staining process and many workers involved in routine diagnostic work evolve rapid procedures which are highly effective and time saving. The reader is encouraged to experiment for himself. A common procedure consists of staining the warm slide immediately after fixation by allowing Crystal Violet to run over the smear, washing it off with iodine solution, decolourizing immediately, washing briefly in water and then counterstaining for 5 sec, washing and drying. With a little practice it is certainly possible to produce a good Gram stain in 15–20 sec by such methods.

4. Ziehl–Neelsen stain

Among the acid-fast bacteria found mainly in the order *Actinomycetales* the tubercle bacillus is particularly important for the diagnostician. This organism resists decolourization by both acid and alcohol and the Ziehl–Neelsen strain is primarily used for the identification of this important pathogen. By omitting the alcohol decolourization step acid-fast organisms can be detected.

The stain is often carried out as a diagnostic procedure on specimens of sputum or pus from patients suspected of tubercular infection. These materials vary widely in their consistency and acid decolourization must be judged by experience from the gross appearance of the smear. Faults usually lie on the side of under decolourization leaving red stain deposit in the background and rendering detection of the slender tubercle bacilli difficult.

Reagents

A.   Basic Fuchsin   1 g
     Ethanol   10 ml
     1 : 20 phenol solution   100 ml

Dissolve the dye in the alcohol and add the phenol solution to it. Or:

B.   Basic Fuchsin   5 g
     Phenol   25 g
     95% Ethanol   50 ml
     Distilled water   500 ml

The Basic Fuchsin is dissolved by shaking with phenol in a boiling water bath for ca. 5 min. The distilled water is then added and the stain filtered.

C.   Sulphuric acid   20%
D.   Ethanol   95%
E.   Counterstain either Loeffler's Methylene Blue (see 1 above) or 1% Aqueous Malachite Green.

Procedure

1. Make smear as usual and fix with gentle heat
2. Flood slide with stain A or B
3. Heat periodically for 5 min, but do not boil, ensuring that the stain does not dry out
4. Wash with water
5. Immerse the slide in 20% sulphuric acid for 1 min
6. Wash with water
7. Repeat 5 and 6 until preparation is pale pink
8. Wash with water
9. Immerse the slide in 95% ethanol for 2 min
10. Wash with water
11. Counterstain with Loeffler's Methylene Blue or 1% Aqueous Malachite Green for 15–20 sec
12. Wash, blot and dry.

N.B. If Malachite Green is used—which is preferable—a deep blue-green filter may be used in the light source to render the counterstain invisible and the acid-fast bacilli black.

Acid-fast organisms stain red, other organisms and background material blue or green according to the counterstain used.

Tubercle bacilli occur as characteristic clumps of slender granular rods.

Fluorescent stain for acid-fast bacteria

Tubercle bacilli are often present in very small numbers in specimens from patients. Searching smears under the oil immersion objective can be time-consuming but it is not easy to identify the bacilli under low powers of magnification. Fluorescent stains are used routinely in a modification of Ziehl–Neelsen stain and smears examined under low power ($\frac{2}{3}$ or $\frac{1}{4}$ objective) using a fluorescence microscope. The identity of fluorescent objects must be checked using a high power oil immersion lens.

Caution: The reader is referred to the Chapter by Walker and Batty (this Volume, p. 219) for a description of this instrument and attention is particularly drawn to the safety precautions mentioned there.

Reagents

A.	Auramine-phenol	
	Aqueous phenol (3%)	100 ml
	Warm to about 30°C and added	
	Auramine	0·3 g
	Shake well and filter	
B.	Acid—alcohol decolourizer	
	Pure hydrochloric acid	0·5 ml
	Sodium chloride	0·5 g
	Methanol	75 ml
	Distilled water	25 ml
C.	Aqueous potassium permanganate solution	0·1%

Method

1. Fix slide by heat
2. Stain with Auramine-phenol for 10 min
3. Wash with water
4. Flood with potassium permanganate solution for 20 sec
5. Wash with water
6. Blot, allow to dry and examine under ultraviolet illumination.

5. Breed's stain for bacteria in milk smears

The detection of bacteria in milk is difficult because of the fat and protein present. Several stains enable bacteria to be detected in milk smears. Breed's stain is one of the most reliable and is used as a method of direct counting of bacteria in milk.

Reagents

A.   Breed's Methylene Blue

Methylene Blue	0·3 g
Ethanol	30 ml
Dissolve and add Phenol (2% in water)	100 ml

B.   Xylene

C.   Ethanol   90%

Method

1. Place a clean slide over a 1 cm square drawn on a cardboard or paper
2. With a pipette place 0·01 ml of milk on the slide and spread with a needle to cover the square
3. Dry with gentle heat
4. Immerse in xylene for a few minutes to remove fat
5. Fix in alcohol for 3 min
6. Flood with Breed's Methylene Blue for 2 min
7. Wash rapidly with 90% alcohol until the smear is only faintly coloured
8. Dry in air and examine

Bacteria are dark blue against a light blue background. They may be counted using a micrometer eyepiece.

F. Cytological stains

The stains described in this section are more complex than the general stains described in E above, in that they are designed to detect specific cell components. It should be recognized that some of them are still little more than crude histochemical tests and the grouping into "general" and "cytological" is merely one of convenience.

1. *Albert's stain*

Several stains have been developed for staining the "metachromatic granules" of *Corynebacterium* spp. In our experience, the following modification is most satisfactory.

Reagents

Solution 1

Toluidine Blue	0·15 g
Malachite Green	0·2 g
Glacial acetic acid	1 ml
Ethanol (95%)	2 ml
Distilled water	100 ml

Solution 2

Iodine	2 g
Potassium iodide	3 g
Distilled water	300 ml

Add the iodine and the iodide to about 25 ml water. When solution is complete, add the remainder of the water.

Method
1. Prepare the smear as usual and fix with gentle heat
2. Stain for 5 min in Albert's solution 1
3. Drain without washing
4. Treat for 1 min with solution 2
5. Wash briefly in tap water
6. Blot dry and examine.

Metachromatic granules stain black in contrast to the light green of the cell body. The bars, or striations, seen in some species are stained dark green to black. The main advantage of this modification of Albert's Stain is that both granules and cells are more intensely stained without lessening the contrast between them.

2. *Capsule stains*

There are many capsule stains published in the literature. Some are positive stains in which the material of the capsule is stained, others are negative stains in which the background is stained and the capsule appears as an unstained halo surrounding the cell. The methods are not easy, neither are they uniformly successful for different organisms—which is hardly surprising in view of the wide range of different capsules and slime layers seen in bacteria.

One of the problems is the staining of the cell itself in such a way as to distinguish it from the capsule without staining the capsule and obscuring the picture. With the general availability of the phase constrast microscope this problem has now disappeared and we strongly favour the negative stains based on India ink. Best of all is a wet India ink preparation examined under phase optics. With a little practice this stain can be made to work effectively on any bacterium with a capsule or slime layer. A 1% aqueous solution of Nigrosin can be substituted for India ink in the following instructions, sometimes with advantage since it tends to give a smoother background.

(a) *Wet India ink preparation*
Reagent
India ink which may be diluted with water if necessary but is better used undiluted.

Method

1. Thoroughly mix a large loopful of India ink and a small quantity of bacterial culture from growth on solid medium.
2. Place a clean coverslip on the ink drop, cover with filter paper and press down FIRMLY. This should give a film some areas of which are sufficiently thin for examination. This thin film will dry out rapidly and the edges should be sealed (see IIB above). REMEMBER. The filter paper will be heavily contaminated with bacteria and should be disposed of accordingly.
3. Observe under phase contrast.

The capsules appear as bright halos demarcating the phase dark cells from a grey background. Often these preparations show remarkable detail in the structure of the larger capsules and slime layers seen with some organisms such as *Azotobacter* for example. It may be necessary to search for an area of satisfactory density.

(b) *Dry India ink film*

Reagents

A.   India ink
B.   6% Aqueous Glucose solution
C.   1% Aqueous Methyl Violet
D.   Methanol

Method

1. Emulsify a small amount of bacterial culture in a loopful of the glucose solution at one end of a well cleaned slide.
2. Add a loopful of India ink and mix
3. With the edge of a second slide spread the mixture into a thin film
4. Allow to air dry
5. Fix film by flooding with methanol. Drain off excess at once
6. Dry by warming over a bunsen flame
7. Stain with Methyl Violet for 1 to 2 min
8. Wash gently with water
9. Dry by warming over a bunsen flame
10. Examine directly under oil immersion using ordinary optics.

Capsules appear as clear, colourless zones, separating the purple cells from a grey background. Not all areas of the preparation will be satisfactory and it will be necessary to search for an area where the background is uniform and of suitable density.

(c) *Hiss's capsule stain*

This is, in our experience, the most uniformly satisfactory of the positive capsule stains. It may be necessary to experiment with the staining conditions—particularly with the differentiation step using copper sulphate solution—in order to achieve the best results.

Reagents

A.   Aqueous solution of Basic Fuchsin 0·15–0·3%
       or
       Aqueous solution of Crystal Violet 0·05–0·1%
B.   Aqueous solution of copper sulphate 20%

Method

1. Grow the bacteria in the presence of serium or mix with a drop of serum to make a thin smear. Air dry and heat fix in the usual way
2. Flood the slide with either stain
3. Heat until steam rises for a few seconds
4. Wash briefly with copper sulphate solution
5. Blot dry and examine.

The pale blue capsules are differentiated from the dark purple cells.

3. *Spore stain*

Bacterial endospores are undoubtedly best observed by means of phase contrast optics when they appear as brilliant refractile objects against phase dark cells. Several spore stains are described in the literature but are rarely used today. They vary in their effectiveness and it is usually necessary to experiment with a particular bacterium until satisfactory staining conditions are achieved. The method described below is the most satisfactory in our experience.

Schaeffer and Fulton spore stain

Reagents

A.	Aqueous solution of Malachite Green	5%
B.	Aqueous Safranin	0·5%
	or	
	Aqueous Basic Fuchsin	0·05%

Method

1. Prepare smear and heat fix in the usual way
2. Flood the slide with Malachite Green and heat for 1 min. Ensure that the stain does not boil or dry
3. Wash thoroughly with cold water

4. Flood the slide with Safranin or Basic Fuchsin
5. Wash and blot dry

Spores are green against the red of the cells.

Problems in interpretation may arise if the microscope optics are not very good, chromatic aberration (see Quesnel, this Volume, p. 1) may result in the spores and cells coming into focus at different settings of the instrument. When the spores are sharply focussed the cells may be virtually invisible and vice versa.

4. *Stain for intracellular lipid*

Lipid inclusion granules are readily seen by phase contrast microscopy but they cannot easily be differentiated from other types of inclusion material. The following stain is easy and reliable.

Reagents

A.   Sudan Black Stain (Alcoholic)
      Sudan Black B   0·3 g
      70% ethanol   100 ml
      Mix by shaking thoroughly at intervals and stand over-
      night before use. Store in a stoppered bottle at room
      temperature. It may be necessary to filter the stain before
      use.
B.   Xylene
C.   9·5% Aqueous Safranin solution.

Method
1. Prepare smear and heat fix in the usual way
2. Flood slide with Sudan Black Stain and replenish as it dries out for 15 min
3. Drain off excess stain and blot dry
4. Rinse with xylene and blot dry
5. Counterstain with Safranin for 10 sec
6. Wash with water, blot dry and examine.

Lipid inclusions are blue, black or bluish-grey in contrast to the pale red cells.

5. *Stain for polysaccharide inclusions* (Hotchkiss, 1948)

(a) *Schiff's stain.* Cell polysaccharides of bacteria appear red with Schiff's fuchsin sulphite after oxidation with periodate to form polyaldehydes.

Reagents

A.   Periodate solution
        4% aqueous periodic acid solution   20 ml
        0·2 M aqueous sodium acetate solution   10 ml
        Ethanol   70 ml
        The solution should be protected from light.

B.   Reducing rinse
        Add 300 ml ethanol and then 5 ml 2 N hydrochloric
        acid to a solution of 10 g potassium iodide and 10 g
        sodium thiosulphate pentahydrate in 200 ml distilled
        water. Stir. The sulphur which precipitates slowly may
        be allowed to settle out.

C.   Fuchsin-sulphite solution
        Dissolve 2 g Basic Fuchsin in 400 ml boiling water,
        cool to 50°C and filter. To the filtrate add 10 ml of 2 N
        hydrochloric acid and 4 g of potassium metabisulphite.
        Stopper and leave in a cool dark place overnight. Add
        approximately 10 ml or more 2 N hydrochloric acid until
        the mixture when dried on glass does not turn pink. The
        stain remains effective if stored in the dark for several
        weeks.

D.   Sulphite Wash
        Add 2 g potassium metabisulphite and 5 ml concen-
        trated hydrochloric acid to 500 ml distilled water.

E.   Malachite Green.
        0·002% aqueous solution.

Method

1. Prepare smear and heat fix in the usual way
2. Flood the slide with periodate solution for 5 min
3. Wash with 70% ethanol
4. Flood the slide with reducing rinse for 5 min
5. Wash with 70% ethanol
6. Stain with Fuchsin-sulphite for 15–45 min
7. Wash several times with sulphite wash solution
8. Wash with water
9. Counterstain with Malachite Green for a few sec
10. Wash immediately. Blot dry and examine.

Polysaccharides stain red in contrast to the green cells.

(b) *Alcian Blue stain.* This is a simple stain for polysaccharides which is
usually satisfactory. Avoidance of over-staining with the counterstain is
critically important.

Reagents

A.　　Alcian Blue　　　　　　　　　　　　　　　　1 gm
　　　95% Ethanol　　　　　　　　　　　　　　　100 ml
　　Dilute with water 1 : 9 for use.
B.　　Carbol Fuchsin (see E.2 above)
　　　Ziehl-Neelsen's carbol fuchsin
　　　dilute 1 : 9 with distilled water.

Method

1. Prepare smear and heat fix in the usual way
2. Stain with Alcian Blue for 1 min
3. Wash with water and air dry
4. Counterstain with Carbol Fuchsin and wash off immediately with water
5. Air dry and examine

Capsules and other bacterial polysaccharides are blue, cellular material is red.

6. *Stain for* Azotobacter *cysts*

This stain is superior to stains formerly used because it shows stages in encystment as well as the mature cyst. Results correlate well with both phase contrast and electron microscopy.

Reagents

　　　　Glacial Acetic acid　　　　　　　　　　　8·5 ml
　　　　Sodium sulphate (anhydrous)　　　　　　3·25 g
　　　　Neutral Red (Difco)　　　　　　　　　　200 mg
　　　　Light Green S.F. Yellowish　　　　　　　200 mg
　　　　Ethanol　　　　　　　　　　　　　　　　50 ml
　　　　Water　　　　　　　　　　　　　　　　100 ml

Add the ingredients to the water with continuous stirring. After 15 min remove the amorphous precipitate by filtering through a 0·5 μ filter membrane.

Method

Bacteria from solid or liquid media are suspended in the stain for wet preparations which must be carefully sealed. (See II.B above). The suspensions may be preserved in this condition for several months without adverse affect.

For aged cysts, the procedure is the same except that the preparations are gently heated by passing through a bunsen flame.

Vegetative cells are a light yellowish-green. Early stages of encystment show a darker green cytoplasm which has receded from the outer cell covering from which it is separated by a brownish-red layer. An early cyst has a similar appearance but a clear unstained area, the intine, is visible surrounding the cytoplasm. In a mature cyst the central body stains dark green and is separated by the unstained intine from the outer, brownish-red, exine.

Although cysts are easily seen using phase contrast optics the earlier stages are often difficult to determine. This stain has the advantage that there is minimal cell distortion and the early stages are clearly differentiated.

7. Cell wall stain

The bacterial cell wall does not stain with most of the staining procedures used to stain cytoplasm, needing the application of a mordant before it will take up stain. There are several variations of the cell wall stain based on the use of tannic acid as a mordant. The following method is rather time-consuming but can be relied on to give good results.

Reagents

A.	Fixative-Bouin's fluid	
	Saturated aqueous solution of picric acid	75 vols
	Formalin	25 vols
	Glacial acetic acid	5 vols
B.	Mordant	
	5–10% aqueous solution of tannic acid	
C.	Stain	
	0·02% aqueous Crystal Violet	

Method

1. Remove an agar slab about 2 mm thick on which the organism is growing
2. Place it face downwards onto a coverslip
3. Immerse in fixative for 1–12 h depending on the thickness of the slab
4. Carefully remove the agar by lifting and wash the coverslip gently for 2 h with water by allowing water to drip into and overflow from a Petri dish containing the coverslip
5. Treat with mordant for 20–30 min
6. Wash in water
7. Stain with Crystal Violet for 1 min
8. Mount in water and seal with nail varnish.

The bacterial cytoplasm is pale purple and the cell wall much darker often appearing brownish-black. Cross walls are readily visible.

Adequate results can often be achieved by a simpler routine. A normal heat-fixed smear is flooded with mordant and heated for 5 min followed by washing and staining with Crystal Violet for 1 min. The slide may be examined directly after washing and drying. The results are not as elegant as those produced by the staining procedure described above but may serve for rapid routine purposes.

8. *Flagella stains*

Bacterial flagella are too slender to be visible under the light microscope. In order to make them visible their thickness must be increased by stain deposition and this is the basis of the various flagella stains described in the literature. The stains are notoriously difficult to perform because of the tendency for stain to deposit on background material and, indeed, onto the surface of the slide. Really clean slides and optimum timing of the staining process are critical.

Slides should be immersed in absolute alcohol, cleaned in chromic acid and then washed with distilled water. Immediately before use they should be passed slowly through a bunsen flame, with the surface on which the smear is to be prepared facing upwards, until the flame shows yellow round the edges. The slides are then allowed to cool on several thicknesses of blotting paper.

Of the many stains described we prefer those based on Fuchsin to those using silver deposition on the grounds that they show less tendency to deposit and that the staining solutions are easier to prepare and store. Most popular in recent years is Leifson's stain, the wide application of which has done much to extend the use of flagellation as a taxonomic character. Whatever method is used, practice is essential and results will improve markedly with experience.

(a) *Leifson's Flagella stain*

Reagents

A.   Tannic acid   10 g
         Sodium chloride   5 g
         Basic Fuchsin   4 g
         To 1·9 g of this mixture add 33 ml of 95% ethanol. When it has dissolved, add distilled water to 100 ml and adjust the pH to 5·0
         or

B.   Tannic acid   3·0% in water +0·2% phenol
         Sodium chloride   1·5% solution in water
         Basic Fuchsin   1·2% in 95% ethanol (pH 5·0)
         Mix solutions in equal volumes for use.

Different workers favour A or B. Results for both are good in experienced hands. The solutions in B keep well and indeed the mixture will keep for some days in the refrigerator. It is perhaps the more convenient method where flagella staining is only used infrequently.

Method

1. Using a sharp wax pencil mark out 4 rectangular areas approximately 1·3 cm by 2·0 cm on the surface of a well cleaned slide. The inside areas of these "moats" should be equal.
2. Prepare a suspension of cells with care to avoid loss of flagella. The preferred method is to add distilled water gently to the surface of on agar slope culture and allow it to stand at room temperature until the liquid has become cloudy.
3. Place a small loopful of suspension at the top of each moat with the slide tilted at about 80°. Allow the liquid to run to the opposite end and immediately remove the excess with a tissue. Allow the slide to air dry.
4. Add 5 drops of stain to the first moat and then 5 sec later to the second moat and so on.
5. Observe carefully for the appearance of a fine precipitate in the stain. When this has occurred in the first and second moat gently wash off the stain with water. Do not pour off the stain before rinsing as this will increase background deposit.
6. Allow the slide to dry in air without blotting and examine directly using an oil immersion objective.

Bacteria and flagella are red, the background should be largely unstained but there is usually some red deposit which should not be sufficient to interfer with interpretation.

(b) *Cesares-Gil flagella stain*

Reagents

A.   Flagella stain
          Tannic acid   20 g
          Alum chloride-hydrated   36 g
          Zinc chloride   20 g
          Rosaniline hydrochloride   3 g
          Ethanol 60%   80 ml

Grind together the solids in a mortar with about 20 ml of the alcohol. Mix and stir in the remaining alcohol. Dilute the stain with an equal volume of distilled water and filter before use.

5a–8

B.   Ziehl-Neelsen Carbol Fuchsin
         (See III.E.4 above.)

Method

1. Make a smear of a young culture (18–24 h). Allow to air dry. Do not fix.
2. Flood the slide with freshly prepared flagella stain for 1 min
3. Wash with distilled water
4. Stain with Carbol Fuchsin for 5 min
5. Wash with distilled water
6. Blot dry and examine

Bacteria and flagella are red normally with a fine granular deposit in the background.

When cells are being examined critically for type of flagellation, it is important to include known organisms as controls with each batch of unknowns.

9. *Nuclear stain–Robinow's method*

The nuclear apparatus of bacteria can be demonstrated by staining with Giemsa's solution after the cells have been hydrolysed with 1 N hydrochloric acid at 60°C to remove ribonucleic acid.

Reagents

A.   2% aqueous osmium tetroxide solution
B.   Schaudin's fluid
      Absolute ethanol   100 ml
      Saturated aqueous solution of mercuric chloride   200 ml
C.   1 N Hydrochloric acid
D.   Giemsa stain
      2–3 drops G.T. Gurr's R 66 Giemsa stain per ml of phosphate buffer
E.   Phosphate buffer
      Na_2HPO_4 (anhydrous)   5·447 g
      KH_2PO_4 4·752 g

Mix together. Add 1 g of mixture to 2 litres of water. This will give pH 7.0

Technique

1. Expose an agar square with thin growth on it to osmic acid vapour for 2–3 min. (5 ml osmium tetroxide placed in a sealed dish, so that it covers three layers of glass balls).
2. Place the square culture side down on a clean coverslip, remove the agar by gently lifting and air dry the film on the coverslip.

3. Fix with Schaudin's fluid for 5 min by floating the coverslip face downwards on the surface of the liquid.
4. Wash with water and store in 70% ethanol
5. Place coverslip in N hydrochloric acid at 60°C
6. Wash with tap water and rinse with distilled water
7. Stain for 30 min with Giemsa stain at 37°C
8. Wash and mount face down in water
9. Examine immediately using an oil immersion objective

Nuclear structures are reddish-purple.

Nuclear structures differ in their stainability in different species and at different ages in a culture. Failure to stain them adequately is usually related to incorrect timing of the acid hydrolysis stage and this should be varied if 10 min proves unsuitable.

10. *Stain for spirochaetes*

Like flagella some of the more delicate spirochaetes—particularly the pathogenic forms—must be thickened by stain deposition before they are properly visible in the light microscope. The stains described for flagella above are often successful in staining spirochaetes. The method described below is usually successful and is preferred by us to others in the literature. (See also under Giemsa stain below.)

As with flagella stains the use of really clean slides is important when staining spirochaetes.

Fontana's stain

Reagents

A. Fixative
Acetic acid	1 ml
Formalin	2 ml
Distilled water	100 ml

B. Mordant
Phenol	1 g
Tannic acid	5 g
Distilled water	100 ml

C. Ammoniated silver nitrate solution
    Add 10% ammonia dropwise to a 0·5% solution of silver nitrate in distilled water until the precipitate formed just re-dissolves. Slowly add more silver nitrate until the precipitate formed just remains as a light opalescence.

D. Ethanol

Method

1. Prepare a smear on a well cleaned slide and let it air dry
2. Treat the film for 30 sec, three times, with the fixative
3. Wash with absolute ethanol allowing it to act for 3 min
4. Drain off excess alcohol and warm the slide gently until the film is dry
5. Pour on the mordant, heat until steam rises for 30 sec
6. Wash in distilled water and dry the slide
7. Treat with ammoniated silver nitrate solution and heat until steam rises and the film becomes brown in colour
8. Wash in water and observe directly or mount in a permanent mountant.

N.B. Unmounted films tend to fade rather rapidly and cannot be stored for more than a day or so.

The spirochaetes are brownish-black against a lighter background.

11. *The Romanowsky stains*

When Eosin and Methylene Blue are mixed in solution a precipitate forms and the staining characteristics of the dye mixture change. Romanowsky introduced this dye mixture for the staining of blood films and parasites with results which were so superior to other stains that the method rapidly became accepted and is today the routine process for examination of blood smears.

Results are dependent on the oxidation or ripening of the Methylene Blue and the technique has undergone a series of developments leading to the range of variants of the method available today. The modification due to Giemsa is most satisfactory for microbiological work and it will be described here. Today it is not necessary to prepare the stain by the former tedious method from its components; proprietary stains such as the Giemsa stain R.66 by G.T. Gurr (136–144 New Kings Road, London S.W.6, or from most laboratory suppliers) are excellent.

Much of the success of these stains depends on the differentiation step. This is carried out by flooding the slide with distilled water and allowing it to remain until the smear is just pink. This usually takes 1–2 min. Experience is needed to judge the timing correctly. A good Romanowsky stain should show three colours; the nuclei of white blood cells should stain purple, red cells should be red and cytoplasm should be blue.

Reagent

Gurr's Improved Giemsa Stain R.66.
The solution is obtained ready prepared. The staining procedures use diluted stain. 4 ml of concentrated stain

added to 100 ml of distilled water gives a concentration of 1 drop per ml.

Methods

(a) Blood films

1. Fix films in methanol or ethanol for 2 min
2. Dilute 10 drops of the concentrated stain with 10 ml of neutral water and gently mix
3. Stain films with the diluted stain for 10–20 min
4. Rinse in distilled water and blot dry.

(b) Thick blood films

1. Place a large drop of blood on a slide and spread rapidly with a thin rod or pin to a circle about $\frac{3}{4}$ in. diameter and air dry for 8–24 h in a horizontal position, protecting from dust
2. Dilute one part of stain with 50 parts of distilled water buffered to pH 7·2
3. Stain with the above mixture for 1 h
4. Flush off the scum from the surface of the stain before removing the slide from the staining jar
5. Differentiate with distilled water
6. Blot dry.

(c) Spirochaetes and Trypanosomes

1. Fix the film with ethanol for 2 min
2. Stain with diluted stain (one drop stain to 1 ml neutral water) for half to 3 h, or overnight if desired
   (Time of staining depends on the thickness and age of the film and may be reduced by warming in an incubator)
3. Rinse with water and blot dry

(d) "Rapid Method"

1. Fix films with ethanol for 2 min
2. Flood films with stain diluted as in method 3 and warm over a flame to steaming for 15 sec; wash off with fresh stain and repeat the warming process four or five times; the last lot should be allowed to steam for 2 min.
3. Wash with water; blot dry.

(e) Exudate, Bacterial smears, etc.

1. Cover unfixed thin films with a few drops of concentrated stain for half to 2 min

2. Add 5–10 volumes of distilled water, mix and continue staining for 2–5 min
3. Wash with distilled water and blot dry

Bacteria stain dark blue, capsules light blue and mucus red.

REFERENCES

Culling, C. F. A. (1969). "Handbook of Histopathological Technique", Butterworth (London).

Hopwood, D. (1969). Fixatives and fixation—a review, *Histochem. J.*, **1**, 323–360.

Hotchkiss, R. D. (1948). A microchemical reaction resulting in the staining of polysaccharide structures in fixed tissue preparations. *Arch. Biochem.*, **16**, 131–141.

CHAPTER III

Techniques Involving Optical Brightening Agents

A. M. PATON AND SUSAN M. JONES

Division of Agricultural Bacteriology, University of Aberdeen, Scotland

I.	Introduction	135
II.	The General Nature and Properties of Optical Brighteners	135
III.	Developments as a Biological Tool	136
IV.	Sources	136
V.	Brighteners Recommended for General Use	137
	A. Type "A"	137
	B. Type "N"	137
VI.	Staining	137
	A. Direct	137
	B. Indirect	141
VII.	Fluorescence Microscopy	143
	References	144

I. INTRODUCTION

In this Chapter it is proposed to introduce the compounds generally known as optical brightening agents to a wider application than has, so far, been considered. The techniques described are intended to demonstrate the potential of the substances in a wide field of microbiology. As they are still in the initial stages of development the methods are open to considerable revision and variation according to the requirements of the user.

II. THE GENERAL NATURE AND PROPERTIES OF OPTICAL BRIGHTENERS

The compounds have been described in some detail by Adams (1959), in relation to their uses in the treatment of textiles and paper, and their incorporation into washing soaps and detergents. Their ability to enhance the "brightness" or "whiteness" of both natural and synthetic fibres is now well recognized, and has stimulated considerable research effort towards producing more useful and effective formulations. Brighteners are strongly fluorescent when excited by ultraviolet radiation and are accepted and retained as dyes by a wide variety of organic structures. As industrial applications have expanded, especially for the treatment of wool, cellulose, and synthetic fibres, many chemical derivatives, largely of unknown properties, have become available for biological investigations at low cost.

Industrial samples are not usually pure and may contain several chemically related components. They are also associated with diluents, such as sodium sulphate, sodium carbonate or urea, often in large proportions. In the absence of purified compounds, these additives must be taken into account in laboratory applications such as are described here. In practice, however, the additives do not interfere with the reactions and commercial preparations of the brighteners are, at present, satisfactory.

III. DEVELOPMENTS AS A BIOLOGICAL TOOL

The application of fluorescent brightening agents to the staining of microorganisms is due largely to Darken (1961a, b, 1962) and Darken and Swift, (1963, 1964). Darken (1962) stated that they appeared to be non-toxic, stable biological markers. She demonstrated that the brighteners were absorbed by growing cultures of bacteria and other micro-organisms. The brighteners were said to be more useful than the well known vital fluors, such as acridine orange, fluoresceins and rhodamine.

It was, therefore, clear that the substances had much to offer in microbiology, particularly at a stage when the fluorescence microscope is becoming a common tool in the laboratory. Several applications were soon considered and found effective. Many applications are involved, such as spore germination (Darken and Swift, 1964), labelling of *Salmonella* (Paton, 1964a), labelling of fungal conidia (Wilson, 1966; Patton and Nichols, 1966) and the labelling of trypanosomes (Herbert *et al.*, 1967).

A study of a large selection of brighteners by the present authors has made it obvious that many further applications are possible, and it is opportune to describe the basic techniques involved.

IV. SOURCES

Fluorescent brightening agents are produced by many concerns throughout the world. As their present uses are remote from biology, examples are rarely found in the usual laboratory catalogues and it is necessary to seek supplies direct from the manufacturers or their associates. The authors are particularly indebted to CIBA Clayton Ltd, Geigy (UK) Ltd and Imperial Chemical Industries Ltd, all of Great Britain, and to the American Cyanamid Company.

V. BRIGHTENERS RECOMMENDED FOR GENERAL USE

A. Type "A"

These are 4,4'–diamino–2,2'–stilbenedisulphonic acid derivatives—

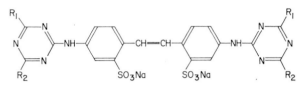

Most brighteners are of this type. Being anionic they are normally pre-cipitated by cationic surface-active agents. They stain the vegetative cells of bacteria, the cytoplasm of leucocytes and the surface of plant cells. They are non-inhibitory to micro-organisms.

B. Type "N"

These are oxacyanine compounds—

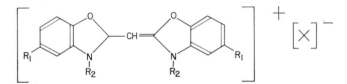

Being cationic, they are precipitated by anionic surface active agents. They stain endospores of bacteria, the nuclei of leucocytes and the starch granules of potato cells. This type tends to be more inhibitory to the growth of micro-organisms than Type "A".

Examples of the above types are: Type "A", Calcofluor White M2R New (American Cyanamid Company Dyes Department, P.O. Box 5670A, Chicago 80, Illinois, U.S.A.); Type "N", Tinopal AN (Geigy (UK) Ltd, Dyestuffs and Textile Chemicals Division, Simonsway, Manchester 22).

VI. STAINING

A. Direct staining

1. Bacteria and fungi

Thin films of bacteria are prepared normally and fixed by heat or methanol. Aqueous solutions of the dyes (1·0% w/v) are applied directly. The staining effect is rapid, but generally a period of contact of approximately 5 min is necessary for an optimum effect. Particular care must be taken to wash films well in water or saline for about 5 min, as residual brightener will cause the background to fluoresce, reducing the necessary contrast.

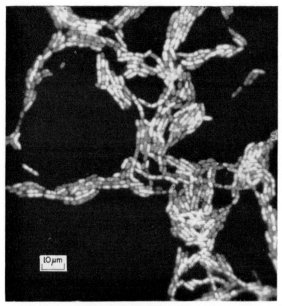

FIG. 1. Film of a young culture of *Bacillus megaterium* treated directly with a brightener of Type "A". Note the variation in the fluorescence of the individual bacteria.

The stain is retained by the organisms and the dried film may be examined immediately or up to some weeks thereafter.

A choice of brightener must be made for each organism, as the staining effects often vary with the species and with the state of the organism. For example, different brighteners will be effective for vegetative cells and for endospores.

The size of most bacteria makes it difficult to distinguish the site of staining beyond that of major morphological features, such as endospores. A critical examination, however, reveals that specific staining within the cell does occur, but its nature is, as yet, undetermined. Organisms derived from a single colony may demonstrate both strong and weak staining reactions (Fig. 1).

In the case of the filamentous fungi, some brighteners of Type "A" are particularly useful for demonstrating septa and cell-wall structures (Fig. 2). The yeasts also show septa readily with the same materials while with other brighteners the cytoplasmic contents stain more readily.

2. *Plant and animal cells*

The nature of the staining effect is more obvious when larger cells are used. Single plant cells are prepared by the agar embedding method (Paton,

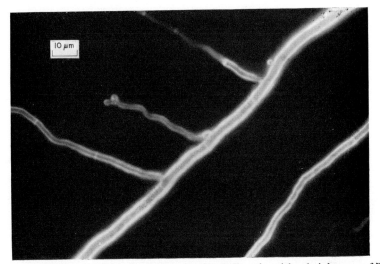

FIG. 2. Hyphae of *Gibberella fujikuroi* stained directly with a brightener of Type "A".

1964b) and suspended in solutions of brightening agents for a few minutes. They are washed free of excess dye and mounted in buffered glycerin, usually at pH 6·5–7·1. Turnip root cells and potato tuber cells can readily be used in this procedure.

In the case of the potato tuber cell, some brighteners (for example, of Type "A") stain the cell wall or membrane and do not affect the contents of the cell, whereas others (for example, of Type "N") do not attach to the outer structures, but strongly associate with the surface of the starch granules. Another demonstration of specificity can be observed with thin films of human blood. When fixed with methanol and treated with brightener solutions, differences are apparent in the staining of the leucocytes. In the case of some brighteners (Type "A") the cytoplasm becomes strongly stained leaving the nucleus non-fluorescent (Fig. 3). With others (Type "N") the nucleus is fluorescent leaving the cytoplasm unstained (Fig. 4). Mature human erythrocytes never become fluorescent. The nucleated forms of erythrocytes will stain with brighteners, such as Type "N".

3. Parasites in blood films

As already stated, the brighteners do not cause non-nucleated erythrocytes to fluoresce. This makes it possible, by treating blood films, to reveal the presence of parasites, such as *Plasmodium*, within such cells. Thin films are advocated, as the presence of fibrin, which absorbs brighteners, may detract from the clarity of the image. A 0·1% (w/v) solution should be

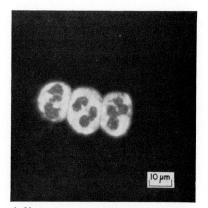

FIG. 3. Human-blood film treated with a brightener of Type "A". The poly-morphonuclear leucocytes show cytoplasmic fluorescence. The background of erythrocytes is scarcely visible. Compare with Fig. 4.

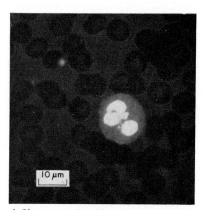

FIG. 4. Human-blood film treated with a brightener of Type "N". The poly-morphonuclear leucocyte show nuclear staining. Compare with Fig. 3.

used and applied for 1 min. As with other direct-staining procedures, excess of stain should be removed carefully by washing in water. This technique is particularly well suited to the examination of blood films for malarial parasites (Fig. 5). Treated smears of infected blood readily show the presence of trypanosomes (Herbert et al., 1967) (Fig. 6) and toxoplasma. The dull non-fluorescent background of red cells make an excellent background for the fluorescent leucocytes and parasites, which can be readily observed and differentiated. The use of various optical brightening agents indicates that, as for other cells, those of the parasites can be stained with a view to demonstrating different portions of the cell structure.

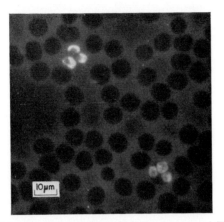

FIG. 5. Monkey-blood film treated with a brightener of Type "N" showing *Plasmodium berghei* associated with erythrocytes.

B. Indirect

1. *Staining of viable cells*

An important feature of the fluorescent brighteners is their general lack of toxicity to micro-organisms and, as far as is known, to other cells. When added to nutrient media, they can impart a fluorescence to the growth of many organisms. This is particularly evident when brighteners are incorporated into solid media. The ensuing colonies, when examined by ultraviolet light can be shown to be largely composed of stained organisms although a proportion, as found with the direct-staining technique, are almost non-fluorescent.

A few brighteners inhibit the growth of some Gram-positive bacteria (as Type "N" at 0·1% w/v) but most have no apparent effect on the rate of growth or other recognizable characteristics of a wide range of organisms.

A noteworthy feature may be a reflection of specificity. Bacteria differ in their ability to produce fluorescent colonies. This may be well demonstrated by plating soil dilutions on the surface of "brightened" nutrient agar. While some colonies are brilliantly fluorescent others are only slightly so, or even non-fluorescent. This observation is significant, as it emphasizes the need for a careful selection of a brightener for the labelling of any particular organism.

The media used for growing fluorescent organisms should not contain materials, such as particulate plant or animal components, as these will take up the brighteners and so reduce the effective concentration available to the cultures. While most common nutrient media are suitable for incorporating

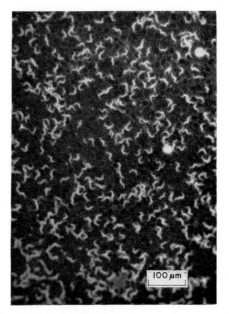

FIG. 6. Mouse-blood infected with *Trypanosome vivax* treated with a brightener of Type "N".

the different brighteners, some are autofluorescent and render it difficult to assess the degree of fluorescence of surface growth. This does not, of course, affect the uptake of dye by the micro-organisms. It is, therefore, found useful on occasions to incorporate water-insoluble nigrosin into the medium sufficient to mask the autofluorescence.

Staining cell while, at the same time, retaining their viability is a technique with many uses. Paton (1964) demonstrated the site of passage of a species of *Salmonella* through egg shells by using organisms labelled in this manner. The technique is simple and is generally applicable. The organisms are grown on nutrient agar with 0·05% brightener, (of, for example, Type "A"). The cells are harvested after suitable incubation and washed free by centrifugation of unabsorbed fluorescent compounds. A difficulty may arise if the fluor has not been correctly selected as it may wash out of the cells leaving them insufficiently fluorescent.

The fluorescent cells are capable of normal division into micro-colonies of progeny that share the fluorescence of the original cell. Such microcolonies can be detected against a non-fluorescent background with an ultraviolet lamp. It would appear that the concentration of brightener on the parent cell is greater than that required for optimum fluorescence.

Other authors (Patton and Nichols 1964; Wilson, 1966) have used brighteners for staining basidiospores and the conidia of fungi. Although the brightener content of an organism is shared by its daughter cells if division is essentially one of binary fision, other forms of growth result in a less effective distribution. For example, the extensions of hyphal structures in the filamentous fungi do not receive sufficient brightener from the older structures to fluoresce.

Little information is available on the vital fluorescent staining of cells other than microbial cells. However, recent work in the authors' laboratory has shown that animal cells (Landschütz tumour cells) can be stained, cultured *in vivo* or *in vitro* and after some normal divisions retain detectable fluorescence.

VII. FLUORESCENCE MICROSCOPY

Persons familiar with the use of the modern fluorescence microscope will find no difficulty in obtaining satisfactory results with brighteners. Most of the wide variety of such microscopes now available are suitable for use with these stains. Several factors must, however, be taken into account. Since a suitable visible emission is only obtained from ultraviolet radiation, sources of illumination, such as quartz–iodine lamps, that are relatively weak in that range, are not recommended. The results described in this Chapter have been obtained with a Reichert Zetopan fluorescence microscope with a high-pressure mercury-vapour lamp. The filters used with normal condensers were UG 1/2·5 mm + BG 12/3 mm (exciter) with 2 × GG 13/1 mm + Wratten foil 2B (absorption). A dark-ground condenser can be used with considerable effect, and in this case the exciter filter should be UG 1/1·5 mm. The "Tiyoda" dark-ground condenser (Sakura Finetechnical Co. Ltd, Tokyo, Japan), designed especially for use with ultraviolet light is particularly useful for the purpose.

For high magnifications, a glycerin-immersion lens is recommended. Glycerin remains non-fluorescent and does not become viscous on ageing or after prolonged exposure to ultraviolet light; moreover, it is readily removed from lenses and from slides. However, a distinct disadvantage applies to its use with this technique. If glycerin is used as immersion fluid directly on a brightener-treated film, it may remove the fluor from the preparation and becomes itself fluorescent. In a short time the image is weakly fluorescent and the necessary contrast with the background is lost. This can, of course, be avoided by using a non-aqueous mountant under a coverglass. An alternative procedure is to apply a 0·1% (w/v) solution of an ionic surface-active agent to the film immediately after "brightening". This renders the brightener less soluble. The choice of surface-active agent depends on the nature of the brightener. The latter is commonly anionic (as Type "A") and may be precipitated by a cationic agent, such as cetyl-trimethylammonium bromide. Brighteners of Type "N" are rendered less soluble by anionic agents, such as the di-octyl ester of sodium sulphosuccinic acid.

Most brighteners fade on prolonged exposure to the intense ultraviolet beam normally applied during observations, but this does not usually cause difficulty except in the photography of weak images.

While using optical brighteners it is difficult to avoid "contaminating" slides and working surfaces. Parts of the microscope, particularly the exposed surfaces of the lenses may also be affected. Moreover, it must be noted that many paper tissues commonly used for cleaning slides and microscope stages are already treated with brighteners by the manufacturers. It is, therefore, advisable by strict precautions to avoid introducing undesirable fluorescent materials at critical stages of the technique.

REFERENCES

Adams, D. A. W. (1959). *J. Soc. Dyers Colourists*, **75**, 22–31.
Darken, M. A. (1961). *Science, N.Y.*, **133**, 1704–1705.
Darken, M. A. (1961b). *Appl. Microbiol.*, **9**, 354–360.
Darken, M. A. (1962). *Appl. Microbiol.*, **10**, 387–393.
Darken, M. A., and Swift, M. E. (1963). *Appl. Microbiol.*, **11**, 154–156.
Darken, M. A. and Swift, M. E. (1964). *Mycologia*, **56**, 158–162.
Herbert, W. J., Lumsden, W. H. R., French, A. McK., and Paton, A. M. (1967). *Vet. Rec.*, **81**, 638.
Paton, A. M. (1964a). *Nature, Lond.*, **204**, 803–804.
Paton, A. M. (1964b). *J. appl. Bact.*, **27**, 237–243.
Patton, R. F., and Nichols, T. H. (1966). *In* "Breeding Pest-Resistant Trees", pp. 153–164. Proceedings of a NATO and NSF Symposium, Pennsylvania State University, August 30th–September 11th, 1964. Pergamon Press, New York.
Wilson, H. M. (1966). *Science, N.Y.*, **151**, 212.

CHAPTER IV

Motility

TETSUO IINO AND MASATOSHI ENOMOTO

National Institute of Genetics, Mishima, Shizuoka-ken, Japan

I.	Introduction	145
II.	Observation and Measurement of Motility	146
	A. Cellular motility	146
	B. Clonal motility	150
III.	Observation of Locomotive Organelles	151
	A. Optical microscopy	151
	B. Electron microscopy	153
IV.	Techniques of Isolating and Typing Motility Mutants	155
	A. Isolation of motility mutants	155
	B. Identification of various motility mutant types	155
	C. Application of motility mutants to genetic analysis	159
V.	Preparation of Media and General Precautions	160
	A. Preparation of motility media	160
	B. Environmental factors	160
	C. Tactic responses	161
	References	162

I. INTRODUCTION

In the microbial world, motile organisms are distributed in most of the phyla. The mode of movement differs in different microbial groups and has been regarded as an important characteristic for the identification of microbes.

Myxomycetes, slime moulds and some unicellular protozoa move over the surface of solid substances as a result of cytoplasmic movement. This kind of movement is commonly called *amaeboid movement*. Another movement which is possible only when organisms are in contact with a solid surface is the *gliding movement*, such as seen with myxobacteria, blue–green algae, and some of the filamentous bacteria such as the Beggiatoales.

Many motile micro-organisms can move actively in a liquid medium. This *swimming movement* is brought about by locomotive organelles known as flagella or cilia. Flagella of eubacteria differ in their fine structure from those found in eucaryotic micro-organisms, such as flagellates or protozoa. The former consists of a single fibril of molecular dimensions, whereas the latter is surrounded by a membrane and encloses a system of very regularly arranged longitudinal fibrils comprising a single central pair and nine outer pairs arranged in a circle. Among the bacteria, spirochetes swim in a liquid medium with the aid of an axial filament.

The techniques described in this Section concern mainly the observation of motility and locomotive organelles of eubacteria. However, the techniques may be applied in various respects to other micro-organisms with appropriate modifications.

II. OBSERVATION AND MEASUREMENT OF MOTILITY

The methods are divided into two groups: the one, involving the use of a microscope, is applied to the movement of individual bacteria (cellular motility); the other, requiring the growth of bacteria, to the movement of bacterial populations (clonal motility).

A. Cellular motility

Bacteria, such as *Proteus*, swarming on a wet solid medium can be observed directly on an agar plate with an ordinary or a phase-contrast microscope (Morrison and Scott, 1966). In order to determine the presence or absence of motility and the mode of locomotion, a "hanging-drop" preparation, is commonly used. However, an ordinary preparation, namely a bacterial suspension on a microscope slide covered by a coverslip and sealed with Vaseline, is also useful, and is indispensable to observation under a dark-field microscope. Ogiuti (1936) calculated the speed of the movement of a bacterium under a dark-field microscope by tracing the pathway of a bacterium in a given time with Abb's drawing apparatus (camera lucida) and computing the true distance from a drawn line.

1. *Hanging-drop method*

A thin film of Vaseline is laid around the concave area of a hollow-ground slide. A drop of culture is placed on the centre of a coverslip with a wire loop, and the slide is placed over it and pressed down gently. The slide is turned over quickly and the drop of culture should remain in the centre of the coverslip in the form of a hanging drop. Microscopic examination should be carried out immediately after the preparation is made. Observation is facilitated by focusing the microscope on the edge of the drop and then moving the slide to centre the drop in the field of the microscope. For the observation of a culture grown on a solid medium, a small amount of the culture should be emulsified in a drop of saline or broth on a coverslip; when many bacterial clones are to be examined quickly, a thick slide glass with a square depression (26 ×26 ×2 mm) is conveniently used. 10–15 drops can be hung on a 24 ×32 mm coverslip. Such a slide is made by sticking two strips of slide glass ca. 2 mm thick on a slide leaving a space 26 mm wide.

Observation by the hanging-drop method is a simple and convenient method suitable for all bacterial species. However, care is needed to distinguish bacterial movement from drifting in water streams and from Brownian movement, and to observe floating cells only, since bacteria sticking to a coverslip often become non-motile. When the speed of movement is too high for observation, the viscosity of the suspension may be increased by the addition of methylcellulose (Pijper, 1947) (0·5–1·5% w/v).

2. Capillary-tube method

The motility of bacteria in a suspension can be measured quantitatively by a simple method in which the number of organisms that pass across a small aperture in a given time are counted (Shoesmith, 1960). A sample of suspension is drawn into a flat capillary tube and observed with an oil-immersion objective by phase-contrast illumination. An opaque disc with a small hole in the centre is fitted in the focal plane of an ocular. This diminishes the field visible under the microscope. Using a 2 mm oil-immersion objective and a ×10 ocular with a 0·8 mm dia hole, the field becomes 10 μm in diam. The number of bacteria traversing this field in a given time is expressed as a "motility count", which is proportional to the average speed of the bacteria and to the bacterial concentration. Several precautions for this procedure are described in the original paper of Shoesmith (1960).

The capillary-tube method is also used in qualitative assay of bacterial movement. An advantage of this technique is that it is suitable for longer periods of observation than are possible with a slide and a coverslip since drying of the preparation is much reduced. This method is also used for studying chemotaxis of bacteria (Sherris et al., 1957; Adler, 1966a).

3. Special chamber method

Motility of individual bacteria submitted to the action of chemical or physical agents can be observed for any length of time in a special chamber invented by De Robertis and Peluffo (1951). This device is made of a Cellophane membrane, a slide and coverslips. First a slide with a groove 3·5 mm wide is made by sticking two coverslip strips on a slide. A thin film of bacterial culture is made at the centre of a sterile coverslip. The coverslip is then inverted over a sterile Cellophane strip (24 ×28 mm), on the centre of which a loopful of melted plain agar has previously been dropped. After setting, the Cellophane flaps are turned up at both ends and the coverslip carrying the Cellophane is placed on the grooved slide with the Cellophane down. The space between the coverslip and the Cellophane on the groove is used as the culture chamber. The periphery of the chamber is sealed with wax leaving both ends of the groove open, the groove becoming a channel for the media or chemical agents to be examined. The feature of this chamber is that prolonged observation of bacteria is possible under the influence of a chemical agent to which the Cellophane membrane is permeable and that several chemical agents can be successively applied to a single chamber.

4. Flying-spot microscopy

The pathway and the speed of locomotion of a bacterium can be recorded by flying-spot microscopy using photoelectric registration (Young and Roberts, 1951). A television tube (cathode-ray oscilloscope), having high brilliance and a very short time constant, is placed in front of the ocular as the source of illumination and a minute phosphorescing spot is focused on a specimen through the objective lens. This scans the specimen as the electron beam of the oscilloscope scans its screen and the transmitted spot is received by a multiplier photocell placed beyond the condenser. The output of the photocell is fed into another cathode-ray oscilloscope and the locomotion of bacteria is depicted on the screen as a discrete pathway of "blips" that are made at the points of collision between a bacterium and a scanning beam. The track of locomotion can be recorded by exposing a standard film to the image. From the number of blips per unit length of the track, the scale of enlargement and the speed of scanning beam, the speed of movement of a bacterium can be calculated (Davenport et al., 1962).

This apparatus, not yet used commonly, has several advantages for measuring motility; a pathway of locomotion is easily recorded as a track; changes in the speed may be estimated by comparing the length of blips under different conditions; heating of a specimen is minimal with the various illumination systems adopted for observing cellular motility, since the illumination of the field is low and intermittent.

5. Stroboscopy and microcinematography

Stroboscopy and microcinematography have been used for examining microbial motility. Photographs taken under a dark-field microscope with stroboscopic illumination are useful in the analysis of the movement of cilia and flagella of eucaryotic cells (Gray, 1955; Brokaw, 1965). This technique is, however, unsuitable for most bacterial species, owing to their small size. It may be applied only to the analysis of body movement of bacteria of relatively large size; for example, Metzner (1920) examined the movement of flagella and bacterial bodies of *Spirillum* species by this technique.

For the same reason, microcinematography has been little used in the analysis of bacterial movement. A successful example of the application of this technique is found in the reports of Pijper (1940, 1953), who used microcinematographs for examining the movement of bacteria and their flagella.

6. *Microelectrophoresis*

Bacterial surfaces are negatively charged in a solution of neutral pH, so that non-motile bacteria move toward the cathode in electrophoresis. Actively motile bacteria, however, move randomly, some toward the anode when examined with low potential gradients. The electrical potential necessary to counteract the movement toward the anode can be measured for individual bacteria (Harris and Kline, 1956).

A suspension of motile bacteria washed with a suitable buffer solution with motility restored by incubation at the appropriate temperature is placed in a flat-type electrophoretic cell made of glass, and observed with an ordinary microscope. The current is applied and a bacterium moving in the direction opposite to the flow of the current is selected for observation. Then the current is increased rapidly until the cell stops and just begins to move in the opposite direction under the force of the electrical potential. The current at that time is measured with a milliammeter. After measuring the specific resistance of the suspending buffer, the value of the electrical potential needed to overcome the movement of the bacterium is calculated from Ohm's law. The values obtained by this technique may be taken as an index of the movement of bacteria. However, the values are influenced not only by the motility of the bacteria but also by the surface charges, including the charge of flagella. Therefore, to use such an electrical potential as the index of motility, one should compare the value obtained with a motile strain to that obtained with a paralysed (non-motile though flagellate) mutant derived from the motile strain.

B. Clonal motility

Clonal motility can be demonstrated and measured by more simple methods than those applied to individual bacteria. It must be remembered that clonal motility is greatly influenced by cultural conditions, because it is usually related to the growth stage of the culture.

1. *Culture-tube method*

The simplest method is to stab the culture into soft nutrient agar or nutrient gelatin agar (NGA; for its preparation, see Section VAI) kept in a culture tube. Both of these media contain a low concentration of agar, so that motile bacteria can move into the medium during their growth. The culture of bacteria is stabbed with a straight platinum wire into the medium in a single straight line. During incubation, motile bacteria grow, spreading from the line of the stab, and non-motile bacteria grow restricted to the line only. When the culture is inoculated on the top of the medium instead of being stabbed, motile bacteria spread through the medium toward the bottom of the tube. This method can be used to measure the speed of movement (Enomoto, 1965) and also to obtain a pure motile culture of bacteria (Lederberg and Edwards, 1953; Skaar et al., 1957). An advantage of NGA as compared with soft nutrient agar, in these experiments is that, because of its high gelatin content, the spreading of the motile bacteria can be stopped at any time by lowering the incubation temperature from 37°C to below 25°C. In order to isolate pure motile bacteria, special culture tubes have been designed, such as a U tube, Rovida's tube and a tube containing a piece of glass tubing cut off obliquely at the bottom (Craigie tube) (Wilson and Miles, 1955).

The culture-tube method is generally suitable for observing the motility for a long time because the medium in a tube can be kept for weeks without drying out.

2. Plate method

A Petri dish containing NGA medium is as suitable for detecting and isolating motile bacteria as the culture tube. On this plate motile bacteria spread and grow below the surface of the medium to give a concentric growth called a "swarm" (Stocker et al., 1953) when incubated at 37°C. A superior feature of this method is that many motile clones spreading as swarms on a plate can be examined separately. Therefore, this method has been extensively used for examining the number and nature of motile clones produced from non-motile bacteria by mutation, transduction, conjugation or transformation. When a small definite volume of different cultures is spotted on, or stabbed into, a plate, the diameters of the resulting swarms are comparable with each other as a rough index of motility of the culture. On handling the NGA plates the following precautions must be taken; plates must be cooled to about 5°C and the culture suspensions used as inocula must be dried from the surface of the plates before incubation.

3. Capillary-tube method

Clowes et al. (1955) measured the speed of movement by using a capillary tube containing soft nutrient agar. The tube is inoculated with motile bacteria at one end, and, after incubation, the soft agar in the capillary is extruded from the uninoculated end, in measured lengths, into aliquots of nutrient broth. The extent of movement of the motile bacteria can be found by the end point of growth in the aliquots after incubation.

A quantitative assay method using capillary tubes has been developed by Adler and Dahl (1967) whose paper describes the procedure in detail. Here it is briefly summarized. Motile bacteria, after washing with a suitable buffer solution, are inoculated with a fine capillary at one end of a capillary tube (10 cm long, 0·8–1·2 mm i.d.) filled with a buffer solution containing an energy source, and the ends of the tube are closed with agar and clay. After incubation the tube is marked into equal segments and the contents of each compartment are drawn out with a fine capillary. Viable bacteria in each compartment are counted by plating (complete assay). Alternatively, the presence or absence of motile bacteria is observed after inoculating the contents of each compartment into broth and incubating them (frontier assay). When a migrating band is produced by motile bacteria, the location of the band can be measured simply with a ruler or scanned with a recording microdensitometer (Adler, 1966b).

III. OBSERVATION OF LOCOMOTIVE ORGANELLES

A. Optical microscopy

Large-sized flagella or cilia present in protozoa or flagellata are visible in hanging drops of the living cells under an ordinary optical microscope, preferably with phase-contrast illumination. In order to observe the mode of movement in detail, the organisms are suspended in a viscous solution, such as methylcellulose solution (0·5–1·5% w/v), which slows down their movement (Marsland, 1943; Stiles and Hawkins, 1947).

The finer flagella carried by eubacteria have thicknesses ranging from 100 to 200 Å. These dimensions are far below the resolving power of the ordinary optical microscope. Therefore, a special staining procedure is required for observing the individual flagellar fibres. Flagellar bundles formed when bacteria are moving can sometimes be seen by dark-field microscopy.

1. Staining of bacterial flagella

The general principle of flagellar staining for optical microscopy is to treat flagella with a colloidal solution of an appropriate mordant, which precipitates on the whole surface of the flagella and causes an increase of the apparent thickness. The most commonly used mordant for this purpose is tannic acid. Since the classical reports by Löffler (1889), various staining procedures for bacterial flagella have been proposed and different techniques have been used in different laboratories. A simple and widely adopted technique is the one perfected by Leifson (1951), whose report describes not only the staining procedure but also various factors affecting the process of specimen preparation.

(a) *Preparation of specimens for staining.* The bacteria must be treated carefully to ensure that flagella are not sheared off and that their natural waves are retained. Care must also be taken that slides are as clean as possible; when an agar surface culture is used, bacteria must be suspended gently in distilled water. With liquid cultures, the bacteria must be washed by centrifugation to free them completely from extraneous material and must then be suspended in distilled water. Centrifugation and re-suspension in distilled water should be carried out as gently as possible. A suitable concentration of bacteria in the final distilled water suspension is 10^5–10^6 cells/ ml.

Addition of formalin (40% formaldehyde) to a concentration of 5–10% (v/v) to a bacterial culture before washing is recommended when the organisms are pathogenic. The formalin treatment has another advantage in that after the fixation, flagella of slime-producing and capsulated bacteria are more readily stained. Attention must be given to the occasional effect of formalin fixation causing a change in the shape of flagella. This change is specific and useful for the characterization of flagellar type (Leifson, 1961; Iino and Mitani, 1967a).

For preparing smears a heavy-lined rectangle is drawn with a wax pencil over a half area of the slide. At one edge of the rectangle on the slide, a loopful of the bacterial suspension is dropped, the slide is then tilted, causing the liquid to flow to the opposite pencil line, and allowed to dry at room temperature.

(b) *Staining.* According to Leifson (1951), the simplest method to prepare the stain is to make the following solutions separately: (1), 1·5% (w/v) NaCl in distilled water; (2), 3·0% (w/v) tannic acid in distilled water; and (3), 1·2% (w/v) basic fuchsin certified for biological staining in 95% (v/v) alcohol. Mix the solutions in exactly equal proportions and store the mixture in a tightly stoppered bottle in a refrigerator. The staining mixture thus stored may be used for several weeks. At higher temperature, however, gradual chemical change on storage results in increasingly fainter staining of flagella.

For flagellar staining, quickly flood 1 ml of the stain over the pencilled rectangle on the slide and leave 5–20 min at room temperature. The time of the staining differs depending on the temperature and humidity of the room, but a standard procedure in our laboratory is 10 min at 23°C in 65% humidity. The staining time must be increased with a decrease of temperature or an increase of humidity. As soon as a precipitate forms throughout the film, rinse the slide with running water without pouring off the stain beforehand. Unless this is done, a heavy deposit settles on the slide obscuring the stained film. Cell bodies may be counter-stained with 1% (w/v) aqueous methylene blue solution for 1 min. Rinse and dry. With these procedures, the bacterial bodies are stained blue and the flagella red. The stained bacteria are observed at a magnification of around 1000 using an oil-immersion objective.

2. Observation of flagellar bundles with dark-field microscopy

In actively moving multi-flagellated bacteria, either peritrichous or polar, the individual flagellar fibres gather and form bundles or so-called "tufts", which can be seen on living bacteria under a dark-field microscope with strong illumination. Pijper (1931, 1938) succeeded in observing flagellar bundles of various bacteria by illuminating them with sunlight. An alternative method is to suspend the bacteria in an appropriate colloidal solution which not only slows down bacterial movement but also precipitates on and coats their flagella, thus improving the visibility of flagellar bundles.

The procedure which has been employed in our laboratory follows principally Pijper's method (1947, 1957) using methylcellulose solution as the colloidal solution. The bacterial sample to be observed is prepared as follows. A bacterial culture is mixed with an equal volume of saline containing 0·5% (w/v) methylcellulose at a final viscosity of 2·8 $\times 10^{-2}$ P (20°C). The mixture is then dropped onto a 1·4 mm thick glass slide and covered with 0·25 mm coverslip. Nail varnish is used for sealing the edges of the coverslip. The sample is observed under the dark-field microscope in combination with an aplanatic paraboloid condenser of numerical aperture 1·4, and an immersion objective $\times 90$ (with iris) with a $\times 10$ ocular. The light source is a high-pressure mercury vapour lamp (150–250 W) with an ultraviolet filter. A low-voltage lamp (6 V, 30 W) with an ordinary-light microscope may be used but the image has less contrast.

Under the dark-field microscope, most of the bacteria are motile when observed immediately after mounting on slides, but the number of motile bacteria decreases gradually with time. When the organisms are moving rapidly, bundled flagella are blurred and have the appearance of a smooth straight tail as reported by Pijper (1957). When the organisms are moving slowly, the spirals of bundled flagella are clearly seen.

B. Electron microscopy

General procedures for electron microscopy are described elsewhere In this Section, comments on the application of various methods for the preparation of specimens for electron microscopy of bacterial flagella will be described.

1. Shadowing methods

A shadowing method gives convincing results for the observation of waves of individual flagella. The method is preferable to a negative staining method for measuring the thickness of the fibres; since it gives more reproducible results (Lowy and Hanson, 1965).

A standard shadowing method for observing bacterial flagella is as follows. Flagellated cells are suspended in distilled water at pH 7·0, ordinarily without any fixative, and droplets of the suspension are placed on collodion-coated grids which are dried in a vacuum and shadowed with chromium at an angle of 20°. For electron microscopy, an initial magnification of ×5000 is appropriate.

The shadowing method is also recommended for the stereoscopic observation of flagellar bundles. With cells carrying short-wavelength flagella (curly phase), flagellar bundles can be seen by using the standard procedure as described above (Iino and Mitani, 1966). With normal flagellar cells, bacteria are suspended in 0·25% (w/v) methylcellulose solution, and kept 3 h at room temperature. Small drops of the suspension are then placed on collodion-supporting films.

2. Negative staining method

For observing the overall shape of flagella, the negative staining method also gives convincing results. This method is especially advantageous for observing the fine structure of the flagellar fibres, that is, the arrangement of flagellin particles in the fibres. Results vary considerably according to the composition of the staining solution. Lowy and Hanson (1965) reported that the most commonly used method of Brenner and Horne (1959), using potassium phosphotungstate at pH 7·4, does not show fine structure in flagella lying on the carbon film. A better method, applied to the staining of several bacterial flagella by Lowy and Hanson (1965), involves the use of uranyl acetate in place of phosphotungstate.

A drop of bacterial suspension is placed on a grid, withdrawn after a few seconds, and replaced with a drop of distilled water in order to dilute any substances that might precipitate with uranyl acetate in the subsequent steps. The washing fluid is withdrawn and immediately replaced with 1% (w/v) uranyl acetate at a pH of about 4·4. After a few seconds the stain is withdrawn and the preparation is allowed to dry. For observing flagellin particles, it is desirable to have a minimum accelerating voltage of the electron microscope of 80 kV. It is worth noting that Grimstone and Klug (1966) reported that the fine structure of flagella of Trichonympha and other flagellates, in contrast to bacterial flagella, is observed less satisfactorily with uranyl acetate negative staining than with phosphotungstate, in that the fibres are poorly preserved and less structural detail is apparent.

The negative staining method has also been applied to the observation of flagellar basal granules on plasmolysed or partially lysed cells (Abram *et al.*, 1965; Hoeniger *et al.*, 1966; Ritchie *et al.*, 1966).

3. Thin sectioning

A thin sectioning technique has been applied to two aspects of the structure of bacterial flagella; the one is the cross-sectional structure of flagella and the other is the observation of basal granules, especially in relation to other cell-membrane structures.

A supplementary comment here is that, for determining the cross-section of the thin fibres of flagella, it is desirable to make a mounted block in which a mass of flagella are clustered and arranged in parallel. Such a sample is obtained as follows (Kerridge *et al.*, 1962). A drop of a dense flagellar suspension is placed onto a silicone-treated glass slide and dried at room temperature, causing the flagellar fibres to align in the drop. When dry, the layer is removed from the slide with a razor blade, cut into small pieces each about 1 cu.mm, and after fixation embedded.

IV. TECHNIQUES OF ISOLATING AND TYPING MOTILITY MUTANTS

For isolating and preliminary typing motility mutants, NGA (nutrient gelatin agar) plates are most conveniently adopted as the selective and indicator medium. A standard procedure that has been used on *Escherichia coli* and *Salmonella* will be described below. For other organisms, the procedure may be modified depending on their growth habits.

A. Isolation of motility mutants

In order to select motile mutants from non-motile strains, bacteria are inoculated in lines on the NGA plates and incubated at 37°C. Motile mutants which are present in the inoculum start to form visible swarms in 10–20 h. The number and size of the swarms continue to increase for 2 days. These swarms are mutant clones which appear during the course of incubation. Further increase is not observed, probably because spreading of the organisms is inhibited by changes in the media.

For the selection of non-motile or less motile mutants from a motile strain, a suspension of the parent organism is streaked to give a cell-concentration gradient on a NGA plate. The plate is incubated at 37°C for 4–8 h and then at 25°C overnight. Non-motile mutant clones, if present in the inoculum, grow as dense patches of various types depending on the type of motility, as will be described in the next Section. When the ratio of the mutant cells in the motile cell culture is low, a mutant clone can be detected only in an area where swarms spread confluently. Consequently, cells of the mutant patches should be picked up and re-streaked on a NGA medium; this procedure enriches the mutant cells and permits the isolation of the pure mutant clones.

Among the bacteriophages, there is a group which is called motility-phage. In order for this bacteriophage to infect the host bacterium, the presence of motile flagella is essential. Motility-phages have been described for *E. coli, Salmonella, Serratia* (Meynell, 1961; Iino and Mitani, 1967b; Schade *et al.*, 1967) and *Bacillus* (Joys, 1965; Frankel and Joys, 1966). For these bacteria, non-motile mutants, including non-flagellated and paralysed types, can be efficiently selected from motile strains using motility-phage as the selective agent (Joys and Stocker, 1965; Iino and Enomoto, 1966). In the *Salmonella*–motility-phage χ system, the following procedure is routinely used.

A nutrient-broth culture of bacteria at a concentration of about 10^8 cells/ ml is mixed with an equal volume of χ-phage suspension at 10^5 particles/ ml. The mixed suspension is incubated for 15 h at 37°C, then a drop of the suspension is plated on NGA medium. After 24 h cultivation, motile cells are mostly lysed, whereas non-motile clones grow to form smooth compact colonies on the medium. Their stability as non-motile mutant clones must be examined by respreading them on phage-free NGA medium.

On the above mentioned screening medium, swarms occasionally appear. They are either temporary χ-resistant clones, probably composed of so-called phage carriers, or motile χ-resistant mutant clones with altered phage receptors on flagella. In order to establish mutant clones with such a screening medium, it is desirable to repeat single-colony isolation at least twice so that the culture is not contaminated with the bacteriophage.

The technique for isolating motility mutants can be applied to the isolation of mutants which differ from a parental type in their flagellar (H) antigen. Addition of an appropriate concentration of H-antiserum to the cultural medium inhibits the motility of the bacteria having homologous flagellar antigen, without affecting their viability and growth (Lederberg and Iino, 1956; Joys and Stocker, 1966). Therefore, when motile cells with specific flagellar antigen are cultivated on NGA medium supplemented with the homologous H-antiserum in the minimal amount necessary to inhibit cellular motility, they grow at the site of inoculation only. If any mutant cells that differ in their antigenicity and therefore escape from the antiserum inhibition are present in the culture, they spread into the medium and form swarms.

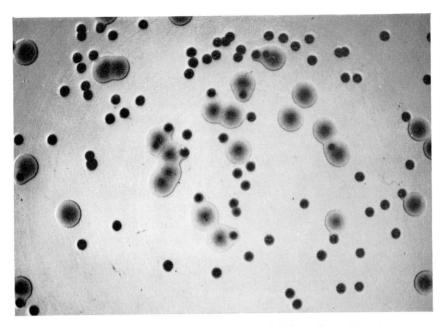

Fig. 1. Colonial types of non-motile mutants of *Salmonella typhimurium* grown on NGA medium. Large colonies are LP-type formed by a non-flagellated mutant, SJ 374. Small ones are SD-type formed by a paralysed mutant, SJ 78.

B. Identification of various motility mutant types

For completly characterizing various types of motility mutants, it is necessary to observe not only clonal motility on NGA medium but also cellular motility in liquid medium and flagellar morphology as well. However, it is possible to identify several different motility types by observing growth characteristics on NGA plates. There is a continuous range of colony types between swarms, formed by motile clones, and compact colonies, formed by non-motile clones. The types also differ in their behaviour depending on the concentration of the agar in the medium and the temperature of incubation.

1. *Non-spreading types*

Bacteria that cannot spread on NGA medium form compact colonies. Among them, there are two distinctive colonial types that have been termed LP and SD (Enomoto and Iino, 1963). LP refers to the large pale-coloured and SD the small dense colonies (Fig. 1). The former are produced by non-flagellate clones. Cells of the latter type have flagella, but these are non-functional. The mutants of this type found in *Salmonella* are represented by the following three sub-types: the paralysed type, which produces flagella

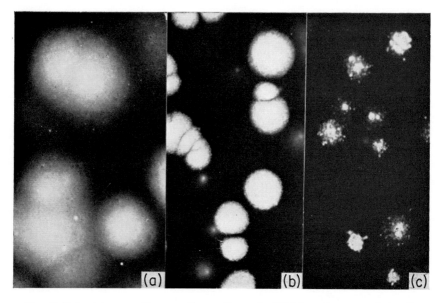

FIG. 2. Colonial types of a normally motile strain, TM 2, and two pauci-flagellate mutants, SJ 781 and SJ 376, of *Salmonella typhimurium* grown on NGA plates: (a) swarms of TM 2; (b) compact swarms of SJ 781; (c) satellite colonies of SJ 376.

morphologically indistinguishable from the normal functional flagella but which are paralysed (Enomoto, 1966a); the straight flagellar type (Iino and Mitani, 1967a); and some of the pauci-flagellated types (Enomoto, 1965). The first two are completely non-motile in liquid medium as well as on NGA medium, whereas the last is slowly motile on NGA medium. The straight flagellar and the paralysed mutants are distinguishable by their response to motility-phage: the former is sensitive whereas the latter is resistant. Not only the genetically non-flagellated bacteria but also pheno-typically deflagellated bacteria, produced by phenol treatment, form LP type colonies. The differences between the SD type and LP type are clearly observed with smooth strains, but less clearly with rough or mucoid ones.

2. Weak-spreading types

Many pauci-flagellated strains and flagellar shape mutants form inter-mediate types of colonies between a swarm and a compact colony on NGA medium. When the component cells of a clone are uniformly weakly motile, the clone forms compact swarms (Fig. 2b); this is observed in a group of pauci-flagellated strains, leaky paralysed strains and strains having small-amplitude or curly flagella (Iino and Mitani, 1966). It is not possible to dis-tinguish these types simply by colonial morphology on NGA medium.

Among them, the cells of the curly flagellar type are markedly limited in their ability to spread on NGA medium containing higher concentrations of agar, and the colonies formed by them are almost like the SD type, but the same is also observed in some pauci-flagellated strains. Furthermore, the curly type is characterized by rotation of individual organisms and their spontaneous aggregation in broth culture (Iino, 1962). With other types, translation and wriggling as well as rotation of cells is generally observed in broth, and these types have no tendency to aggregate except when the cells are of the extremely rough type. The final identification of these types can be made only after direct observation of their flagella by microscopy.

In a group of leaky motility mutants, motile cells appear randomly in only a fraction of the population, and their motility lasts only during one to three cell generations. The colonies formed by such mutants on NGA medium are surrounded by minute colonies, each of which is composed of progeny cells from a motile cell appearing in the clone (Fig. 2c). The colonies surrounded by such minute colonies are called "satellite" colonies.

An extreme case of production of minute colonies as a result of segregation of motile and non-motile cells is observed in abortive transduction of flagellation genes in *Salmonella* (Stocker *et al.*, 1953), where the motile genetic trait is inherited linearly through successive cell generations and the clone forms an array of minute colonies which is termed a "trail".

C. Application of motility mutants to genetic analysis

In principle, the procedures described for selecting motile mutants are used for the screening and scoring of recombinants in conjugation, transduction and transformation experiments. The practical procedures differ to some extent in different organisms and in different recombination systems. Readers are advised to consult references from the following literature, depending on the organism and recombination system of their interest—

Salmonella—transduction
   Stocker *et al.* (1953); Lederberg and Iino (1956); Iino and Enomoto (1966).
Salmonella—recombination
   Enomoto (1966b).
Escherichia coli—transduction
   Armstrong and Adler (1966).
E. coli—recombination
   Ørskov and Ørskov (1962).
E. coli–Salmonella—hybridization
   Mäkelä (1964).
Bacillus anthracis—transduction
   Brown *et al.* (1955).

B. subtilis—transformation

Stocker (1963); Joys and Frankel (1967).

V. PREPARATION OF MEDIA AND GENERAL PRECAUTIONS

A. Preparation of motility media

1. *Nutrient gelatin agar* (*NGA*) *medium* (*semi-solid motility medium*)

The composition of NGA medium is—

NGA medium (*Edwards and Ewing, 1955*)

Peptone	10 g
Meat extract	3 g
NaCl	5 g
Gelatin	8 g
Agar	4 g
Distilled water	1 litre

Heat the mixture in a steamer until it has dissolved, adjust the pH to 7·2 with an alkaline solution (more alkali is required than in preparation of broth because a solution of gelatin is acidic), and sterilize the medium at 121°C for 15 min. Immediately after sterilization shake the medium vigorously to dissolve the gelatin completely. The solidity of the medium may be modified by changing the concentration of agar.

2. *Chemically defined media*

A minimal medium, such as phosphate-buffered salts medium (Lederberg, 1950; Weinberg and Brooks, 1963) can be used for the cultivation of motile *Salmonella* and *Bacillus*. Analytical grade (special grade) reagents should be used in the preparation. A medium suitable for the cultivation of motile *E. coli* has been devised by Adler and Dahl (1967). In this medium, instead of glucose, which prevents flagellar synthesis in *E. coli* (Adler and Templeton, 1967), the 20 common amino-acids are contained in phosphate buffer. A defined medium for *Proteus* has been devised by Jones and Park (1967). In the preparation of synthetic medium one should take care to avoid contamination with heavy metal ions, which are known to prevent flagellar synthesis (Weinberg and Brooks, 1963; Sokolski and Stapert, 1963) and motility (Adler and Templeton, 1967).

B. Environmental factors

Besides metal ions, several environmental factors influence the synthesis of flagella and hence motility. These are described in the following references—

Escherichia coli—temperature, pH, nutrients, chelating agents, oxygen
Adler and Templeton (1967).
Salmonella typhimurium, Proteus vulgaris—temperature
Bisset and Pease (1957).
Proteus hauseri—temperature
Coetznee and Sacks (1960).
Proteus—temperature
Coetznee and Klerk (1964).
P. vulgaris—chemical agents, metabolic inhibitors
De Robertis and Peluffo (1951).
Proteus mirabilis—pH
Hoeninger (1965).
S. typhimurium—temperature, nutrients, analogues of amino-acids,
purines and pyramidines
Kerridge (1961).
A review of surface antigens influenced by environmental factors
Lacey (1961).
E. coli—temperature
Morrison and McCapra (1961).
Salmonella, Escherichia,Vibrio, Spirillum spp.—temperature, pH, nutrients,
chemical agents etc.
Ogiuti (1936).
Eberthella typhosa, etc.—methylcellulose
Pijper (1947).
Pasteurella pseudotuberculosis—temperature
Preston and Maitland (1952).
S. paratyphi C—temperature
Quadling (1958).
S. typhimurium—temperature
Quadling and Stocker (1962).
Pseudomonas—oxygen, arginine
Sherris *et al.* (1957).
Pseudomonas viscosa, Bacillus brevis, E. coli—pH, viscosity of the medium
Shoesmith (1960).

C. Tactic responses

When observing and measuring motility, one should always take into account the tactic response, that is, the movement of motile bacteria toward or away from a source of stimulation. On this subject the reader is referred to a review article (Weible, 1960) and to the extensive studies on *E. coli* (Adler, 1965, 1966a, 1966b; Armstrong *et al.*, 1967).

5a–9

162 T. IINO AND M. ENOMOTO

REFERENCES

Abram, D., Koffler, H., and Vatter, A. E. (1965). *J. Bact.*, **90**, 1337–1354.
Adler, J. (1965). *Cold Spring Harb. Symp. quant. Biol.*, **xxx**, 289–291.
Adler, J. (1966a). *J. Bact.*, **92**, 121–129.
Adler, J. (1966b). *Science, N.Y.*, **153**, 708–716.
Adler, J., and Dahl, M. M. (1967). *J. gen. Microbiol.*, **46**, 161–173.
Adler, J., and Templeton, B. (1967). *J. gen. Microbiol.*, **46**, 175–184.
Armstrong, J. B., and Adler, J. (1966). *Genetics*, **56**, 363–373.
Armstrong, J. B., Adler, J., and Dahl, M. M. (1967). *J. Bact.*, **93**, 390–398.
Bisset, R. A., and Pease, P. (1957). *J. gen. Microbiol.*, **16**, 382–384.
Brenner, S., and Horne, R. W. (1959). *Biochim. biophys. Acta*, **34**, 103–110.
Brokaw, C. J. (1965). *J. exp. Biol.*, **43**, 155–169.
Brown, E. R., Cherry, W. B., Moody, M. D., and Gordon, M. A. (1955). *J. Bact.*, **69**, 590–602.
Clowes, R. C., Furness, G., and Rowley, D. (1955). *J. gen. Microbiol.*, **13**, i–ii.
Coetznee, J. N., and De Klerk, H. C. (1964). *Nature, Lond.*, **202**, 211–212.
Coetznee, J. N., and Sacks, T. G. (1960). *J. gen. Microbiol.*, **23**, 209–216.
Davenport, D., Wright, C. A., and Causley, D. (1962). *Science, N.Y.*, **135**, 1059–1060.
De Robertis, E., and Peluffo, C. A. (1951). *Proc. Soc. exp. Biol. Med.*, **78**, 584–589.
Edwards, P. R., and Ewing, W. H. (1955). "Identification of Enterobacteriacea", p. 179. Burgess, Mineapolis.
Enomoto, M. (1965). *J. Bact.*, **90**, 1696–1702.
Enomoto, M. (1966a). *Genetics*, **54**, 715–726.
Enomoto, M. (1966b). *Genetics*, **54**, 1069–1076.
Enomoto, M., and Iino, T. (1963). *J. Bact.*, **86**, 473–477.
Frankel, R. W., and Joys, T. M. (1966). *J. Bact.*, **92**, 388–389.
Gray, J. (1955). *J. exp. Biol.*, **32**, 775–801.
Grimstone, A. V., and Klug, A. (1966). *J. Cell Sci.*, **1**, 351–362.
Harris, J. O., and Kline, R. M. (1956). *J. Bact.*, **72**, 530–532.
Hoeniger, J. F. M. (1965). *J. Bact.*, **90**, 275–277.
Hoeniger, J. F. M., van Iterson, W., and van Zanten, E. N. (1966). *J. Cell Biol.*, **31**, 603–618.
Iino, T. (1962). *J. gen. Microbiol.*, **27**, 167–175.
Iino, T., and Enomoto, M. (1966). *J. gen. Microbiol.*, **43**, 315–327.
Iino, T., and Mitani, M. (1966). *J. gen. Microbiol.*, **44**, 27–40.
Iino, T., and Mitani, M. (1967a). *J. gen. Microbiol.*, **49**, 81–88.
Iino, T., and Mitani, M. (1967b). *J.Virol.*, **1**, 445–447.
Jones, H. E., and Park, R. W. (1967). *J. gen. Microbiol.*, **47**, 369–378.
Joys, T. M. (1965). *J. Bact.*, **90**, 1575–1577.
Joys, T. M., and Frankel, R. W. (1967). *J. Bact.*, **94**, 32–37.
Joys, T. M., and Stocker, B. A. D. (1965). *J. gen. Microbiol.*, **41**, 47–55.
Joys, T. M., and Stocker, B. A. D. (1966). *J. gen. Microbiol.*, **44**, 121–138.
Kerridge, D. (1961). *In* "Microbial Reaction to Environment". (Ed. G. G. Meynell and H. Gooder), pp. 41–68. Cambridge University Press, London.
Kerridge, D., Horne, R. W., and Glauert, A. M. (1962). *J. molec. Biol.*, **4**, 227–238.
Lacey, B. W. (1961). *In* "Microbial Reaction to Environment". (Ed. G. G. Meynell and H. Gooder), p. 343. Cambridge University Press, London.
Lederberg, J. (1950). *Meth. med. Res.*, **3**, 5–22.
Lederberg, J., and Edwards, P. R. (1953). *J. Immun.*, **71**, 232–240.

Lederberg, J., and Iino, T. (1956). *Genetics*, **41**, 743–757.
Leifson, E. (1951). *J. Bact.*, **62**, 377–389.
Leifson, E. (1961). *J. gen. Microbiol.*, **25**, 131–133.
Löffler, F. (1889). *Zentbl. Bakt. ParasitKde, Abt. I. Orig.*, **6**, 209–224.
Lowy, J., and Hanson, J. (1965). *J. molec. Biol.*, **11**, 293–313.
Mäkelä, H. (1964). *J. gen. Microbiol.*, **35**, 503–510.
Marsland, D. A. (1943). *Science, N.Y.*, **98**, 414.
Metzner, P. (1920). *Jb. wiss. Bot.*, **59**, 325–412.
Meynell, E. W. (1961). *J. gen. Microbiol.*, **25**, 253–290.
Morrison, R. B., and McCapra, J. (1961). *Nature, Lond.*, **192**, 774–776.
Morrison, R. B., and Scott, A. (1966). *Nature, Lond.*, **211**, 255–257.
Ogiuti, K. (1936). *Jap. J. exp. Med.*, **14**, 19–28.
Ørskov, F., and Ørskov, I. (1962). *Acta path. microbiol. scand.*, **55**, 99–109.
Pijper, A. (1931). *Zentbl. Bakt. ParasitKde, Abt. I. Orig.*, **123**, 195–201.
Pijper, A. (1938). *J. Path. Bact.*, **47**, 1–17.
Pijper, A. (1940). *J. biol. photogr. Ass.*, **8**, 158–164.
Pijper, A. (1947). *J. Bact.*, **53**, 257–269.
Pijper, A. (1953). *J. Bact.*, **65**, 628–635.
Pijper, A. (1957). *Ergebn. Mikrobiol. ImmunForsch. exp. Ther.*, **30**, 37–91.
Preston, N. W., and Maitland, H. B. (1952). *J. gen. Microbiol.*, **7**, 117–128.
Quadling, C. (1958). *J. gen. Microbiol.*, **18**, 227–237.
Quadling, C., and Stocker, B. A. D. (1962). *J. gen. Microbiol.*, **28**, 257–270.
Ritchie, A. E., Keeler, R. F., and Bryner, J. H. (1966). *J. gen. Microbiol.*, **43**, 427–438.
Schade, S. Z., Adler, J., and Ris, H. (1967). *J. Virol.*, **1**, 599–609.
Sherris, J. C., Preston, N. W., and Shoesmith, J. G. (1957). *J. gen. Microbiol.*, **16**, 86–96.
Shoesmith, J. G. (1960). *J. gen. Microbiol.*, **22**, 528–535.
Skaar, P. D., Richter, A., and Lederberg, J. (1957). *Proc. Nat. Acad. Sci., U.S.A.*, **43**, 329–333.
Sokolski, W. T., and Stapert, E. M. (1963). *J. Bact.*, **85**, 718–719.
Stiles, K. A., and Hawkins, D. A. (1947). *Science, N.Y.*, **105**, 101–102.
Stocker, B. A. D. (1963). *J. Bact.*, **86**, 797–804.
Stocker, B. A. D., Zinder, N. D., and Lederberg, J. (1953). *J. gen. Microbiol.*, **9**, 410–433.
Weible, C. (1960). *In* "The Bacteria" (eds., I. C. Gunsalus and R. Y. Stanier), vol. 1, pp. 153–205. Academic Press, New York.
Weinberg, E. D., and Brooks, J. I. (1963). *Nature, Lond.*, **199**, 717–718.
Wilson, G. S., and Miles, A. A. (1955). "Principles of Bacteriology and Immunity", p. 436. Edward Arnold, London.
Young, J. Z., and Roberts, F. (1951). *Nature, Lond.*, **167**, 231.

CHAPTER V

Production and Isolation of Flagella

R. W. SMITH AND HENRY KOFFLER

Department of Biological Sciences, Purdue University, Lafayette, Indiana, U.S.A.

I.   Introduction   165
II. Isolation of Flagella   166
III. Purification of Flagella   167
IV. Purification of Flagella Protein   168
References   170

I. INTRODUCTION

Individual bacterial flagella are too small to be observed with the ordinary light microscope. They do associate in a side-by-side fashion, however, which renders bundles or aggregates of flagella visible in the dark-field microscope (Weibull, 1950a, b). Early descriptions of bacterial flagella, i.e. observations made by ordinary microscopy, probably represented bundles of these organelles and not separate, individual structures. By electron microscopy the flagellum can be seen to consist of at least three morphologically distinct structures, a basal region, the hook and the filament. When observable, the basal region (Abram et al., 1965; van Iterson et al., 1966; Hoeniger et al., 1966; Abram et al., 1966; Cohen-Bazire and London, 1967; Abram, 1968) generally appears as a spherical vesicle, 100 to 500 Å in diameter, surrounded by membrane material. This structure is located just within the cell membrane and is most easily observed when attached to the second differentiated region, the flagellar hook (Houwink and van Iterson, 1950; Houwink, 1953; Rinker, 1957; Kerridge et al., 1962; Glauert et al., 1963; Rogers and Filshie, 1963; Abram et al., 1964; Hoeniger et al., 1966; van Iterson et al., 1966; Ritchie et al., 1966; Betz, 1967; Cohen-Bazire and London, 1967; Mitchen and Koffler, 1968). The hook passes through the cell membrane and wall and connects the basal region to the most familiar portion of bacterial flagella, the filament. The hook regions of flagella of Bacillus pumilus are 100 to 120 Å wide and approximately 660 Å in length (Abram et al., 1967). They differ from the filaments in fine structure and are more stable to acid, alcohol, and heat than are filaments (Abram et al., 1966). These differences have been exploited to isolate and purify flagellar hooks, which sediment in the centrifuge after a three step procedure which consists of treatment with 10^{-4} M HCl in 50% ethanol, 10^{-3} M aqueous HCl, and 3% (w/v) deoxycholate at pH 9·0 or 0·3% (w/v) saponin at pH 7 (Abram and Koffler, 1964; Mitchen, 1969).

The main portion of the flagellum is the filament. In suspension, filaments occur as helices, but collapse upon drying to form a sine wave with a wavelength of 2 to 3 μm and an amplitude of 0·2 to 0·6 μm. The filament is a tube composed of fibres of the globular protein, flagellin (Weibull, 1948; De Robertis and Franchi, 1951; Glauert et al., 1963; Abram et al., 1966) probably coiled about each other. Filaments of B. pumilus appear to consist of six of these fibres (Weibull et al., 1945; Abram et al., 1966; Betz, 1967). Models for the packing of subunits into the filaments of other organisms have been presented in the literature (De Robertis and Franchi, 1952; van Iterson, 1953; Glauert et al., 1963).

Sheath-like structures are occasionally associated with portions of the flagellum (De Robertis and Franchi, 1951; Starr and Williams, 1952; Braun, 1956; Gordon and Follett, 1962; Follet and Gordon, 1963; Martinez, 1963; Lowy, 1965; Lowy and Hanson, 1965; Das and Chatterjee, 1966; Ritchie et al., 1966; Abram and Shilo, 1967; Nauman, 1967; Seidler and Starr, 1967, 1968). These differ in composition from other flagellar substructures since they have distinct physical and chemical properties. In general, sheaths are easily removed by agitation or washing and probably do not occur in significant amounts in preparations of highly purified flagellar filaments.

II. ISOLATION OF FLAGELLA

Before attempting isolation of flagella, one needs to optimize the culture selected and growth conditions. If possible, an organism should be chosen that grows readily on a solid medium, is actively motile by means of many flagella per cell, forms no or only little slime layer or capsular material, is resistant to clumping, and forms colonies that are easily dispersed in liquid. If more intensive studies are contemplated, it may be desirable eventually to understand the genetic basis for other findings. For that reason, an organism with known genetics or one probably amenable to genetic analysis should be considered. After selection of the species to be used, optimal growth conditions should be determined and used during further manipulations of the organism. Depending on the organism, the number of flagella per cell may be affected by medium, growth temperature, and length of incubation. A medium should be selected that permits satisfactory growth, minimizes capsule formation, and, if applicable, postpones sporulation. In general, one finds that cells in the late logarithmic or early stationary phase of growth are the most heavily flagellated. This is probably due to a slowing of the rate of cell division and not to an increased rate of production of flagella. These cells also are probably more resistant to lysis by the physical forces used to remove flagella than are rapidly dividing cells. Caution should be exercised, as in the case of *Bacillus* cultures, however, to prevent lysis due to sporulation or the spontaneous induction of lysogenic bacteriophage.

The temperature of incubation may also be critical. An increase of only a few degrees may completely prevent the formation of flagella. Specific adverse effects on motility are more commonly observed during an increase, rather than a decrease, in the growth temperature.

Within a population the number of flagella per cell may vary among individual cells. Yields of flagella can be improved if the more motile, i.e. the more heavily flagellated cells are selected. This is most simply achieved by selecting cells from the leading edge of an isolated growing colony. A better method, however, is to inoculate cells into one end of a U-tube filled with a semi-solid medium, and after an appropriate period of incubation, isolate cells from the opposite end. The assumption of this procedure, which can be repeated several times, is that the cells isolated have a more rapid motility due to a larger number of flagella per cell.

Flagellar filaments are readily broken from intact cells by vigorous shaking. Breakage frequently occurs at the juncture of the hook and filament leaving the hook and basal region attached to the cell body. In our hands, approximately one-fifth of the broken filaments of *B. pumilus* possess hooks (Abram *et al.*, 1967). A smaller number has visible basal regions and membrane material attached. After shaking and differential centrifugation a preparation of flagellar filaments may be obtained which is 95 to 98% flagellin on a dry weight basis.

More specifically, isolation of flagella involves the harvesting of cells from a solid growth medium by suspension in water or buffer. Cells are then sedimented by centrifugation at 3300 to 6000 g for 30 min. The pellets are suspended in water to a concentration, based on wet weight, of 30 to 60 gm/litre and are agitated to break filaments from the cells. In this laboratory, suspensions are shaken on a Miracle Paint Rejuvenator for 10 min at about six hundred $1\frac{1}{2}$ in. strokes per min. Intact cells and debris are sedimented by centrifugation at 3300 to 6000 g for 30 min followed by treatment at 16,000 g for 15 to 20 min. Flagellar filaments are then pelleted by centrifugation at 40,000 g for 2 to 3 h.

III. PURIFICATION OF FLAGELLA

Purified filaments sediment into a water-clear gelatinous pellet. The presence of opaque material in the pellet indicates contamination with other cell fragments. For the most part, these may be removed by the following procedure: A suspension of filaments at 0·3 to 0·7 mg/ml is centrifuged at 12,500 g for 20 to 30 min in the No. 30 rotor of a Spinco model L centrifuge. The centrifugal force is then increased to 78,000 g for 90 min. The opaque contaminants become centred and hard-packed at the bottom of the pellet and may be removed by cutting through the cellulose nitrate tube with a razor blade. The purified filaments are then resuspended in distilled water and the procedure is repeated until completely water-clear pellets are obtained. This procedure is capable of yielding preparations of native flagellar filaments that are 95 to 99% flagellin on a dry weight basis. One obvious disadvantage is that many filaments are removed along with the contaminants during the cutting step. As an alternative, highly fragmented filaments have been purified on columns of diethylaminoethyl cellulose (Erlander et al., 1960).

IV. PURIFICATION OF FLAGELLA PROTEIN

Greater purification with little loss of filament protein is possible if native filaments are not required as a final product, i.e. if one is interested in purified flagellin or in reassembled filaments. Fractional solubilization has been found to be very effective in further purification of the protein. Many agents and physical treatments disintegrate filaments into soluble proteins. These include acid (Weibull, 1949, 1950; Mallett, 1956; Ambler and Rees, 1959; Kobayashi et al., 1959; Stenesh and Koffler, 1962; McDonough, 1965; Vegotsky et al., 1965; Farquhar, 1966), alkali (Weibull, 1949; Koffler et al., 1957; Kobayashi et al., 1959; Stenesh and Koffler, 1962; Lowy et al., 1966), heat (Ambler and Rees, 1959; Kobayashi et al., 1959; Stenesh and Koffler, 1962; Glauert et al., 1963; Vegotsky et al., 1965), sonication (Stenesh and Koffler, 1962; Glauert et al., 1963), alcohols (Weibull and Tiselius, 1945, Abram and Koffler, 1964; Asakura et al., 1964; Lowy et al., 1966), dioxane (Asakura et al., 1964; Vegotsky et al., 1965), acetone (Mallett and Koffler, 1957), sodium dodecyl sulphate (Mallett and Koffler, 1957; Kobayashi et al., 1959; Stenesh and Koffler, 1962; Glauert et al., 1963; Roberts and Doetsch, 1966), cetyl pyridinium chloride (Lowy et al., 1966), and guanidine hydrochloride, urea, and acetamide (Kobayashi et al., 1959; Stenesh and Koffler, 1962; Gaertner, 1966; Lowy et al., 1966). Of these, treatment with acid has been used most often. Filaments disintegrate below about pH 3·8. Non-flagellin components such as hooks, basal material, membrane and wall fragments, and other cell debris largely remain insoluble when mild conditions for the disintegration of the filaments are used, and may be removed by sedimentation. Filaments suspended in water at a concentration of 1 to 2·5 mg/ml are acidified with 0·1 to 1·0 N HCl to a final concentration of 0·01 to 0·001 N HCl. After 30 min to 1 h at room temperature, the insoluble material is removed by centrifugation at 104,000 *g* for 1 h. Flagellin remains in the supernatant liquid and may be further purified by techniques commonly used in general protein chemistry. These include electrophoresis with cellulose acetate (Uchida *et al.*, 1952), disc gel (Suzuki and Iino, 1966; Sullivan, 1969) or starch gel (Enomoto and Iino, 1966), preparative poly-acrylamide electrophoresis (Suzuki and Iino, 1966), and column chromato-graphy on diethyl-aminoethyl cellulose (Smith and Koffler, 1965; Farquhar, 1966; Sullivan *et al.*, 1969), polysulphonic acid resins (Ada *et al.*, 1963), and hydroxyapatite (Ada *et al.*, 1964). The flagellins may be removed from solution and concentrated by precipitation with ammonium sulphate (Weibull, 1949; Uchida *et al.*, 1952; Koffler and Kobayashi, 1956, 1957; Farquhar and Koffler, 1965) or ethanol (Koffler and Kobayashi, 1956). In general, flagellins from various species range in molecular weight from 20,000 to 50,000 (Ambler and Rees, 1959; Kobayashi *et al.*, 1959; Weibull, 1960; Abram and Koffler, 1962; Stenesh and Koffler, 1962; Glauert *et al.*, 1963; Ada *et al.*, 1964; McDonough, 1965; Abron, 1966; Farquhar, 1966) and have a pronounced tendency to associate with one another to form dimers and larger polymers (Stenesh and Koffler, 1962). Association becomes increasingly marked near the isoelectric point which usually lies between pH 5 and 6. The pH at which association occurs may be decreased by addition of salts or by lowering the temperature (Stenesh and Koffler, 1962). The amino-acid composition of flagellins, in general, is note-worthy in that cysteine, histidine, proline, tryptophan, and tryosine residues are either absent or present in very small amounts (Ambler and Rees, 1959; Kobayashi *et al.*, 1959; Weibull, 1960; Abram and Koffler, 1962; Ada *et al.*, 1964; McDonough, 1965; Abron, 1966; Farquhar, 1966). Although individual filaments of a given organism may consist of a single type of protein, various cultures of *Salmonella* (Smith and Stocker, 1962; Iino and Lederberg, 1964; McDonough, 1965; Asakura, 1966; Sullivan *et al.*, 1969) and *B. pumilus* (Smith and Koffler, 1965; Suzuki and Iino, 1966) (and probably many other organisms) produce more than one species of flagellin molecules. These may be similar structurally with only a few amino-acid differences. In *Salmonella*, a given cell produces only one species of flagellin at a time.

In the cases examined, the different molecules produced by a given species have retained the chemical ability to interassociate and are capable of copolymerization (Wakabayashi *et al.*, 1969). In *B. pumilus*, questions concerning the simultaneous synthesis of the different flagellins within a cell and the occurrence of filaments containing several or only one type of protein have not yet been answered. The molecules are similar in several respects since results obtained with mixtures on the analytical centrifuge (Ada *et al.*, 1964) and the presence of a single N-terminal amino-acid (Lederberg and Edwards, 1953; Sala *et al.*, 1968) indicate homogeneity. Apparently they do have slightly different isoelectric points, however, since separation has been accomplished by electrophoresis with disc gels (Sullivan, 1969) or cellulose acetate, preparative polyacrylamide electrophoresis (Suzuki and Iino, 1966), chromatography on diethylaminoethyl cellulose columns (Smith and Koffler, 1965), and specific isoelectric precipitation (Sala and Koffler, 1967). Caution is advised in future work since many observations now need to be reinterpreted in view of these results.

Conceivably some contaminants may also be soluble under the conditions used to disintegrate flagellar filaments. Reassembly of the soluble flagellin followed by sedimentation and washing generally allows removal of these materials. Under the proper conditions of pH, temperature, flagellin concentration and ionic environment, soluble flagellin molecules reassociate to form structures morphologically similar to native flagellar filaments (Mallett and Koffler, 1955; Elek *et al.*, 1964; McDonough, 1965; Abram *et al.*, 1966; Wakabayashi *et al.*, 1969). As an example, the procedure used to reassemble acid-disintegrated filaments of *B. pumilus* is as follows: The flagellin solution at pH 2–3 is carefully adjusted to pH 3·9. Between pH 4 and 5 an amorphous precipitate rapidly forms and this pH range should be avoided. The solution is then mixed with 0·4 M KH_2PO_4–K_2HPO_4 buffer, pH 5·4, to give a final phosphate concentration of 0·02 M and a final pH of 5·3 to 5·5. The rate of assembly is dependent upon protein concentration, pH, and temperature. At pH 5·4, 23°C, and 2 mg/ml of flagellin, assembly is essentially complete after 4 h; at 5 mg/ml the reaction is practically instantaneous. Optimal procedure and conditions for reassembly, of course, depend upon the particular flagellin under investigation. Repeated disintegration, sedimentation, reassembly and sedimentation results in preparations that are essentially 100% flagellin.

REFERENCES

Abram, D. (1968). *Bact. Proc.*, 30.
Ada, G. L., Nossal, G. J. V., Pye, J., and Abbot, A. (1964). *Aust. J. exp. Biol. Med. Sci.*, **42**, 267.
Abram, D., and Koffler, H. (1962). *Abstr. 8th Intern. Congr. Microbiol.*, Montreal, p. 21, August.
Abram, D., and Koffler, H. (1964). *J. molec. Biol.*, **9**, 168.
Abram, D., Koffler, H., Mitchen, J. R., and Vatter, A. E. (1967). *Bact. Proc.*, 39.
Abram, D., Koffler, H., and Vatter, A. E. (1965). *J. Bact.*, **90**, 1337.
Abram, D., Koffler, H., and Vatter, A. E. (1966). *Abstr. 2nd Intern. Biophys. Congr.*, September, Vienna, Austria, Abstr. No. 45.
Abram, D., Vatter, A. E., and Koffler, H. (1964). *Bact. Proc.*, 25.

Abram, D., Vatter, A. E., and Koffler, H. (1966). *J. Bact.*, **91**, 2045.
Abram, D., and Shilo, M. (1967). *Bact. Proc.*, 40.
Abron, H. E. (1966). M. S. Thesis, Purdue University, Lafayette, Indiana.
Ada, G. L., Nossal, G. J. V., Pye, J., and Abbot, A. (1963). *Nature, Lond.*, **199**, 1257.
Ada, G. L., Nossal, G. J. V., Pye, J., and Abbot, A. (1964). *Australian J. Exptl Biol. Med. Sci.*, **42**, 267.
Ambler, R. P., and Rees, M. W. (1959). *Nature, Lond.*, **184**, 56.
Asakura, S., Eguchi, G., and Iino, T. (1964). *J. molec. Biol.*, **10**, 42.
Asakura, S., Eguchi, G., and Iino, T. (1966). *J. molec. Biol.*, **16**, 302.
Betz, J. V. (1967). *Bact. Proc.*, 40.
Braun, H. (1956). *Arch. Mikrobiol.*, **24**, 1.
Burge, R. E. (1961). *Proc. Roy. Soc.*, **A260**, 558.
Champness, J. N., and Lowy, J. (1968). *Symp. on Fibrous Proteins*, Canberra.
Cohen-Bazire, G., and London, J. (1967). *J. Bact.*, **94**, 458.
Das, J., and Chatterjee, S. N. (1966). *Ind. J. med. Res.*, **54**, 330.
DeRobertis, E., and Franchi, C. M. (1951). *Exp. cell Res.*, **2**, 295.
DeRobertis, E., and Franchi, C. M. (1952). *J. appl. Phys.*, **23**, 161.
Elek, S. D., Kingsley-Smith, B. V., and Highman, W. (1964). *Immunology*, **7**, 570.
Enomoto, M., and Iino, T. (1966). *Japan. J. Genetics*, **41**, 131.
Erlander, S., Koffler, H., and Foster, J. F. (1960). *Archs Biochem. Biophys.*, **90**, 134.
Farquhar, M. N. (1966). M. S. Thesis, Purdue University, Lafayette, Indiana.
Farquhar, M., and Koffler, H. (1965). Unpublished observations.
Follett, E. A. C., and Gordon, J. (1963). *J. gen. Microbiol.*, **32**, 235.
Forslind, B., and Swanbeck, G. (1963). *Expl. cell Res.*, **32**, 179.
Gaertner, F. (1966). Ph.D. Thesis, Purdue University, Lafayette, Indiana.
Glauert, A. M., Kerridge, D., and Horne, R. W. (1963). *J. cell. Biol.*, **18**, 327.
Gordon, J., and Follett, E. A. C. (1962). *Proc. 5th Int. Congr. Elec. Microscopy*, New York, **2**, M-5.
Hoeniger, J. F. M., van Iterson, W., and van Zanten, E. N. (1966). *J. cell Biol.*, **31**, 603.
Houwink, A. L. (1953). *Biochim. Biophys. Acta.*, **10**, 360.
Houwink, A. L., and van Iterson, W. (1950). *Biochim. biophys. Acta.*, **5**, 10.
Iino, T., and Lederberg, J. (1964). In "The World Problem of Salmonellosis" (Ed. E. Van Oye). p. 111, Dr. W. Junk, publisher, The Hague.
van Iterson, W. (1953). In "Bacterial Cytology" p. 24, *Symp. 6th Congr. Int. Microbiol.*, Oxford, Blackwell.
van Iterson, W., Hoeniger, J. F. M., and van Zanten, E. N. (1966). *J. cell Biol.*, **31**, 585.
Kerridge, D., Horne, R. W., and Glauert, A. M. (1962). *J. molec. Biol.*, **4**, 227.
Kobayashi, T., Rinker, J. N., and Koffler, H. (1959). *Archs Biochem. Biophys.*, **84**, 342.
Koffler, H. (1957). *Bacteriol. Rev.*, **21**, 227.
Koffler, H., and Kobayashi, T. (1956). *Bact. Proc.*, 39.
Koffler, H., and Kobayashi, T. (1957). *Archs Biochem. Biophys.*, **67**, 246.
Koffler, H., Mallett, G. E., and Adye, J. (1957). *Proc. Nat'l. Acad. Sci., U.S.*, **43**, 464.
Labaw, L. W., and Mosley, V. M. (1955). *Biochim. biophys. Acta.*, **17**, 322.
Lederberg, J., and Edwards, P. R. (1953). *J. Immun.*, **71**, 232.
Lowy, J. (1965). *J. molec. Biol.*, **14**, 297.
Lowy, J., and Hanson, J. (1964). *Nature, Lond.*, **202**, 538.

Lowy, J., and Hanson, J. (1965). *J. molec. Biol.*, **11**, 293.
Lowy, J., Hanson, J., Elliott, G. F., Millman, B. M., and McDonough, M. W. (1966). *In* "Principles of Biomolecular Organization" (Eds. G. E. W. Wolstenholme and M. O'Connor), p. 229, Little, Brown, and Co. Boston.
Lowy, J., and McDonough, M. W. (1964). *Nature, Lond.*, **204**, 125.
Mallett, G. E. (1956). Ph.D. Thesis, Purdue University, Lafayette, Indiana.
Mallett, G. E., and Koffler, H. (1955). *Proc. Am. chem. Soc.*, (128th) 20C.
Mallett, G. E., and Koffler, H. (1957). *Archs Biochem. Biophys.*, **67**, 254.
Martinez, R. J. (1963). *J. gen. Microbiol.*, **33**, 115.
Marx, R., and Heumann, W. (1962). *Arch. Mikrobiol.*, **43**, 245.
McDonough, M. W. (1965). *J. molec. Biol.*, **12**, 342.
Mitchen, J. R. (1969). M.S. Thesis, Purdue University, Lafayette, Indiana.
Mitchen, J. R., and Koffler, H. (1968). Unpublished observations.
Nauman, R. K. (1967). *Bact. Proc.*, 40.
Rinker, J. N. (1957). Ph.D. Thesis, Purdue University, Lafayette, Indiana.
Ritchie, A. E., Keeler, R. F., and Bryner, J. H. (1966). *J. gen. Microbiol.*, **43**, 427.
Roberts, Jr, F. F., and Doetsch, R. N. (1966). *J. Bact.*, **91**, 414.
Rogers, G. E., and Filshie, B. K. (1963). *In* "Ultrastructure of Protein Fibres" (Ed. R. Borasky), p. 123. Academic Press, New York.
Sala, F., Gaertner, F. H., and Koffler, H. (1968). *Gior. Bot. Ital.*, **102**, 327.
Sala, F., and Koffler, H. (1967). *Fed. Proc.*, **26**, 612.
Seidler, R. J., and Starr, M. P. (1967). *Bact. Proc.*, 42.
Seidler, R. J., and Starr, M. P. (1968). *J. Bact.*, **95**, 1952.
Smith, R. W., and Koffler, H. (1965). Unpublished results.
Smith, S. M., and Stocker, B. A. D. (1962). *Br. med. Bull.*, **18**, 46.
Starr, M. P., and Williams, R. C. (1952). *J. Bact.*, **63**, 701.
Stenesh, J., and Koffler, H. (1962). *Fed. Proc.*, **21**, 406.
Stenesh, J., and Koffler, H. (1962). Unpublished results.
Sullivan, Ann (1969). M.S. Thesis, Purdue University, Lafayette, Indiana.
Sullivan, A., Bui, J., Suzuki, H., Smith, R. W., and Koffler, H. (1969). *Bact. Proc.*, 30.
Suzuki, H., and Iino, T. (1966). *Biochim. biophys Acta.*, **124**, 212.
Suzuki, H., and Koffler, H. (1968). Unpublished observations.
Swanbeck, G., and Forslind, B. (1964). *Biochim. biophys. Acta.*, **88**, 422.
Uchida, H., Sunakawa, S., and Fukumi, H. (1952). *Jap. J. Med. Sci. Biol.*, **5**, 351.
Vegotsky, A., Lim, F., Foster, J. F., and Koffler, H. (1965). *Archs Biochem. Biophys*, **111**, 296.
Wakabayashi, K., Hotani, H., and Asakura, S. (1969). *Biochim. biophys. Acta.*, **175**, 195.
Weibull, C. (1948). *Biochim. biophys. Acta.*, **2**, 351.
Weibull, C. (1949). *Biochim. biophys. Acta.*, **3**, 378.
Weibull, C. (1950). *Acta Chem. Scand.*, **4**, 268.
Weibull, C. (1950a). *Ark. Kemi*, **1**, 573.
Weibull, C. (1950b). *Ark. Kemi*, **1**, 21.
Weibull, C. (1951). *Discussions Faraday Soc.*, **11**, 195.
Weibull, C. (1951). *Acta Chem. Scand.*, **5**, 529.
Weibull, C. (1953). *Acta Chem. Scand.*, **7**, 335.
Weibull, C. (1960). *In* "The Bacteria" (Eds. I. C. Gunsalus and R. Y. Stanier), Vol. I, p. 153, Academic Press, New York.
Weibull, C., and Tiselius, A. (1945). *Arkiv Kemi Mineral. Geol.*, **20B**, No. 3.

CHAPTER VI

Antigen-antibody Reactions in Microbiology

C. L. OAKLEY

Department of Bacteriology, The Medical School, Leeds, England

I.	General Considerations	174
	A. The nature of antibodies	174
	B. The nature of antigens	175
	C. The use of standard sera	176
	D. Antigen-antibody reactions	177
II.	The Precipitation Reaction	177
	A. Techniques of precipitation	177
	B. Purification of antibody	181
	C. Separation of macroglobulin from microglublin activity	182
	D. Extensions of the precipitin reaction	182
III.	Agglutination	193
	A. Types of agglutinogen	194
	B. The defects of agglutination	196
	C. Tanned red cell agglutination	200
IV.	Cytophilic Antibodies	201
V.	Complement-fixation Reactions	202
	A. The technique of complement fixation	204
	B. The Wassermann reaction	207
VI.	Antitoxic Antibodies	208
VII.	Antiviral Antibodies	208
	A. Virus neutralizing antibodies	208
	B. Other virus antigen-antibody reactions	210
VIII.	Other Antigen-antibody Reactions	210
IX.	Use of Passive Transfer of Antibodies	211
X.	The Production of Antisera	212
	A. Agglutinating antisera	213
	B. Precipitating antibodies	213
	C. Complement-fixing antibodies	214
	D. Antitoxic antibodies	215
	E. Anti-viral sera	216
	F. Anti-mycoplasmal sera	216
References		217

I. GENERAL CONSIDERATIONS

The reactions between antigens and antibodies are of several different types and have many applications in the biological sciences. It is the object of this Chapter to provide a discussion of the background to the use of antigen-antibody reactions at the research level, with particular emphasis on microbiological applications. Some details of appropriate injection schedules for the preparation of antisera of various types are also included. Experimental details are given where appropriate, and full details of many of the techniques described will be found in the Chapters by Walker *et al.* and Batty in this volume.

A. The nature of antibodies

Antibodies (immunoglobulins) are produced by warm-blooded animals, and by some poikilothermous vertebrates maintained at temperatures of 19°C and above (e.g. Bissett, 1947a, b, 1949–50; Maung, 1963; Alcock, 1965) in response to parenteral experiences, natural or artificial, of antigens—substances of high molecular weight. These are, with few exceptions, foreign to the producer's structure. The antibodies produced in response to an antigen combine almost specifically with that antigen. They are all, as far as is known, members of the globulin class of proteins, but antibodies to a particular antigen may be divisible into immunoglobulins of very different molecular weight. In the early stages of immunization of mammals, for example, the antibody produced is mainly macroglobulin (19S antibody, IgM) of molecular weight 900,000; in the later stages microglobulin (7S antibody IgG) is the main antibody produced, though some macroglobulins may always be present. In certain cases other immunoglobulins (IgA, IgE) may be produced.

If an antibody-producing animal of species X has parenteral experience of an antigen A, it will therefore produce a group of immunoglobulins capable of combining with antigen A. All these immunoglobulins have, beside their antibody specificity, a specificity due to their immunoglobulin class (IgM, IgG, IgA, IgE) in species X—a specificity that they share with macro- or micro-globulins of species X. In the plasma, serum, tissue fluid or tissue extracts of individuals of species X immunized with antigen A we therefore find the antibodies with which we are concerned mixed with a great deal of material similar in nature but incapable of combining with antigen A.

Assay of antibodies to antigen A cannot therefore depend on their nature as macro- or micro-glublins, but must be based on their capacity to combine with antigen A; but we may often find it useful that antibodies share the properties of the general classes of serum proteins to which they belong. In particular, the macro- and micro-globulins of an animal of species X will be antigenic in animals of species Y, which will produce antibodies against them; such antibodies will combine not only with the normal macro- and micro-globulins of species X, but also with antibodies of the corresponding globulin classes, produced in animals of species X.

This is particularly useful when one is concerned to show that antibody to antigen A has combined with that antigen, when, for one reason or another, the usual accompaniments of antigen-antibody reaction have not occurred or cannot be observed. If the antigen A-antibody complex can be washed free of uncombined antibody of species X origin, addition to the supposed antigen-antibody complex of anti-X-globulins prepared in species Y may make it clear that the complex does contain an X-globulin, and therefore probably an antibody. This is the basis of the Coombs antiglobulin test and of many fluorescent antibody techniques.

B. The nature of antigens

So far we have dealt with antigens as if they were single molecular species, and as if the antibody were uniform apart from the class of immunoglobulin to which it belonged. We find, however, that even when the antigen is a single molecular species, the antibody is seldom directed against the entire antigen molecule, only against specific electrically charged groups on the molecule, and that the antibody molecules vary in the particular area of the antigen with which they will combine. All the antibody molecules, however much they may differ in this respect, will combine with the whole antigen, and so will be assayed together as "antibody" in any test in which the whole antigen is used. Distinction can be made only in tests against separated "fragments" of the antigen.

If the "antigen" is embedded in a cell surface, similar variations in antibody specificity occur. For instance in a bacterial surface particular antigenic areas are repeated at intervals, and antibodies are produced against all the varieties of them. It is usual to say that the antibodies are produced against particular antigens, but it seems that the molecular structure of the bacterial surface forms a continuous spiral, and that the antibodies are directed against parts of it very much as they would be against parts of a pure single antigen. All the various antibody molecules will combine with the whole antigen, in this case the bacterial surface. They can be tested for their specificity for a particular area only against bacteria that possess only one of the areas present in the bacterium used to produce the antibody.

Antigens *within* a cell may be exposed by making homogenates of the cell, or by cutting thin sections of it, and antibodies against cell components may be assayed by their effect on the homogenate, or on cell sections, or on cells on whose surface the components of the cell homogenate have been adsorbed. It is worth remembering however that cells or structures in different organs may have many antigens in common, that sera that react with homogenates of many organs may be reacting with the same antigen in all of them, and that values ascribed to antibodies as anti-liver, anti-kidney, anti-spleen antibodies may be only reflections of the same specificity if the tests go no further than this (Thewaini Ali and Oakley, 1967).

Until, therefore, we are in a position to define our antigens and antibodies very thoroughly indeed—and this means not only their chemical composition but their molecular arrangement—figures for assays of antibody are likely to be averages over a fair range of specificity. This does not mean that they are of no value, only that we should use them only for purposes for which their defects are of little moment. Lastly, every bacterial filtrate that has been thoroughly investigated has proved to be an exceedingly complex mixture; in many instances particular activities of the filtrate, e.g. lethal activity or haemolysis, have been shown to be shared by several of the filtrate components. When animals are immunized against these filtrates or against unpurified materials derived from them, the antisera (a convenient term for sera containing antibodies) so obtained are likely to contain antibodies against some or all of the antigenic filtrate components. Attempts to assay an antiserum by determining the degree to which it reduces, e.g. the lethal activity of such a lethal bacterial filtrate, are likely to give misleading results unless a great deal is known about the composition of the filtrate, or relatively pure antigens have been made from it.

C. The use of standard sera

It is convenient, especially when it is difficult to standardize the conditions of an experiment involving the assay of antibodies with any great exactness, to include a standard antiserum, and to assay the test antisera or other antibody-containing material in terms of it. In testing antitoxins this is constant practice (cf. Batty, this Vol. p. 255), but it is not perhaps so easy to see its value in, for example, agglutination tests. Since, however, the results of an agglutination test will depend on the number of organisms used and on the exact antigenic composition of the bacterial surface, which may have been changed by repeated subculture of the organism, it is extremely useful to include in the test a comparison with a reagent that will allow for some of these differences, and may even draw attention to their occurrence.

Last, but by no means least, it is easy to make simple errors in antibody testing, and the inclusion of a standard antiserum may sometimes make it evident what the errors were.

D. Antigen-antibody reactions

All antibody assays depend on the facts that antibody combines almost specifically with its antigen, and that this antigen-antibody combination produces detectable changes in the system in which it is observed. The degree of change will depend on the concentrations of the components; if they are sufficiently high, macroscopic changes such as precipitation or agglutination may occur, and even at very low concentrations it may be possible to show that antibody-containing fluids reduce the activity (enzymic, toxic or infective) of an antigen or of the cell of which it forms part. If the concentrations are too low for detection of direct changes of this kind, it is often possible to determine antibody effects indirectly by more sensitive methods such as complement fixation, or by the use of fluorescent antibodies.

II. THE PRECIPITATION REACTION

When the antigen is a single molecular species which is usually multivalent, addition to it of specific antibody, which is usually divalent, leads to the formation of a lattice of antigen-antibody molecules in which the antibody forms bridges between antigen molecules. In a short time the lattice becomes visible, and with sufficiently high concentrations of antigen and antibody, a precipitate forms. If the antiserum is derived from a rabbit, the precipitate is soluble in excess of antigen only, and in consequence, precipitation occurs over a wide range of antigen concentrations; if it is derived from a horse, the precipitate is soluble in excess of both antigen and antibody, and if the antibody is in some fixed concentration, precipitation occurs over only a narrow range of antigen concentration—a form of precipitation called flocculation.

A. Techniques of precipitation

1. *Ring test*

The usual method of performing precipitation tests is to place standard volumes of antiserum in capillary tubes closed at the bottom, and to layer logarithmic dilutions of antigen over them. Placing either component in fine capillary tubes is often difficult for the inexperienced; a convenient method is to pull a fine capillary Pasteur pipette that will readily enter the entire length of the capillary tube, provide it with a rubber teat, and mount it in one clamp of a retort stand; the rubber teat is then fixed in another clamp that can be varied in width by a screw. The teat is first compressed by tightening the screw, and the Pasteur pipette filled with a convenient volume of serum by placing its tip well under the serum level in a tube or

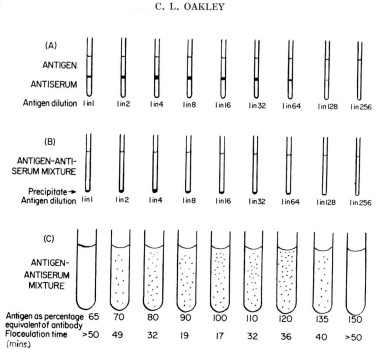

FIG. 1. Comparison of precipitin test (A and B) and flocculation test (C). In A the precipitin test is as it appears about 15 min after the antigen dilutions have been layered on top of the antiserum; in B the antigen-antibody *mixture* has been incubated at 37°C for 30 min and centrifuged. In C the antigen dilutions have been mixed with a constant amount of antibody and incubated for 50 min at 50°C. The floccules are illustrated much as they appear when fully developed. Antigen: diphtheria toxin. Antibody in A, B, rabbit diphtheria antitoxin; in C, horse diphtheria antitoxin.

bottle and loosening the screw. Serum can then be expelled into the capillary tubes with great ease, regularity and freedom from bubbles by inserting the tip of the Pasteur pipette deep into the bottom of the capillary tubes and tightening the screw. It is convenient to have all the serum levels equal.

Dilutions of the antigen two-fold, four-fold or ten-fold, or other convenient logarithmic dilutions are then made in saline, and layered over the serum by the same capillary method, starting at the lowest concentration of antigen. The tip of the loading pipette should be just above the serum meniscus. Precipitate forms on standing at the interface between antigen and antiserum and is best detected by examining the tubes against a back light. (Fig. 1A and B).

If is is necessary to determine the neutral region it is better to mix standard volumes of serum with constant volumes of antigen dilutions, and incubate the mixture at 37°C. for half-an-hour. The mixtures are then cooled to room temperature and the precipitate spun off. The supernatants are then tested by the ring test, as described above, or mixture test for antibody by adding a convenient dilution of antigen, and those negative for antibody are tested for antigen by adding a convenient amount of antibody. The tubes in the neutral zone contain neither antigen nor antibody in the supernatant.

The determination of the neutral zone makes it possible to assay the antibody by a chemical method. The amount of nitrogen in the antigen is determined by a sensitive chemical method, such for instance as the micro-Kjeldahl method; a quantity of antigen is then allowed to precipitate with a little more antibody than will just give a neutral mixture. The precipitate is spun down, washed in ice-cold saline by centrifugation until the supernatant contains no detectable nitrogenous material, and then is tested for its nitrogen content. The difference between this value and that for the antigen alone is due to the antibody, and is multiplied by 6·25 to give the antibody content of the volume of antiserum used.

Obviously the chemical method requires the antigen to be relatively pure; nitrogenous non-antigenic impurities in it will reduce the accuracy of the antibody assay. The most reliable results are clearly those in which the antigen contains no nitrogen (e.g. some pneumococcal polysaccharides); antigens that contain some nitrogen will necessarily be determined with some error variance, and the variance of the antibody nitrogen result will necessarily be the sum of the variances of the estimate of the nitrogen in the antigen and that in the precipitate. The lower the amount of antigen added, within useful limits, the better.

2. Flocculation

The methods given above are usable for rabbit antisera. For horse sera the particular kind of precipitation called flocculation can be used, and antisera can in some cases be tested against a standard and assayed in terms of it.

To a logarithmic range of volumes of antigen in 75 × 8 mm tubes add a constant amount of the standard horse antiserum. Make the volume up to a constant level (usually that of the tube containing the maximum amount of antigen) with saline and incubate in a water bath at some convenient temperature between 50° and 55°C, with the meniscus well above the water level to encourage convection currents. The approach of flocculation is usually heralded by generalised turbulence in the mixture, and the flocculation time is read for the first and second, and sometimes the third tube to flocculate, i.e. the time at which a fine granularity first appears. This time varies slightly for different people examining the same mixture, but each observer is usually consistent.

A first time of flocculation of about 20 min is usually convenient as this allows a sufficiently large difference between the flocculation time of the first and subsequent tubes. The first tube that flocculates is the nearest to neutral, and a more exact determination of the neutral point can be made by interpolation; as the tests can in some cases be done at 10% differences, results can be read to about 2·5% and checked by repetition (Fig. 1C).

The amount of antigen so determined is called the test dose, and this amount of antigen is used in subsequent tests to assay antibody in test antisera. This amount of antigen is placed in the tubes, and a logarithmic range of volumes of the test antiserum is added. The volume is made constant with saline, and the procedure repeated as above. The volume of antiserum in the first flocculating tube, or that given by appropriate interpolation, contains the same amount of antibody as the volume of standard antiserum used in the determination of the test dose.

The test obviously requires that either the antigen shall be pure, or that other antigens present shall flocculate with their antibodies at relatively very different concentrations or at very different times. Technicians who read tetanus toxin-antitoxin flocculation tests, in which, if the toxin is not pure, several flocculation zones occur, seldom have difficulty in deciding which is the correct toxin-antitoxin zone, as the amount of precipitate in the first flocculating tube has to be what they ordinarily experience at that point. The matter can, however, be checked by using a different sample of antigen produced in such a way as to make the relations in amount between the various antigens present very different, determine its test dose and use it in parallel tests; the answer for a given serum should be the same within the errors of the test.

I have not made any reference so far to the method devised by Pappen-heimer and Robinson (1937), who showed that some samples of diphtheria antitoxin, when allowed to flocculate with diphtheria toxin, gave over part of the range of addition of antitoxin floccules showing a nitrogen content linearly related to the amount of antitoxin added. Extrapolation to no antitoxin added would clearly give the amount of nitrogen due to the toxin precipitated, and substraction of this from the floccule nitrogen at any point would give the nitrogen derived from the antitoxin, and consequently the concentration of antitoxin in the serum. Pappenheimer and Robinson (1937) therefore recommended that diphtheria antitoxin should be assayed by adding it to a constant amount of diphtheria toxin, whose flocculating nitrogen content was known, so as to lie in the flocculating range, and by estimating the nitrogen in the washed floccules estimating the antitoxin content of the serum. Unfortunately, though some sera behave in such a way as to make the test applicable, the majority of commercial diphtheria antitoxic sera do not, and the method cannot be recommended for them.

B. Purification of antibody

So far we have been dealing with pure antigens (however complex the antibody system) and relying on this to give us specific precipitation. Flocculation with horse antisera, with its characteristically narrow zone, will help in avoiding co-precipitation of different antigens with their antibodies, but with rabbit antibody the wide zone of precipitation makes separation difficult. To avoid this it is necessary to absorb or precipitate out unwanted antibodies (better still, of course, not to have them in the first place, but anyone may be unlucky).

Thus, for instance, in the early stage of immunization of rabbits against the sera of other vertebrates, the rabbit antisera show little cross-precipitation with sera other than that against which they were immunized, but if immunization is continued for a long period, or adjuvants are used after the early stages, the antisera commonly show some non-specific cross-precipitation. This non-specific activity can usually be removed, without much loss of specific activity, by treating the antisera with suitable antigens. Thus if the antiserum is made against human serum, it may, if the rabbit is immunized for a long period, show some capacity to precipitate with ox serum. The best way to get rid of this activity is to do a straightforward precipitation test with ox serum, centrifuge out the precipitate, and test the supernatant for neutrality. Then to a larger bulk of rabbit anti-human serum add a little more ox serum than will remove all that non-specific antibody, incubate at 37°C for 30 min and centrifuge out the precipitate. Do not forget that in doing so you will dilute the antiserum, and that many absorptions will dilute it to a degree that may seriously affect its precipitating properties, quite apart from the fact that it will contain traces of sera from other species that may raise inconvenient problems if the sera are used for purposes other than those for which they were made. Besides this non-specific cross-precipitation you will, if you are making, e.g. rabbit antihuman sera, have the difficulties inherent in immunizing animals against mixtures of antigens.

The amount of antibody produced often bears little relation to the amount of antigen injected, and you will clearly get a mixture of antibodies, of which the one you want is not necessarily in the highest concentration; indeed a simple capillary precipitation test with rabbit antihuman serum and human serum will often show two precipitation rings on careful examination. If you need precipitating sera against a particular serum fraction, it is far better to purify that serum fraction by electrophoresis or by Sephadex filtration and to immunize the rabbit with the purified product. Another method is to subject the serum to immuno-electrophoresis, in which the products of electrophoresis in agar gel and a rabbit antiserum against the whole serum are allowed to interact by diffusion (Grabar and Williams, 1955) (Fig. 9, p. 191). Precipitation arcs appear in positions from which the probable nature of the antigen can be determined; the arc corresponding to the antigen required is cut cleanly from the gel, washed thoroughly to remove excess serum, ground gently and injected directly into the popliteal gland of an anaesthetized rabbit. Relatively pure antisera (in the sense that they precipitate almost exclusively with the one antigen) can be prepared in this way.

C. Separation of macroglobulin from microglobulin activity

If you need to determine the macroglobulin and microglobulin precipitating activities of a serum, the easiest way is to determine the total amount of antibody chemically, then to destroy the macroglobulin activity by treating the serum with 2-mercaptoethanol, dialysing the treated serum exhaustively against a phosphate buffer plus iodoacetamide, concentrating the dialysed material, and retesting it chemically for 7S precipitating antibodies. The difference between the two figures will give the macroglobulin (19S) antibody level.

D. Extensions of the precipitin reaction

In systems where the numerous antigens cannot readily be separated, precipitation in agar gels can be employed. Several methods are available.

1. *The Ouchterlony or immuno-diffusion methods*

The principle of these is that the antigen mixture is placed in a basin cut or moulded in an agar plate (the basin may be circular, but Ouchterlony recommends a lozenge shape); the antiserum is placed in another similar basin, either with a fixed distance between its centre and that of the antigen mixture if the basins are circular, or at a convenient distance between the nearest sides if the basins are lozenge-shaped. Antigens and antibodies diffuse into the *arena* between the basins, and precipitation occurs, when the appropriate concentrations are reached, as lines normal to the line joining the centres of circular basins, thickest on that line and fading off away from it, or parallel to the approximated sides of lozenge-shaped basins. With reasonable luck the concentrations of each antigen and antibody will be such as to give separate lines for each system, though some may overlap.

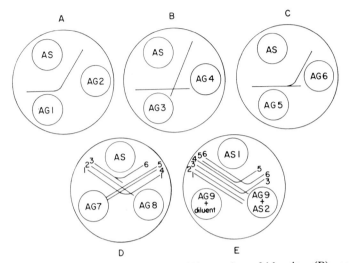

FIG. 2. Ouchterlony plates, illustrating (A) reaction of identity; (B) reaction of non-identity; (C) reaction of partial identity; (D) complex system, showing all three reactions shown in (A), (B), (C); (E) use of these reactions to indicate partial and complete neutralization of the antigens in a filtrate by antiserum.

Another filtrate may then be compared with the first or sera may be titrated against the antigen mixture by adding a mixture of serum and antigen in another basin, which if circular will have its centre so placed as to form an equilateral triangle with those of the other two basins, or if lozenge-shaped will lie symmetrically at the same distance from the antiserum basin as the antigen basin does. If the antiserum under test has antibody against any of the components of the antigen mixture, the corresponding lines will be formed nearer the antigen-serum mixture basin than they will to the antigen basin. Sera can then be assayed against a standard antiserum for their capacity to move the lines to a standard position, each line, however moved, being identified by its fusion with the line between the antigen and reference antiserum basins.

It may be convenient at this point to refer in some detail to the various reactions that may occur when two or more filtrates diffuse against the same antiserum. For convenience let us take two filtrates; the argument is readily generalizable. If there is an antigen common to filtrates AG1 and AG2, to which the antiserum possesses antibody, and the concentrations of the antigen are much the same in the two filtrates, lines of precipitation will appear between the AG1 and the antiserum basins, and between the antiserum basin and filtrate AG2, and these will fuse completely in the "reaction of identity" (Fig. 2A). If the two filtrates (AG3 and 4) have no antigen in common, but there is antibody to both in the antiserum, the precipitation lines will cross one another in the "reaction of non-identity" (Fig. 2B). When an antigen to which antiserum AS possesses antibody is present in filtrate AG5 and a related but not identical antigen in AG6, the lines of precipitation may fuse, but one line may be continued as a spur in the "reaction of partial identity" (Fig. 2C). If we imagine that antigen AG5 has active antigenic patches a, b, c, d, e and therefore excites the production of antibodies A, B, C, D and E, and antigen AG6 has active patches a, c, d, g, h, then during diffusion AG6 will remove antibodies A, C and D from the antiserum and give a line of precipitation, but antibodies B and C will diffuse away to precipitate with AG5, and thus continue the AG5 precipitation line as a spur.

Figure 2D shows all three types of reaction. Antigen 1 in AG7 and AG8 shows a reaction of identity; antigen 2 in AG7 shows non-identity with all antigens in AG8, and antigens 4 and 5 in AG8 show non-identity with all antigens in AG7. Antigen 3 in AG7 shows a reaction of partial identity with antigen 6 in AG8.

Reactions of identity and non-identity are easy to understand; partial identity is more difficult, as some spurs may be due to differences in concentration of the same antigen in two filtrates. Thus if two filtrates containing the same antigen are diffused against the same antiserum the line corresponding to the higher concentration of the antigen will appear first, nearer the antiserum basin, and may have extended a considerable distance before the line corresponding to the lower concentration of antigen appears further away from the antiserum basin; fusion of the lines may then leave a redundant spur. If spurs appear on lines nearer the antiserum basin, a further test should be done in which the filtrate containing the higher concentration of antigen is diluted to give a line in much the same relative position as that given by the weaker filtrate; if a spur is then produced, partial identity of the two antigens is more probable.

Evidently if the basins are arranged in a circle round the central well, numerous antigenic mixtures can be tested against a single serum, or numerous antisera can be tested against a single antigenic mixture (Fig. 3). Six basins in the outer ring is a convenient number; care should be taken that the distances between centres of the inner well and each of the outside wells is constant.

The method is useful, but very rough, and wasteful in that a large percentage of non-productive diffusion occurs. All that matters to the experiment is the arena between the basins, but most of the diffusion takes place outside this area. To get over this difficulty Oakley (unpublished) some years ago devised a diffusion box (Fig. 4), in which a T-shaped cavity is constructed by sawing out of a Perspex plate about 5 mm thick and

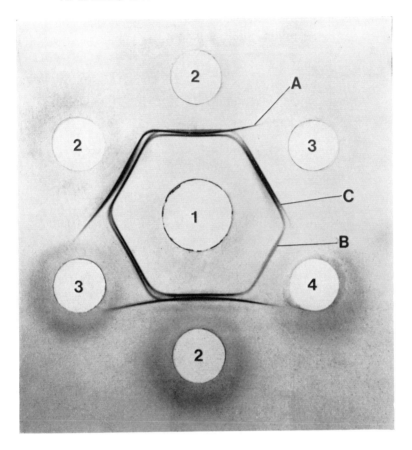

FIG. 3. Photograph of an Ouchterlony plate showing reactions of identity and non-identity. (From Norris, 1969.)

Oxoid ionagar No. 2 (1%) in saline poured to give a 2 mm deep layer on a 3 in. square glass plate. Wells cut using a Shandon (Shandon Scientific Co. Ltd., 65 Pound Lane, London, N.W.10) gel cutter.

1. Solution of *Bacillus thuringiensis* crystal protein (1 mg/ml) (Norris, 1969).

2. Rabbit antiserum prepared against whole protein solution.

3. Specific antiserum prepared by injecting into rabbits precipitin lines B and C cut from Oakley and Fulthorpe (1953) columns.

4. Antiserum produced by injecting crystal protein from a different strain of *Bacillus thuringiensis* into rabbits.

A, B and C: Specific preciptin lines.

Basin, contains antigens a, b and c, basin 3 contains antigens b and c, and basin 4 contains antigen b.

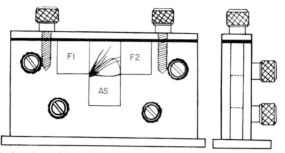

Fɪɢ. 4. Diffusion box. Two filtrates (F1, F2) diffuse against a single antiserum (AS). If the antigens producing lines are numbered from below on the left, filtrate F1 contains antigens 1, 2, 6, 7, 8, 9, all absent from F2 ,antigens 3 and 4, much weaker in F2, and antigen 5, weaker in F2. Similar results might be obtained if F2 was a filtrate plus a definite volume of antiserum. In this event the antiserum would have considerable neutralizing power against antigens 1, 2, 6, 7, 8, 9, less against antigens 3 and 4, and less still against antigen 5.

cementing a rectangular plate of perspex about 10×5 cm on each side of the plate containing the T, whose dimensions are 3 cm wide by 1 cm deep for the horizontal arm of the T, and 1×1 cm for the vertical arm. Pieces sawn out of the Perspex are smoothed to form two $1 \cdot 5 \times 1 \cdot 0 \times 0 \cdot 5$ cm blocks, through each of which is drilled a 2 mm cylindrical hole parallel to the $1 \cdot 5 \times 1$ cm face and the $1 \cdot 5 \times 0 \cdot 5$ cm side. Similar holes are drilled through the T-shape containing block, beyond the T, on either side, and a projecting headless screw is driven into each of these. A rubber gasket is cut to cover the top of the box, and it and a piece of Perspex of similar shape are perforated so that the projecting screw passes through them, and the lid may be screwed on firmly with a milled nut over the projecting screw.

Fill the vertical arm of the T with a mixture of equal volumes of 2% agar and reference antiserum made up at 50°C and allow it to set; allow for the contraction of agar on setting by filling to a little above the top of the vertical arm (the surface tension will usually prevent the serum-agar mixture from running over). Then place two blocks each perforated with a vertical cylindrical hole, one in the left and the other in the right parts of the horizontal arm of the T, so that the 1·5 cm lengths are vertical. Overfill the centre space (1×1 cm) with 1% agar and allow to set. Remove one of the blocks, and cut the agar flush with the top of the box with a razor blade; remove the other block. Warm the antigen mixture and antigen-serum mixture to 37°C and place one in the left and the other in the right of the horizontal arm of the T; insert the gasket and screw down the lid. Incubate at 37°C. In this system all the diffusion is useful and the mixtures cannot dry up.

Obviously the position of the lines in the left-hand side of the arena

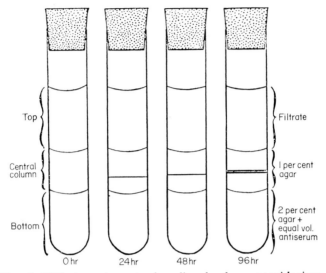

FIG. 5. Diffusion column to show line development with time.

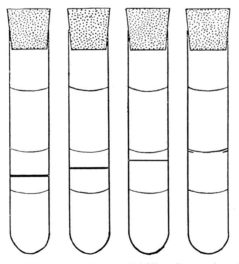

Figs. 5 and 6 from Oakley and Fulthorpe (1953)— Increasing dilution of filtrate, or increasing addition of antiserum to it.

FIG. 6. Diffusion column, to show effect on line position of dilution of filtrate or addition of antiserum to it.

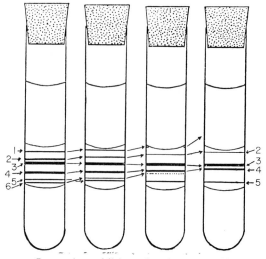

— Increasing addition of antiserum to filtrate.

FIG. 7. Diffusion column, to show effect on a complex system of lines of increasing addition of antiserum to the filtrate. The lines are numbered 1–6, and then movement may be followed by means of the arrows. The antiserum added to the top evidently has no antibody to antigens responsible for lines 3 and 5; it possesses some antibody to all the other antigens. (From Oakley and Fulthorpe, 1953).

will depend on the relative proportions of the antigens in the left-hand basin and the antibodies in the lower one. The more concentrated the antigens compared with their antibodies the more acute will be the angle made by the line and the horizontal; the weaker the antigen in proportion to the antibody the higher up the arena the line will appear, the less acute the angle it will make with the horizontal, and the more slowly will it develop across the arena. If no antibodies are added to the antigens on the right, the diffusion pattern will be completely bilaterally symmetrical, each line joining in the centre vertical line of the arena with its opposite number. Addition of antibodies to the antigens on the right will reduce the quantity of their antigens available for diffusion, raise the origins of their lines and displace the junction point to the right. At some stage of increasing addition of antibodies, some particular antigen will be so reduced on the right as not to produce any diffusion line at all. In the event the line produced on the left of the arena will continue undeviated across it. This method can clearly be used to compare sera against a standard antiserum— probably the best such standard is the reference serum used in the base of the T, since it will remove the lines in their own order—using as indicator complete failure to produce a right-hand diffusion line.

After use the agar blocks can be removed from the box with a jet of cold water directed through a 1 mm-bore Pasteur pipette; the boxes cannot be autoclaved, so cleaning must be very thorough. The boxes slowly discolour with age and repeated use.

The method suggested above can be made very accurate, but it is somewhat clumsy, and numerous boxes are required to obtain a small amount of information; besides which, bubbles of gas coming out of solution from the filtrates may seriously impede diffusion. A rather more convenient method involves the use of double diffusion into an agar gel (Oakley and Fulthorpe, 1953) (Figs 5–7). In this, constant volumes of a mixture of equal amounts of 2% agar and antiserum is placed in 75×8 mm tubes and allowed to set. Above this is placed a layer of 1% agar of uniform height; as agar contracts on setting, it is necessary to add the agar to a standard height while it is still fluid, and not to add it to the same height as in the preceding tube, in which it may already have solidified. When the intermediate layer of agar has set, filtrates or their dilutions or filtrates plus antisera are poured on top of it, and the tubes are closed with rubber bungs and incubated at 37°C.

The height of the layer of antiserum plus agar at the bottom of the tube does not seem to matter very much, as long as its meniscus is well clear of the "shoulder" of the tube; the height of the layer of agar will depend on the nature of the materials and the information one needs. The less its height, the denser the precipitation lines formed, but the closer they are together; the greater the height, the better the separation of the lines, but the fainter they tend to be, and the longer they take to form.

Oakley and Fulthorpe found 8 mm a very convenient height for many purposes, but with very concentrated antigen mixtures agar columns of much greater length may be appropriate.

Dilution of antigens in tests of this kind displaces the line (disc) of precipitate upwards towards the upper meniscus of the agar column, and antigens may be assayed by determining the dilution of antigen that, diffused against an antiserum, produces a line in the upper meniscus. Another method is to determine the amount of antigen that when mixed with a standard number of units of antiserum and allowed to diffuse against an antiserum produces a line at some specified level in the agar, preferably in the upper meniscus. Sera may then be assayed against this amount of antigen in comparison with a standard serum. If the antigenic material is complex, multiple lines may be produced in the agar, and special methods must be used to distinguish them (see Oakley and Fulthorpe, 1953, for these; and Preer, 1956, for a modified capillary tube method). The chief defect of this method is that direct comparison of antigen mixtures for line fusion is not possible.

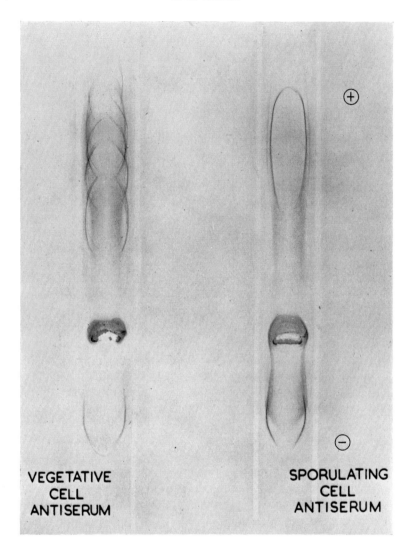

FIG. 8. Immuno-electrophoresis of a mixture of bacterial antigens. Sporulating cells of *Bacillus cereus* were disintegrated by ultrasonic irradiation. The disintegrate was electrophoresed from wells in a 2 mm thick layer of agar (Oxoid ionagar No. 2 (1% w/v), veronal/acetate buffer pH 8·6) on $3\frac{1}{4}$ in. square glass plates at a voltage of 6 v/cm for 120 min. At the end of electrophoresis trenches were cut from the agar and filled with rabbit antiserum produced by injection of disintegrated vegetative cells (left) and disintegrated spores (right) and incubated at 30°C for 24 h. The relationship between spore and vegetative cell antigens in the sporulating cells is clearly seen. (From J. R. Norris; unpublished.)

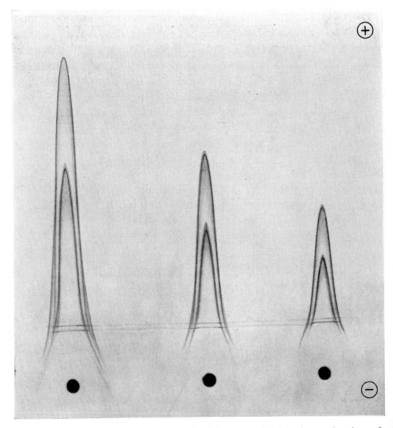

Fig. 9. Photograph showing the method for quantitative determination of anti-gen solutions by electrophoresis into agar containing antiserum. Doubling dilutions of an antigen mixture (2 components) were placed in wells cut into 2 mm thick sheets of agar (Oxoid ionagar No. 2 1% (w/v), Oxoid Veronal-acetate buffer pH 8·6, Rabbit antiserum 7% (v/v)) on $3\frac{1}{4}$ in. square glass plates and the preparation subjected to a voltage of 8 V for 24 h. The amounts of antigens in the solutions are determined by comparison with standards using a calibration curve based on measurement of the area below the precipitation curve. (J. R. Norris; unpublished.)

2. *Immuno-electrophoresis*

Another way in which precipitation techniques may be exploited is in immuno-electrophoresis (Grabar and Williams, 1955) (Fig. 8). In this process a layer of agar dissolved in an appropriate buffer is allowed to set on a glass plate so that a wick is fixed in the agar at each end; each of these wicks is then led into a trough of buffer, connected by an inverted U-tube filled with buffer to another trough of buffer into which one of the electrodes is placed.

A basin is cut out of the agar, and filled with the "substance" to be subjected to electrophoresis, and narrow troughs are cut in the agar parallel to the line of current flow. A fixed potential difference is then established between the two electrodes, and molecular species in the material in the basin then move into and through the agar in a direction and to a degree determined by the character and size of the residual charges on their molecules and by the effects of endosmosis. After a time of exposure to the electric field determined by experience or by the movement of a known substance stained with a convenient dye, the agar plate is removed from the field and the lateral troughs are filled with antisera. After a convenient diffusion period, arcs of precipitation develop where antigen is present in the agar and the corresponding antibody in the trough, and the antigens may be identified by the position of the corresponding arcs in the gel. The use of different antisera in the troughs may provide extra information.

This method of performing immuno-electrophoresis cannot be made quantitative, but two other methods can. The first (Afonso, 1964, improved method, 1965; objections, Bogdanikowa et al., 1966, answered by Afonso, 1967) involves electrophoresis of 2 μl of material in a uniform thin layer of agar at pH 8·6 in a potential gradient of 10 V per cm; the plate is allowed to stand for $4\frac{1}{2}$ h, and is then sprayed with the antiserum (4–5 μl per cm^2 of trimmed agar) and incubated in a moist atmosphere at 30°C for 1 h (after which the same dose of antiserum is reapplied), and then for a further 10–12 h. Arcs of precipitation are developed with 0·5% acetic acid or stained with naphthalene black (saturated solution in methyl alcohol containing 10% acetic acid and 10% water). The diameter of the precipitation circles is proportional to the concentration of the antigen, provided the antibodies are present in much the same concentrations.

A rather neater method is described by Laurell (1966) who suggests subjecting the substance under test to electrophoresis in agar containing the antiserum (Fig. 9). (This had also been considered by Afonso, but rejected because the patterns of precipitation in complex mixtures were so distorted.) If this method is used both the antigens in the test substance and the antibodies in the agar move in the electric field, and the precipitation arcs resemble "up-and-down" rocket trajectories. If an antiserum against only one component of the test substance is included in the agar, only the arc corresponding to this component develops, and the concentration of the component can be determined by comparison with the arcs produced in the same sheet of agar from standard concentrations of pure component. Scolari et al. (1968) showed that a slight modification of Laurell's method gave excellent results with pure serum proteins. For a detailed account of this kind of technique the reader is referred to Clarke and Freeman (1966).

If antibodies to a few components are present in the serum, determination of the concentrations of the corresponding antigens may be possible, but the method is far more suitable for single antigen-antibody systems; if the antigen is pure, it does not matter how complex the antiserum is.

III. AGGLUTINATION

When antigen and antibody are of comparable size, the complex of antigen and antibody normally forms a lattice that separates from solution either as a precipitate or as a mass of floccules. When the antigen is fixed in the surface of a bacterial or other cell, antibody may form bridges between similar antigenic areas on the bacterial or cellular surface, and so lead, if sufficient bridges are present, to clumping of the bacteria or cells into masses as large as the forces between antigen and antibody will support. This clumping is called **agglutination**, the antigens to which it is due are called **agglutinogens** and the antibodies **agglutinins**. It is a natural consequence of the fact that antisera are closely specific that antisera produced against a particular bacterial strain will agglutinate all bacteria sharing antigens with the homologous strain; not only will bacteria sharing *all* agglutinogens with the homologous strain be agglutinated, but also those sharing only some of them.

If for instance we produce an agglutinating serum by injecting into a rabbit killed suspensions of organism A possessing agglutinogens 1, 2, 3, 4, that serum will agglutinate not only bacteria possessing agglutinogens 1, 2, 3, 4 (i.e., organisms of the same strain) but also organisms B, C, D, E, F, G, H, J, K, L, M, N, P and Q, possessing antigens 1, 2, 3, 4, 1 and 2, 1 and 3, 1 and 4, 2 and 3, 2 and 4, 3 and 4, 1, 2 and 3, 1, 2 and 4, 1, 3 and 4 and 2, 3 and 4 respectively. It is true that the titre of the serum (i.e., the dilution of the serum that will just produce agglutination of a standard suspension of a bacterium) may differ in agglutination tests against different bacteria; but exact ideas on the degree of relationship between bacteria must depend on absorption tests.

If a serum is prepared against bacterium A, it will contain antibodies against agglutinogens 1, 2, 3, 4; and, if it is mixed with a suitable suspension of bacterium A, it will agglutinate it. The bridge-formation between bacteria necessarily involves removal of the antibodies from the solution by fixation on the bacteria; and if the bacteria are centrifuged out, the supernatant will not agglutinate organism A. (This process may involve dilution of the antiserum, and several repetitions of this process, usually called agglutinin-absorption.)

If, however, this serum against bacterium A is mixed with a suitable suspension of organism H, possessing agglutinogens 1 and 4, it will agglutinate it, but the bacteria will remove from the serum only the agglutinins against antigens 1 and 4; and so the supernatant from the mixture will still agglutinate bacterium A, though the titre may be somewhat reduced. Obviously if we examine a number of bacterial strains by this technique against a single antiserum, we can divide them into those that will remove all the agglutinins from the serum, and those that will not do so, even after repeated absorptions, and therefore lack some of the agglutinogens in bacterium A, though they may of course have others that it does not possess.

The next stage (mirror agglutinin-absorption tests) involves the production of antiserum against the organism H, with antigens 1 and 4. Obviously bacterium A will remove all the agglutinins against bacterium H from it, and so naturally will bacterium H itself; and a comparison between the effects of bacterium A and H on the two sera shows that bacterium A possesses at least one antigen absent in H, and that H does not possess any antigen absent in A. If other strains, such for instance as B(1), C(2), D(3), E(4), F(1, 2), G(1, 3) are available, further comparisons are possible. Anti-A serum absorbed with bacterium H will clearly agglutinate C(2), D(3) and G(1, 3), but not B(1) or E(4), and mirror agglutinin-absorption tests between these will readily demonstrate the differences between them and A and H.

A. Types of agglutinogen

So far we have been considering agglutination as if agglutinogens were all of the same kind. In fact they are of at least three kinds—**somatic (O)**, **flagellar (H)** and **capsular** (sometimes of more than one kind). The somatic antigens are those of the body of the bacterium, the flagellar antigens are those of the flagella; the first are complex lipoproteinpolysaccharide complexes, the second are proteins. The capsular antigens are those of the polysaccharide or polypeptide capsular envelope of the bacterium. Though flagella are fairly readily detached from the bacterial body, flagellar and somatic agglutination tests are usually done with intact organisms; the capsular material may be removable from the organism, in which case capsular agglutination tests on the whole organism may be supplemented by precipitation tests with rabbit antisera and the separated capsular material.

If the organism is naturally non-flagellate or is a non-flagellate mutant of a normally flagellate species, agglutination by a specific antiserum, whether against a flagellate or non-flagellate form, will necessarily be of somatic (O) type, in which the antibody bridges are between the antigens in the bacterial walls, which are therefore brought close together. Such agglutination therefore produces small tight agglutinated masses of bacteria, very difficult to disperse by shaking. Flagellar (H) agglutination, on the other hand, is produced by antibody bridges between adjacent flagella on different organisms, so that though the flagella are brought nearer together the bacteria are not. This type of agglutination tends, therefore, to produce a larger fluffier agglutinum, readily redispersable by shaking. Capsular agglutination, since the capsule is over the somatic antigens, tends to produce a small agglutinum like that of O-agglutination.

As the antigens of flagella are commonly different from the somatic antigens of the same organism, and the capsular antigens different from either, determination of all these antigens may effectively specify the whole organism, or rather the antigenic composition of its surface. Classification of *Salmonella* "species" has depended on O-antigens, by which the genus has been divided into groups, and "H"-antigens, by which the groups have been divided into "species". In some of these species (e.g. *S. gallinarum*, *S. pullorum*) there are no flagella, and no H-antigens are available for examination; in others there are two phases in flagellar antigenic composition. A knowledge of such structural characters may be invaluable for recognition of particular organisms, and a collection of single-factor antisera (i.e., sera capable of reacting with only one major antigenic group on a bacterial body, flagellar surface or capsule) may make this recognition even easier.

Cooper (1956) made use of the fact that flagellar agglutination abolishes the motility of flagellate organisms to identify members of the Enterobacteriaceae. The organisms are sown on a suitable medium, and drops of flagellar antisera (active against all flagellar phases of each species) are placed on the area where the lawn will develop. Plaques of inhibition of growth occur over the area over which *specific* flagellar antiserum has been applied.

For detailed considerations of agglutination methods in an organism with somatic, flagellar and capsular antigens (*Escherichia coli*) see Bettelheim and Taylor (1969).

It has been assumed so far that all bacterial suspensions used in agglutination tests, whether for the detection of antibodies in sera, or for the immunization of animals against particular groups of antigens, or for examination for particular antigens, are non-fimbriate.

Fimbriae (Duguid *et al.*, 1955; sometimes unnecessarily called pili: Brinton, 1959, 1965; but see Duguid, 1964) are non-flagellar filamentous appendages of many species of Enterobacteriaceae; they are shorter than flagella, and the commonest type (type 1) confer on the bacteria that bear them the ability to agglutinate suspensions of suitable red cells (e.g., those of guinea-pigs); this haemagglutinating property is mannose-sensitive (Duguid, 1959, 1964). They have also a specific antigenic constitution, which bears little or no relation to the O-, H-, or capsular antigenic structure of the bacteria from which they are derived; in consequence, fimbriate strains of bacteria may show fimbrial agglutination by appropriate antisera.

Thus all fimbriate species of *Salmonella* examined by agglutination and agglutinin-absorption tests (Duguid and Campbell, 1969) shared one fimbrial antigen, which was also shared with the fimbriae of strains of *Arizona arizonae*, *Citrobacter ballerupensis* and *C. freundii*, but not with those of *Shigella flexneri*, *Escherichia coli*, *Klebsiella aerogenes* or *Enterobacter cloacae*. A minor fimbrial antigen is shared by *Shigella flexneri* and all type-1 fimbriate strains of *E. coli* (Gillies and Duguid, 1968); and some species of *Salmonella*, but not all, may share other fimbrial antigens. It is clear from this that if fimbriate organisms are used to immunize animals, the antisera obtained may give very misleading results if the possibility of common fimbrial antigens is not borne in mind.

The obvious remedy is to use non-fimbriate strains for all such purposes; there are a number of methods for obtaining them. Strains grown for short periods (e.g., 6–12 h) in static broth are usually non-fimbriate. Non-fimbriate variants of normally fimbriate forms may also be employed, or one may remove the fimbriae by suspending washed bacteria in 0.005 N HCl at 37°C for 5 min and then neutralizing the acid with 0.05 N NaOH, or remove both fimbriae and flagella by heating the bacterial suspension at 100°C for 30 min. Such defimbriated bacteria do not agglutinate 2% suspensions of guinea-pig red cells (Duguid and Campbell, 1969).

Viruses, if sufficiently large, may show agglutination by appropriate antisera (Ledingham, 1932, 1933), besides producing precipitating and other antigens.

Electrolytes are essential for agglutination, so the sera are best diluted and the organisms best suspended, in saline (1% NaCl) (but see below). Since Gram-negative organisms readily undergo lysis if fresh, complement-containing antisera are added to them, the complement in the sera should be inactivated by heating to 55°C for 30 min (see below).

B. The defects of agglutination

Many bacteria, especially the members of the genus *Clostridium*, show auto-agglutination—i.e., suspensions of these bacteria in 1% saline show spontaneous agglutination in the absence of specific antisera. Reduction of the concentration of salt in the suspending and diluting fluids to 0·45% or less will often help, but some suspensions, of *Cl. welchii* for example, auto-agglutinate in tap-water; in cases like this a fluorescent antibody method is essential (see Walker *et al.*, this Volume, p. 219).

Agglutination necessarily involves only surface antigens and the homologous antibodies, whereas immunization against a particular bacterium will provoke in the animal antibodies against not only the superficial antigens, but also against the deeper ones. Thus for instance virulent *Salmonella typhi* possess beside the usual antigenic complement of these organisms, an antigen Vi, so distributed over the bacterial surface as to cover up completely the deeper surface antigens, which in non-virulent *S. typhi* are freely exposed. Antisera produced against non-virulent strains of *S. typhi* will therefore not agglutinate the virulent strains of this organism, and conversely.

1. *Prozone formation*

As an agglutinating serum is diluted out, it normally produces less and less agglutination of the homologous organism until finally a state of affairs is reached in which the bacterium-serum dilution mixture contains too little antibody to provide enough bridges to produce visible or even microscopic agglutination. This is simple enough, but some organisms, especially *Brucella* spp. show a much less readily explainable phenomen, the prozone. These organisms may show no agglutination when they are mixed with antiserum diluted 1 in 20, 1 in 50, 1 in 100 or even 1 in 200, but are readily agglutinated by serum diluted 1 in 400, 1 in 800, 1 in 1600, or 1 in 3200. Moreover it is often observed that the sera of patients suffering from brucellosis taken in the early acute stages of the disease may show no prozone; those taken later commonly do so.

The reason, at any rate in brucellosis, is that the 19S macroglobulin, produced as usual in the earlier stages of immunization or in the earlier stage of the disease, readily agglutinates *Brucella abortus*; the 7S and IgA microglobulins produced in the later stages are commonly univalent, and do not therefore agglutinate the homologous organism (Kerr *et al.*, 1967). 7S antibody is preferentially taken up by the antigenic sites on the bacterial body, so preventing the action of 19S antibody; no agglutination is in consequence produced until 7S antibody is diluted out sufficiently to allow 19S antibody to be taken up by the bacteria in sufficient amounts to cause agglutination.

We can show that this is true by treating *Brucella abortus* suspensions with *Brucella* antiserum at a high concentration that will not cause agglutination, washing the organisms with saline by repeated centrifugation to remove all unabsorbed antibody, and then treating the resuspended organisms with antibody against the globulins of the animal species in which the *Brucella* antiserum was produced. A general anti-globulin serum will do, but it is clearly tidier to use an anti-microglobulin antiserum for the test. Since the molecules of the anti-microglobulin serum will form bridges between the molecules of the antibody (7S microglobulin) attached to the surface antigens of the bacteria, the bacteria will be secondarily agglutinated by this reaction if they have taken up 7S non-agglutinating antibody. That the agglutinating antibody at lower concentration of *Brucella* antiserum is 19S antibody can readily be demonstrated by treating the antiserum with 6-mercaptoethanol, when it will cease to agglutinate directly at these lower concentrations, whereas, since its 7S antibody is unaffected, it will still show a positive antiglobulin test at lower concentrations. (See Kerr *et al.*, 1968, for a thorough consideration of these points, and details of methods.)

Most of the arguments applied to bacteria apply also to other cells, which may be agglutinated when antibodies against their surface antigens are added to cell suspensions. These antisera should be heated to 55°C for half-an-hour to prevent lysis of the cells. Thus red cells of species X may readily be agglutinated by inactivated antiserum produced by injecting X red cells into an animal of species Y and lymphocytes are agglutinated by antilymphocytic sera.

2. Univalent antibody

I have already said that antibody is *usually* divalent (p. 174) and it is clear that this property is essential to lattice formation and so also to precipitation and agglutination. It may be convenient now to deal with the case where the antibody is unusual in being univalent, or, as some say, incomplete. This condition occurs in foetal incompatibility of the Rh type, in which as a result of the mating of Rh-positive men with Rh-negative women, foetuses whose red cells are antigenically distinct from those of their mothers develop. Unluckily, though the maternal and foetal circulations are separate in mammals, haemorrhages occasionally take place from the foetal into the maternal circulation, so that the mother is exposed to foreign (i.e., antigenically distinct) red cells. She seldom produces such antibody in her first pregnancy, but if she subsequently becomes pregnant of another Rh-positive child (as she must do if her husband is homozygous for Rh-positiveness) she is likely to produce antibody against the foetus's red cells. Both divalent and univalent antibody are produced, but only univalent antibody passes from maternal to foetal circulation. There the univalent antibody combines with the foetal red cell surfaces, but does not agglutinate the red cells; it does, however, alter the red cells in such a way that they are more rapidly removed from the circulation, and the foetus becomes anaemic. Proof that the foetal red cells are coated with antibody can be provided by washing them free from serum, and then treating them with anti-human-globulin sera, by which, as they are coated with human globulin in the form of antibody, they are readily agglutinated.

Electron microscopy has now reached such a level that it is possible to demonstrate antibody molecules fixed to the surface of spirochaetes and even to determine whether they are micro- or macro-globulins (Osechinsky et al., 1969).

3. Mixed agglutination

Cellular antigens effective in agglutination are necessarily fixed in surfaces, and if the surface forms a continuous sheet, as for instance in a tissue culture growing on a coverslip, antibody may be absorbed on the antigenic areas without producing anything resembling agglutination, merely because no bridges are formed; indeed one might regard the antibody as fixed by one combining site only. This leaves its other combining site free to combine with the same antigen. If therefore inactivated antibody against human red cells is applied to a tissue culture, it will be absorbed if the tissue-culture surface contains antigens common to the tissue culture and human red cells. If then the unabsorbed antibody is washed away with tissue-culture medium, and human red cells are added to the washed tissue culture, they should be absorbed by the antibody fixed to the culture surface and so attach themselves to it. If this occurs, and so the tissue culture contains antigens in common with red cells, the chances are that the tissue culture is of human origin; if no fixation of the red cells to the culture surface occurs, it is unlikely that the culture consists of human cells. An even neater method is to use incomplete antibody against red cells, add it to the tissue culture, wash off the excess antibody and add a mixture of red cells and incomplete antibody washed to remove any antibody excess. Then, if the tissue culture and the antibody-coated red cells floating just above it are treated with an antiglobulin antibody against globulin of the species in which the antibody was produced, some of the antiglobulin will agglutinate the red cells, and if some of the antigens on the tissue culture surface are the same as those on the red cell surface, the agglutinated masses of red cells will be fixed by antiglobulin bridges to the surface of the tissue culture. If there are no antigens in common between red cell and tissue culture, the agglutinated red cells will float readily in the liquid, and are easily detached from the tissue-culture surface if they sink on to it.

Coombs et al. (1961) and Coombs (1964) have used these methods to identify the species of origin of a number of tissue cultures, and have shown with disconcerting clarity how often the cells of these cultures are no longer antigenically related to the cells from which they were believed to have started. (See also Steele and Coombs, 1964.)

A kind of marked antibody method has recently been applied to the demonstration of antigens in electron-microscope sections. In this either the packed organisms or the sections are treated with antibody (against somatic, spore or flagellar antigens) coupled with ferritin by means of toluene 2-4 di-*iso*-cyanate. Ferritin shows up in a very characteristic form in electron micrographs, so that the distribution of certain antigens within or on the surface of organisms can readily be made out (Thomson *et al.*, 1966; Walker *et al.*, 1967; Walker *et al.*, this Volume, p. 219).

It would be interesting to find out whether a mixed antibody method was practical, i.e., one in which, say, rabbit antibody to bacterial antigens was applied to the sections, which were then treated with goat anti-rabbit-globulin, and then with rabbit anti-ferritin, and then with ferritin. The control problem would be considerable.

C. Tanned red cell agglutination

The red cell is such a convenient object for agglutination that it is not surprising that Boyden (1951) used it as a carrier for antigens, in the hope that red cells so coated would be agglutinated by the homologous anti-body. He first treated the cells, washed free from serum, with 0·005% tannic acid by slow-speed centrifugation and then treated them with the antigen under standard conditions. Subsequent addition of homologous antibody frequently caused agglutination of the coated red cells, even when it gave no precipitation or other visible effect when mixed with the antigen.

The method given above is a direct agglutination method. It is possible to obtain results of greater accuracy by a blocking or back-titration method. In this the cells are coated with antigen as above, and it is shown that a certain dilution of antibody-containing material will agglutinate them. This dilution of antibody is mixed with known amounts of antigen; the mixture is then allowed to stand for a convenient time for combination and added to the suspension of antigen-coated red cells; the mixture containing the smallest amount of antigen that just prevents red cell agglutination is regarded as neutral. Evidently this amount of antigen can be regarded as a measure of the amount of antibody present in the dilution used; alternatively in appropriate cases, a test-dose of an antigen can be determined in this way against a standard serum, and other sera may be assayed against the standard by determining the minimum dilution of the test serum that, when mixed with the test-dose of antigen, just fails to agglutinate the antigen-coated red cells.

Obviously the reliability of these tests depends on the purity of the antigen used. If the material used for coating the red cells contains several antigens, the antigen that is preferentially absorbed on the tanned red cells may vary according to the concentration of the various antigens in the mixture, or according to the circumstances of absorption. The antibody that leads to agglutination of the coated red cells may therefore not be the same in different tests, and the results of assay may therefore be misleading. (See Fulthorpe, 1957, 1958a, b, 1959, for a very thorough investigation of the method for the assay of tetanus toxin and antitoxin, and Fulthorpe, 1962 for diphtheria antigens, in which the importance of purified antigens is shown, and the conditions for successful testing are determined.)

IV. CYTOPHILIC ANTIBODIES

These were first described by Boyden and Sorkin (1960, 1961) who showed that sera from rabbits immunized with human serum albumin contained a 7S antibody (Sorkin, 1963) that could be absorbed on normal rabbit spleen cells *in vitro*, and that normal cells so treated and washed could absorb specific antigen. Very little (0·1–1%) of the total serum antibody was so absorbed.

Later Boyden (1964) showed that the serum of guinea-pigs immunized with sheep red cells mixed with Freund's *complete* adjuvant contained a similar antibody, and that normal guinea-pig peritoneal cells, obtained by injecting 20 ml of Hanks' solution* containing 10 units of heparin per ml and immediately draining the peritoneum through a size-17 hypodermic needle, exposed to the antiserum *in vitro* would absorb the antibody and so acquire the capacity to agglutinate sheep red cells on their surface. No cytophilic antibody was produced in guinea-pigs immunized with red cells plus *incomplete* adjuvant, though the haemagglutinating titres of the sera of the two groups were similar.

Three tests for cytophilic antibody are available. In the first, the antigen is treated with radioactive ^{131}I, and the amount of radioactivity of cells treated with antibody and then with radioactive antigen is determined and compared with that of cells treated with normal serum and radioactive antigen. This method is obviously more useful for antigens like serum proteins.

The other two are more readily applicable to red cells as antigen. 0·1 ml of washed normal guinea-pig peritoneal cells (2×10^5 per ml) is treated with 0·1 ml of appropriately diluted serum containing cytophilic antibody for 1 h at 0°–4°C; the cells are then washed thoroughly in Hanks' solution, in which they are resuspended to a volume of 0·2 ml. After addition of

* *Hanks' solution*: $NaHCO_3$, 1·4 g; 0·2% phenol red, 5 ml; Distilled water 95 ml. Saturate with CO_2 until orange in colour. Sterilize by autoclaving at 8–9 lb/sq. in. for 10 min.

0·8 ml of 2% sheep red cells the mixture is allowed to stand overnight at 0°–4°C. One drop of the mixture is placed on a slide on which a few glass ballotini (diameter 0·2 mm) have been scattered to prevent the cover-slip breaking up the clumps of cells, covered with a coverslip and examined by phase-contrast microscopy. The result is expressed as the percentage of peritoneal cells to which five or more red cells adhere. Controls with normal serum are used for comparison.

The other method is preferable (Boyden, 1964). Two 1 cm-diameter holes are cut in Perspex slides $7·5 \times 2·5 \times 0·25$ cm. A coverslip is luted over each hole with melted paraffin wax and 2×10^5 peritoneal cells are placed in each chamber so formed, incubated for 40 min at 37°C in a moist atmosphere containing 5% CO_2, and then washed with Hanks' solution. Appropriate dilutions of serum in a volume of 0·1 ml are added, and the slides allowed to stand at 0°–4°C for 1 h, after which excess serum is washed off with five changes of Hanks' solution; 0·1 ml of 1% sheep red cells in Hanks' solution are added to each well, and the slides left on the bench for 1 h. The slides are then inverted in a petri dish containing Hanks' solution to allow free red cells to fall away from the coverslip; the slide is dried, the well covered with another coverslip, and the slide turned so that the macrophage-covered coverslip is uppermost, and the macrophages examined by phase-contrast microscopy to determine the number of red cells adherent to 100 macrophages. Appropriate controls are set up with normal serum.

V. COMPLEMENT-FIXATION REACTIONS

When guinea-pigs or rabbits are immunized with whole cells, such as Gram-negative bacteria or foreign red cells, their sera after a time develop the power of lysing the corresponding antigenic cells. Bordet (1895, 1900) and Bordet and Gay (1906) showed that this activity required two substances, one the antibody (immune body, amboceptor) against the surface cellular antigens, the other a substance present in the normal serum, and therefore also the immune serum, of rabbits, guinea-pigs and men (complement, alexin), which is inactivated by ageing and by heating at 55°C for 30 min. Immune sera that have been heated at 55°C for 30 min retain their capacity to agglutinate the cells against which they were prepared, but usually to a lower titre than that to which they will lyse them in the presence of complement.

Later work (see Mayer, 1965, for review) has shown that complement is a complex mixture of proteins, and that lysis of red cells in its presence involves a cascade of enzyme reactions activated by attachment of the antibody to the red cell, with final perforation and fragmentation of the red cell and release of its contents.

Obviously the lytic activity of antibody in the presence of complement is advantageous in infections, at any rate with Gram-negative organisms. But the reaction is of far greater general use in complement-fixation (or complement-deviation) reactions, for though complement is seldom necessary for antigen-antibody reactions, when antigen and antibody combine some of the components of complement are adsorbed on the complex. Removal of active complement may therefore be used as evidence of antigen-antibody combination and therefore of antigen in a mixture containing antibody and complement, or of antibody in a mixture containing antigen and complement.

As complement is necessarily for haemolysis of red cells treated with anti-red-cell antibody heated at 55°C for 30 min, a suspension of red cells treated with a dose of anti-red-cell antibody, heated at 55°C, just inadequate to agglutinate the red cells but more than is sufficient to haemolyse them in the presence of excess complement—a mixture usually called "sensitized red cells"—is a very sensitive indicator for complement. A complement-fixation test is carried out as follows.

If we need to demonstrate the presence of an antigen in a fluid or in a tissue, we mix the test material with a convenient amount of the antibody specific to the antigen and $2\frac{1}{2}$ effective doses of complement (usually as guinea-pig serum, which is unusually rich in complement). The antibody should have been "inactivated" by heating for 30 min at 55°C, and if the test material may contain complement, it should be treated similarly. The mixture of test material, antibody and complement is allowed to stand for an appropriate period (depending on what is known about the system), and a suitable volume of sensitized red cells is added, and the whole mixture incubated at 37°C for half an hour. Three results may be obtained—complete haemolysis, partial haemolysis, or no haemolysis. If no haemolysis occurs, there cannot have been any complement in the mixture, so complement must have been removed from the test substance-antibody-complement mixture; so it is probably that an antigen-antibody complex has been formed, on which some of the components of complement have been absorbed. The test mixture probably, therefore contains the corresponding antigen.

If complete haemolysis occurs, there must have been complement in the final mixture, so none can have been absorbed, so no antigen-antibody complex has been formed, so it is unlikely that there is any antigen in the test material.

Partial haemolysis will mean the presence of insufficient antigen to form enough complex to absorb all the complement.

Evidently the same arguments can be applied to systems in which the antigen is known and sera are examined for antibody. In either case the test material may be diluted to find the dilution that will just remove all the complement activity in the presence of the other members of the complex. If antigen is being tested for, this dilution will give an idea of the concentration of antigen present. More important, the greatest effective dilution of an antiserum will give an idea of the amount of complement-fixing antibody in it; an increase in the amount of antibody during an illness may be used as evidence of infection with the organism or its products used as antigen in the test.

The test is a sensitive one, but adequate controls are essential. It must be shown that the antigen will not remove complement in the absence of antibody, and that the antibody will not remove complement in the absence of antigen: antigens and antibodies that have this unfortunate property are said to be anticomplementary, and may need to be diluted below the anticomplementary level before they can be used. Some sera have the opposite property and increase the activity of complement, and may in consequence mask slight degrees of complement fixation (pro-complementary sera). Besides this, the sensitized cells must not haemolyse alone at 37°C, and they must haemolyse completely in the presence of the amount of complement used in the test, when incubated at 37°C. Preliminary titrations of immune body and complement are essential.

Not all antisera will fix complement, even under apparently optimum conditions; rabbit, guinea-pig and human sera do so, but bovine and some horse antisera do not, especially if they have been heated to inactivate them. It is possible, in these cases, to perform an indirect complement-fixation test. In this the serum under test is added to the antigen in the presence of guinea-pig complement, and then a rabbit antiserum to the antigen is added. If the test antiserum contains antibody, it will combine with the antigen, and prevent the rabbit antiserum from doing so. Complement will therefore not be fixed, and the sensitized red cells added will be haemolysed (Rice, 1948).

Complement-fixation may be demonstrated in antigen-antibody reactions *in vivo* by fluorescent antibody tests with anti-complement sera on sections (Goldwasser and Shepard, 1958; Burkholder, 1963), and also, if antigen is injected intravenously, by the reduction in the serum complement titre that follows.

A. The technique of complement fixation

The complement-fixation test set out below is only one of many in use, and is given only as a guide; much smaller volumes may be used if little material is available.

1. Titration of haemolytic antibody

Prepare a 1 in 1000 dilution of rabbit or horse anti-sheep-red-cell antiserum previously heated to 55°C for 30 min (i.e., "inactivated" as far as complement is concerned) by diluting 0·1 ml of serum in 100 ml of saline; from this dilution set up a series of two-fold dilutions from 1 in 2000 to 1 in 16,000 either by diluting 1 vol of the 1 in 1000 dilution with 1, 3, 7, or 15 vol of saline, or by adding 1 vol of 1 in 1000 to 1 vol of saline to give 1 in 2000, and by repeating this process with the 1 in 2000 dilution and so on.

Place 0·25 ml of each dilution of antibody in a convenient tube and add to each 0·25 ml of saline, 0.25 ml of guinea-pig serum diluted 1 in 10, and 0·25 ml of a 5% suspension of washed sheep red cells. Mix well, and incubate for 30 min in a water-bath at 37°C. The tube containing the lowest concentration of haemolytic antibody that produces complete haemolysis contains 1 minimum haemolytic dose (MHD) of the antibody.

2. Sensitized red cells

To 1 vol of 10% washed sheep red cells add 1 vol of haemolytic antibody diluted to contain 12 MHD per ml; mix thoroughly and incubate at 37°C for 30 min, agitating at intervals.

3. Titration of complement

Dilute fresh guinea-pig serum 1 in 10, 20, 30, 40, 50, 60 and 70 with saline, and place 0·25 ml of each dilution in a convenient tube; to each tube add 0·5 ml of saline and 0·25 ml of 5% sensitized red cells. The tube containing the lowest concentration of guinea-pig serum that produces complete haemolysis contains 1 MHD of complement.

If commercial freeze-dried complement is used, the manufacturer's instructions should be strictly followed. If you make it yourself, add distilled water to make the sample up to the original volume of the serum from which it was made, and make your dilutions in saline from this.

4. Tests for anti-complementarity

When you are dealing with systems that have been in use for a long time, you may not need to do these tests for the antigen, as the information may already be available. If you are dealing with something new, or something with which you are not familiar, it is wise to do the tests suggested here.

In the first you are finding the dilution of antigen necessary to avoid complement fixation by antigen alone. Make up two-fold dilutions of the antigen, and place in a tube with 0·25 ml of saline and 0·25 ml of a dilution of guinea-pig serum containing $2\frac{1}{2}$ MHD of complement. Allow the mixtures to stand at room temperature, at 4°C or at 37°C, whichever previous experience has suggested is best, for the time found to be most convenient, and then add 0·25 ml of 5% sensitized sheep red cells. Choose as the dilution of antigen to use the one containing the highest concentration of antigen that has no anticomplementary effect, i.e. the highest concentration in which complete haemolysis occurs.

It is often useful to be sure that this dilution is not anticomplementary in the presence of serum; in this case, 0·25 ml of inactivated normal serum is substituted for 0·25 ml of saline in the test described in the last paragraph.

5. *Titration of antigen*

It is also necessary to show that the antigen, used at a concentration at which it is not anticomplementary, will fix complement in the presence of positive sera. To determine this a set of two-fold dilutions of the antigen is made up, and 0·25 ml of each dilution is placed in a number of tubes corresponding to the number of dilutions of the positive serum you expect to use. 0·25 ml of the dilutions of the inactivated positive serum is then added, so that each dilution of antigen is mixed with each dilution of serum. The easy way to do this is to arrange a series of racks so that each rack contains tubes with all the dilutions of antigen in order, and add 0·25 ml of the first dilution of positive serum to all the tubes of the first rack, 0·25 ml of the second dilution to all the tubes of the second rack, and so on to produce a "chessboard" titration; $2\frac{1}{2}$ MHD of complement in a volume of 0·25 ml is then added to each tube, and the racks are allowed to stand as before. After this 0·25 ml of 5% sensitized sheep red cells is added to each tube, and all the mixtures are incubated at 37°C for 30 min. The optimum concentration of antigen is that which gives complete complement fixation with the lowest concentration of positive serum. If, as is often the case, several concentrations of antigen share this property, a mean concentration is chosen.

6. *The complement-fixation test on sera*

This is then carried out with mixtures of 0·25 ml of optimal antigen dilution, 0·25 ml of test serum dilution, and 0·25 ml of saline containing $2\frac{1}{2}$ MHD of complement, with the same steps as before. If only a positive or negative result is required, only one dilution of serum need be tested; but it is far better to assay the serum by testing a number of two-fold dilutions to obtain an end-point dilution at which all the complement is fixed. It is obviously convenient to know this when one is comparing sera for antibody content against the same antigen, and just as important, against related antigens.

Each set of complement-fixation tests should be thoroughly controlled to make sure (1) that the test sera are not anticomplementary; (2) that the antigen is not anticomplementary in the concentration used; (3) that the sensitized red cells are haemolysed when complement is added to them; and (4) that normal sheep red cells are not haemolysed by complement. Some of the controls may be more necessary in some instances than in others.

In (1) the tubes contain 0·25 ml of saline, of the highest concentration of inactivated serum used, and of complement; if complement is fixed, the test should be repeated with a series of two-fold dilutions of inactivated serum, to find out whether the serum fixes complement at lower concentrations in the presence of antigen than in its absence. If it does, the test is useful; if it does not, no conclusions can be drawn.

Controls (2), (3) and (4) are obvious from what has previously been described. The only important point is that the final volume should in all cases be the same as that used in the main test.

For some sera (particularly anti-viral sera) it may be necessary to test a range of antigen dilutions in complement-fixation tests, or to set up a chessboard pattern of such tests. Differences in the amount of antigen producing maximum complement-fixation in such tests may make it possible to decide whether antibody is IgG or IgM (though there is a good deal of doubt about this), and more interestingly, whether in infants recovering from respiratory syncytial virus infection, the antibody is passively acquired maternal antibody, or antibody actively produced by the infant (Jacobs and Peacock, 1970).

B. The Wassermann reaction

Normal serum contains minute traces of an auto-antibody reacting with cardiolipins as antigens. During infection with syphilis and yaws, and in a number of other conditions, like disseminated lupus erythematosus, in which connective tissues undergo extensive damage, these antibodies increase considerably in concentration in serum. They may then be detected by their capacity to fix guinea-pig complement in the presence of liver extracts, or alcoholic extracts of heart muscle, or more recently cardiolipins, "fortified" if necessary with cholesterol. The watery emulsions of these alcoholic extracts are stabilized by the cholesterol, but addition of syphilitic sera may lead to visible precipitation of the antigen. Clearly the Wassermann reaction for syphilis is non-specific, detecting a consequence of syphilitic infection rather than the infecting organism itself, so that more specific tests for syphilitic antibody, such as Nelson and Mayer's (1949) *Treponema* immobilization test and indirect fluorescent antibody tests are now preferred.

VI. ANTITOXIC ANTIBODIES

During the course of infection with toxigenic bacteria, or of immunization of man or animals with toxin derivatives, antitoxic antibodies may be
produced by the patient or immunized animal. These may be assayed for
their capacity to neutralize the toxin or toxins produced by the organism,
and consequently to prevent their *in-vivo* or *in-vitro* activities (see Batty,
this Volume, p. 255). Suitable antisera may also be used to identify particular toxins or to investigate the complexity of bacterial filtrates
(Oakley, 1954).

VII. ANTIVIRAL ANTIBODIES

A. Virus neutralizing antibodies

During infections with viruses, or during immunization with viruses
or their derivatives, virus-neutralizing antibodies may appear in the
plasma of the infected or immunized animal. Since viruses can grow only
by entering living cells, and can show their presence and growth there
only by producing changes in the cells, we can test for the infectivity of
viruses only in whole animals and in organ and tissue cultures.

The whole animals that viruses can infect may be adults—e.g., adult
mice, which when anaesthetized with ether are readily infected with mouse-
adapted influenza virus given intranasally; or newborn animals—newborn mice are readily infected with Coxsackie viruses; or foetal animals—
foetal guinea-pigs can be infected with influenza till late in gestation, though
the adults cannot; or fertile eggs, whose membranes or embryo are often
susceptible to viruses that are unable to attack adult birds or even chicks.

The effects produced in whole animals may be death in a certain time,
or severe illness, or the development of visible lesions. In some cases, e.g.,
in the membranes of the fertile egg, the number of these lesions (pocks,
for instance) may be proportional to the number of virus particles instilled
(Burnet, 1926).

Antisera, therefore may be tested, preferably against a standard, by
determining the amount of serum that will reduce the activity of the virus
suspension to a small fixed amount. (See Casals, 1967). Thus a suspension
of mouse-adapted influenza virus will, if injected intranasally into mice
anaesthetized with ether in suitable amounts, kill the mice within four days
with 100% pneumonic consolidation of the lungs; addition of appropriate
amounts of anti-influenzal serum to the virus, and intranasal injection of
the mixture into anaesthetized mice after it has stood for 30 min or so to
combine, will lead to a reduction in the amount of consolidation of the
lungs and in the number of deaths. If the indicating effect is chosen as the

amount of antiserum that will reduce the effect of the virus to an average 50% consolidation of mouse lungs after four days, antisera can be assayed against the standard by determining the relative amounts of test serum and standard necessary to reduce virus activity to this level. As, in general, a certain amount of antiserum will neutralize a fixed percentage of the virus, however much or little is present, it is best to work at as high a concentration of virus as is safe and practicable.

If pocks are formed, as with fowlpox in the fertile hen egg, antisera can be assayed against a standard by determining the relative amounts of test serum and standard that will reduce the number of pocks produced by the virus alone to a half.

In one particular case the virus, a bacteriophage, is parasitic in a whole organism, the bacterium. If the susceptible bacterium is inoculated on a suitable medium, the plate is dried, and a suitable volume of bacteriophage suspension run on to a limited area of the plate, incubation of the plate will show a lawn of growth with a number of clear plaques where the organisms have been destroyed by the bacteriophage. Antisera can then be assayed by determining the amount of antiserum that when mixed with the same quantity of virus used alone will reduce the number of plaques to one-half.

In many cases the reaction between antiserum and virus need only be very roughly quantitative, as all that is necessary is the identification of the virus; but quantitative estimation of antibody is nearly always useful if several antibodies to different antigens are present in the same antiserum. If a serum for instance contains antibodies to viruses A and B, virus B may be inaccurately identified as A if quantitative work, preferably with several different sera, is not done.

Whole animals are troublesome and expensive, and methods involving organ and tissue culture have in consequence become very popular, especially since antibiotics have made tissue and organ culture so much easier.

In tissue culture (Porterfield, 1967), the effects of virus infection may be death of the cultured cells forming plaques (Porterfield, 1964), with or without specific cytopathic changes, or interference with the growth of other viruses in them. Some cytopathic changes, e.g. the giant-cell syncytia characteristic of the growth of measles virus in tissue culture, may make the identification of the virus group possible. But for exact identification neutralization with appropriate antisera is necessary; mixtures of virus and antiserum may show fewer plaques of dead cells than the virus alone, or show no cytopathic effect when the virus itself does. Alternatively, though the virus may not itself produce any cytopathic effect on the cell it enters and parasitises, its growth may render the cell unable to allow the entry or growth of another virus that does produce a visible effect.

Let us suppose, for instance that a virus C can enter and grow in a particular tissue culture, but does not produce any cytopathic changes; the virus alone will therefore not be directly detectable. But if its growth in the tissue culture prevents the entry and growth of Virus D, which does produce a cytopathic change in normal tissue cells, the growth of an interfering virus may be assumed, especially if a control experimental culture treated with virus C and its antibody still allows the growth of virus D and the development of its cytopathic changes, itself preventable with D-antiserum.

If virus D is not cytopathic, but possesses some other detectable property, e.g. haemadsorption (Vogel and Shelokov, 1957) when it is growing in tissue culture, this property may be used to identify virus C, or to assay antibody against it. If a suspension of virus C is added to the tissue culture, the virus will enter and grow; if the culture is washed after an appropriate time subsequent addition to the culture of virus D will not lead to infection with virus D; and consequently if the culture is rewashed and a suspension of suitable red cells added to it, no haemadsorption will occur. But if there is little virus C in the sample, or it is neutralized with antiserum, and the neutralized mixture added to the culture, no growth of virus C occurs, or it is very much reduced; if the culture is washed, it can readily be infected with virus D. After the culture has, after a suitable time, been washed free from virus D, addition to it of a suspension of suitable red cells will lead to haemadsorption, and so to evidence that virus C is not present in the original material, or, in appropriate cases, that the antiserum used contains antibody to virus C (Marcus and Carver, 1965).

Small fragments of pharyngeal mucosa may be grown as organ cultures, and show vigorous ciliary movement (Porterfield, 1967); some viruses added to the fluid surrounding the cultures may in time stop the ciliary movement, a cytopathic activity that is prevented by appropriate amounts of specific antiviral sera.

B. Other virus antigen-antibody reactions

Appropriate extracts of viruses may be examined by the complement-fixation technique; sometimes the virus itself may be agglutinable by antisera; and fluorescent antibody tests may be used to identify viruses in tissue cultures.

VIII. OTHER ANTIGEN-ANTIBODY REACTIONS

In addition to the applications discussed by Walker et al. (1967) fluorescent antibody methods may also be used to demonstrate the presence of antibody to organisms where no other specific test, e.g., precipitation, agglutination or complement-fixation is available. Thus the spirochaete of syphilis has probably never been grown *in vitro*; but other non-pathogenic strains of *Treponema* can be grown, such as Reiter's and Nicol's strain, that are closely antigenically related to it. If carefully dried films of these spirochaetes are made, treated with supposedly syphilitic human serum, thoroughly washed with phosphate-buffered saline, then treated with fluorescent anti-human globulin, washed and mounted in 90% glycerol, and examined in a dark-ground fluorescence microscope, the spirochaetes will fluoresce in ultraviolet light if the serum contains any syphilitic antibody. It is usually necessary to dilute the antibody-containing serum considerably (e.g. to 1 in 200) to avoid non-specific effects.

IX. USE OF PASSIVE TRANSFER OF ANTIBODIES

If antibodies are free in the serum they and their properties may be transferred passively with the serum to other animals that have had no experience of the antigen against which the antibody is directed. Thus diphtheria antitoxin prepared in one animal may be injected into another, not necessarily even of the same species and will, for a time depending on the rate at which it is destroyed, protect the recipient against diphtheria toxin.

In a similar way sera from animals sensitized to anaphylactic shock by a particular antigen may be injected into other animals, which will, after a suitable period during which the antibody becomes fixed in their tissues, become sensitive to the antigen.

Some antibodies readily become fixed in the skin. Thus sera from animals made sensitive to a particular antigen may in the early stages of that sensitization contain antibodies that when injected into the skin of another animal of the same species may become fixed in the skin, often for months. Such fixed antibodies may render the area of skin so injected sensitive to the antigen, so that injection of the antigen into the treated skin area leads to the production of local swelling, wheal formation and arteriolar flare (Prausnitz and Küstner, 1921). In the later stages of sensitization "blocking" antibodies may be produced. These do not themselves sensitize an animal passively, but may prevent any reaction to an antigen in an animal actively or passively sensitized to the antigen (Cooke, 1947).

Antibodies may in some animals pass from mother to young via the placenta or foetal membranes, or be absorbed by the alimentary canal from the colostrum. This transfer may make it possible to test the immunizing power of vaccines in animals where direct active immunization is impossible.

Thus newborn rabbits are sensitive to cholera toxin only for the first ten days or so of life; during this period it is impossible to immunize them actively. This difficulty is got over by immunizing pregnant rabbits with appropriate antigens so that the immune response is at its peak when the young are born. Antibody then passes from mother to foetus through the yolk-sac splanchnopleur and appears in the foetal serum at concentrations about one-tenth of that in the mother. The newborn rabbit may then be tested for immunity against cholera organisms or cholera toxin (Dutta and Habbu, 1955; Panse *et al.*, 1964; Dutta, Dohadwalla and Chakravarti, 1966).

In a similar way, lambs may be immunized against lamb dysentery either by injecting lamb dysentery antiserum prepared in the horse into the lamb very soon after birth, or by immunizing the pregnant ewe with lamb dysentery toxoid, so that when the lamb is born it becomes passively immunized by absorption of antitoxins from the ewe's colostrum. The efficiency of vaccines can clearly be tested by immunizing ewes, and therefore their lambs, in this way.

Some antibodies are, however, never free in the serum, and cannot be transferred passively with it. They can, however, be transferred with cells, particularly lymphnode cells, and are therefore called cell-bound antibodies. Such are the antibodies responsible for the tuberculin reaction, for sensitivity to many simple chemical substances, for delayed hypersensitivity reactions in general, and for the destruction of homografts. The cells may be transferred by intravenous or subcutaneous injection.

X. THE PRODUCTION OF ANTISERA

The ways in which sera are produced depend very much on the purpose for which they are needed. Most laboratory antisera are made in rabbits, most commercial antisera in horses. Rabbits are perhaps a little small, and not over costly to keep; horses are large, sometimes difficult to control, and costly to maintain. Rabbits produce precipitating antisera, horses flocculating antisera. By careful bleeding it is possible to bleed rabbits 40–50 ml at intervals of about a week; 8 litres of blood can be obtained from a horse three times over eight days. On the whole one can produce more concentrated antitoxic antibodies in horses than in rabbits, stronger agglutinating antisera in rabbits than in horses. It is possible to produce antitoxic antisera in rabbits, rats and guinea-pigs, but the relatively short lives of these animals make it difficult to "rest" them for a long time, and so high value antitoxins are not readily obtained.

A. Agglutinating antisera

For the production of agglutinating antisera, organisms are grown in or on a suitable medium, and gently washed by centrifugation in appropriate cases (e.g. where toxins are likely to be present). The concentration of organisms is adjusted to about 3×10^8 per ml, and rabbits are given intravenous injections of the suspension. The first dose is about 0·1 ml, and further injections, doubling in volume on each occasion, are given on alternate days for a week. The animal is bled about a week after the last injection; boosting injections may be given on the day after the bleeding and the animal bled a week later.

If the organism has no fimbriae, flagella or capsule, this may be entirely satisfactory; but if it has any of these appendages, it may be convenient to obtain antisera against them, or free from antibody against them.

Thus to obtain flagellar antibody we use a flagellated strain of the organism killed with 0·12% formaldehyde. Antisera from animals given injections of this suspension will contain flagellar antibody and some somatic antibody; the latter may be removed by absorbing the antiserum with a non-flagellate variant of the organism, or with flagellated cells heated to 100°C for $\frac{1}{2}$–2 h.

The absorbing suspensions just referred to may also be used to produce anti-O-antibodies, as the boiling denatures the flagellar protein.

Capsular antibodies may be produced by injecting capsulated organisms killed with mercuric iodide in buffered formol-saline into rabbits. Sera so produced contain some somatic antibodies, which can be removed by absorption with non-capsulated variants of the strains. In some cases (e.g., the pneumococcus) the capsular material can be removed from the organism (see, for instance, Heidelberger *et al.*, 1936) and injected with dead bacteria or with adjuvants to provoke the formation of almost pure capsular antibody.

If the strain is fimbriate, organisms without fimbriae can usually be obtained from it by short (6–12 h) culture in static broth; cultures 24 to 48 h old are usually heavily fimbriate. Purely fimbrial antisera can then be obtained by injecting long-term cultures into rabbits, and absorbing out all but the fimbrial antibodies with organisms from short-term cultures, or with non-fimbriate variants of the same species or strain.

B. Precipitating antibodies

These are best produced by intramuscular injection into rabbits of crude or purified samples of the antigen mixed with Freund's complete or incomplete adjuvant.* The antigen is best dissolved and mixed with an equal

* I have used a mixture (v/v) of 10% Arlacel A (Honeywell and Stein, Ltd, Devonshire House, Mayfair Place, Piccadilly, London W.1) and 90% Technical

volume of adjuvant. The volume of the initial injection will depend on the amount of material available; 0·1 ml is a satisfactory volume, and doubling amounts may be given on alternate days for a week. The animal is bled about a week after the fourth injection; a few days later a booster dose is given, and the rabbit bled again a week later. This process may be repeated many times, but later bleedings tend to be less specific for the antigen than earlier ones.

If sera are to be prepared against a mixture of proteins, e.g., whole human serum, e.g., for immuno-electrophoresis, crude serum may be used. If an antiserum is to be prepared against a single component of such a mixture, the procedure is to subject the crude material to immuno-electrophoresis in agar gel, and cut from the agar the precipitation arc corresponding to the antigen required; this is washed free of antigens and antibody, ground gently, and injected into the exposed popliteal glands of rabbits. Very "pure" antisera have been produced in this way (Goudie, Horne and Wilkinson, 1966). Discs of precipitate may be cut from well washed column diffusions (cf. page 188) and injected intramuscularly in admixture with Freund's adjuvant when highly active mono-specific antisera are sometimes produced (see Fig. 3).

It is not unusual at immunological meetings to heat a good deal of grumbling about the perversity of rabbits in producing antisera in which the antibody content is inconsistent from rabbit to rabbit, especially when the animals are immunized with mixtures of antigens. Unfortunately, rabbits appear to have little interest in pleasing immunologists and one rabbit given a mixture of antigens A and B will produce a great deal of anti-A antibody, and very little antibody against B; another will do the exact opposite, another may produce considerable amounts of antibody against both, and another still may not produce antibody against either. Animals given a large amount of laboriously purified antigen A may perversely produce mostly antibody against the trace contaminant antigen B.

It always seems to me that indignation of this kind is misplaced, and that much more use might be made of the variable responses of immunized animals. In examining bacterial filtrates for antigenic components, for example, extreme variety in the relative amounts of antibody to the various components is essential to the analysis (Oakley, 1954).

C. Complement-fixing antibodies

Agglutinating antibodies may also fix complement, and so may precipitating antibody, anti-viral and anti-mycoplasmal antibodies. The antibody

white oil light grade 23 (Esso Petroleum Co. Ltd., Specialty Department, London N.W.1). This incomplete adjuvant may be completed by adding 1 mg heat-killed tubercle bacilli per ml.

to sheep red cells may be made by injecting increasing doses of sheep red cells, washed free from serum by repeated centrifugation with saline, intravenously into rabbits or horses. Increasing doses may be given on alternate days starting with the equivalent of 0·5 ml of blood, and doubling the doses up to 8 ml. The animal is bled when a sample bleeding shows a satisfactory titre (1 in 2000 for horse antisera and 1 in 10,000 for rabbit antisera).

Complement is best obtained from healthy adult male guinea-pigs, which may be bled out from the heart, or less cleanly, from the throat, under anaesthetic. The separated serum, which should show only the faintest tinge of haemoglobin, is either freeze-dried or preserved with sodium azide.

D. Antitoxic antibodies

No doubt it is true that diphtheria antitoxin of high value can be produced in animals by a single injection of diphtheria toxoid emulsified in Freund's complete adjuvant (Freund et al., 1948); but little is said about the avidity of such antitoxic sera. It is usual to find, however, that when an animal is pushed to produce antibody very rapidly, the antibody so produced is commonly non-avid, that is, it readily dissociates from its antigen on dilution. If the antibody is used at high concentrations (as, for instance, in flocculation tests), this is of relatively little moment; but, if low concentrations of antibody are used (as, for instance, in necrotizing tests in skin), the results may be very difficult to read, and vary with the level of test (Glenny and Barr, 1932a, b).

In order to produce avid antisera we need either (1) animals with natural antibody or (2) plenty of time to rest the animal.

Horses may possess natural antibody to diphtheria toxin, *Clostridium welchii* alpha-toxin and *Staphylococcus aureus* alpha-toxin; sheep often have natural antibody to *Cl. welchii* epsilon-toxin. Such animals do not need any particular preparation or period of rest before the main immunization. Generally speaking they may be given injections, preferably intramuscularly, of increasing volumes of toxin (*Cl. welchii* alpha- and epsilon-toxin, *Staphylococcus aureus* alpha-toxin) or toxoid (diphtheria toxoid), three times a week. An experimental bleeding is taken a week after the first injection, and weekly thereafter, and if the animal is responding well, it is bled a week after the rate of increase of antitoxin units shows signs of falling off. (Sera with rather poor antitoxin levels may be quite useful, but they are naturally used up sooner than high value sera, so that high value sera are more valuable if they can be obtained.)

The immunizing process can then be continued with injections of toxin plus alum or toxin with Freund's adjuvants. If alum is used, five injections are usually needed; if the adjuvant is Freund's, one or two may be sufficient. The animal is bled ten days after the first injection. These short immunizations may be repeated as long as the animals remain fit or provide serum with high antitoxin concentrations.

Animals without natural antibody require very different treatment. The usual practice is to give them two injections of alum-precipitated toxin or toxoid at an interval of a month. The animal is bled a small amount ten days after the second injection, to see whether it is responding well. If it is, it is rested for a convenient period, which will vary with circumstances from 1–9 months—the longer the better. At the end of this resting period the first immunization begins, with injection of toxin or toxoid two or three times a week. The first experimental bleeding is taken four weeks after the first injection following the rest, and after that the immunizing process follows that for animals with natural antibody, as it does in short-term subsequent immunizations.

It should be remembered that possession of natural antibody to one component of a bacterial filtrate does not guarantee the possession of natural antibody to any other component, so that antisera produced against, say, *Cl. welchii* alpha-toxin in an animal possessing natural antibody to that antigen, may contain avid *Cl. welchii* alpha-antitoxin, but very non-avid *Cl. welchii* kappa- and mu-antibody. It is often, for this reason, preferable to treat horses with natural antibody, at any rate for research purposes, as if they had none, if minor antigens (so-called) are to be investigated.

To obtain antisera with very different ratios of antibodies to the various components of a filtrate one may use filtrates from strains grown in various media, at different temperatures or for different times; or use filtrates from different toxin types; or use filtrates from degraded strains, or filtrates treated with trypsin so as to leave undamaged only the trypsin-resistant components (e.g., *Cl. welchii* epsilon- and iota-toxins (Oakley, 1954)).

E. Anti-viral sera

These may be obtained from animals infected with non-lethal doses of virus, from animals given injections by an unusual route (e.g., ferrets given an intraperitoneal injection of influenza virus), or from animals immunized with formolized virus, with or without Freund's incomplete adjuvant.

F. Anti-mycoplasmal sera

These may be obtained from infected animals, or from animals immunized with live or formalin-treated mycoplasms, with or without adjuvant.

REFERENCES

Afonso, E. (1964). *Clin. chim. Acta*, **10**, 114.
Afonso, E. (1965). *Clin. chim. Acta.*, **13**, 107.
Afonso, E. (1967). *Clin. chim. Acta*, **17**, 138.
Alcock, Doris (1965). *J. Path. Bact.*, **90**, 31.
Bettelheim, K. A., and Taylor, Joan (1969). *J. Med. Microbiol*, **2**, 225.
Bissett, K. A. (1947a). *J. Path. Bact.*, **59**, 301.
Bissett, K. A. (1947b). *J. Hyg., Camb.*, **45**, 128.
Bissett, K. A. (1949–50). *J. Endocr.*, **6**, 99.
Bogdanikowa, Beata, Drozd, Jadwiga, Dubińska, Lidia, and Bogdanik, T. (1966). *Clin. chim. Acta*, **14**, 807.
Bordet, J. (1895). *Ann. Inst. Pasteur*, **9**, 462.
Bordet, J. (1900). *Ann. Inst. Pasteur*, **14**, 257.
Bordet, J., and Gay, F. P. (1906). *Ann. Inst. Pasteur*, **20**, 467.
Boyden, S. V. (1951). *J. Exp. Med.*, **93**, 107.
Boyden, S. V. (1963). *In* "Cell Bound Antibodies" (Eds. B. Amos and H. Koprowski), Philadelphia, 1964. *Immunology*, **4**.
Boyden, S. V. (1964). *Immunology*, **7**, 474.
Boyden, S. V., and Sorkin, E. (1960). *Immunology*, **3**, 272.
Boyden, S. V., and Sorkin, E. (1961). *Immunology*, **4**, 244.
Burkholder, P. M. (1963). *Amer. J. Path.*, **42**, 201.
Burnet, F. M. (1926). *MRC Spec. Rep. Ser.*, no. 220.
Casals, J. (1967). *In* "Methods in Virology" (Eds. K. Maramorosch and H. Koprowski), New York and London, vol. III, pp. 113–198.
Clarke, H. G. M., and Freeman, T. (1966). *Protides of Biol. Fluids*, **14**, 503.
Cooke, R. A. (1947). "Allergy in Theory and Practice", Philadelphia and London.
Coombs, R. R. A. (1964). *In* "Immunological Methods" (CIOMS Symposium) (Eds. J. F. Ackroyd and J. L. Turk), Oxford.
Coombs, R. R. A., Daniel, Mary R., Gurner, B. W., and Kelus, A. S. (1961). *Nature, Lond.*, **189**, 503.
Cooper, G. N. (1956). *J. Path. Bact.*, **72**, 39.
Cooper, P. D. (1967). *In* "Methods in Virology" (Ed. K. Maramorosch and H. Koprowski), New York and London, vol. III, pp. 244–311.
Deacon, W. E., Falcone, V. H., and Harris, A. (1957). *Proc. Soc. exp. Biol. Med.*, **96**, 477.
Duguid, J. P. (1959). *J. Gen. Microbiol.*, **21**, 271.
Duguid, J. P. (1964). *Rev. lat.-amer. Microbiol.*, **7**, Suppl. 13–14, p. 1.
Duguid, J. P., and Campbell, I. (1969). *J. med. Microbiol.*, **2**, 535.
Duguid, J. P., and Gillies, R. R. (1958). *J. path. Bact.*, **75**, 519.
Duguid, J. P., Smith, Isabel W., Dempster, G., and Edmunds, P. N. (1955). *J. Path. Bact.*, **70**, 335.
Freund, J., Thompson, K. J., Hough, H. B., Sommer, H. E., and Pisani, T. M. (1948). *J. Immunology.*, **60**, 383.
Fulthorpe, A. J. (1957). *J. Hyg., Camb.*, **55**, 382.
Fulthorpe, A. J. (1958a). *J. Hyg., Camb.*, **56**, 183.
Fulthorpe, A. J. (1958b). *Immunology*, **1**, 365.
Fulthorpe, A. J. (1959). *Immunology*, **2**, 104.
Fulthorpe, A. J. (1962). *Immunology*, **5**, 30.
Gillies, R. R., and Duguid, J. P. (1968). *J. Hyg., Camb.*, **56**, 303.

Glenny, A. T., and Barr, Mollie (1932a). *J. Path. Bact.*, **35**, 91.
Glenny, A. T., and Barr, Mollie (1932a). *J. Path. Bact.*, **35**, 142.
Goudie, R. B., Horne, C. H., and Wilkinson, P. C. (1966). *Lancet*, ii, 1224.
Grabar, P., and Williams, C. A. (1955). *Biochim. biophys. Acta*, **17**, 67.
Holt, L. B. (1968). *J. med. Microbiol.*, **1**, 169.
Holt, L. B., and Spasojević, Vera (1968). *J. med. Microbiol.*, **1**, 119.
Jacobs, J. W., and Peacock, D. B. (1970) *J. med. Microbiol.*, **3**, 313.
Jones, D. M., and Sequeira, P. J. L. (1966). *J. Hyg., Camb.*, **64**, 441.
Kabat, E. A., and Mayer, M. M. (1961). "Experimental Immunochemistry", 2nd ed., Springfield, Ill.
Kerr, W. R., McGaughey, W. J., Coghlan, Joyce D., Payne, D. J. H., Quaife, R. A., Robertson, L., and Farrell, I. D. (1968). *J. med. Microbiol.*, **1**, 181.
Kerr, W. R., Payne, D. J. H., Robertson, L., and Coombes, R. R. A. (1967). *Immunology*, **13**, 223.
Laurell, C-B. (1966). *Analyt. Biochem.*, **15**, 45.
Ledingham, J. C. G. (1932). *J. Path. Bact.*, **35**, 140.
Ledingham, J. C. G. (1933). *J. Path. Bact.*, **36**, 425.
Marcus, P. I., and Carver, D. H. (1965). *Science, N.Y.*, **149**, 983.
Maung, R. T. (1963). *J. Path. Bact.*, **85**, 51.
Mayer, M. M. (1965). *In* "Complement" (Ciba Fdn Symp.) (Eds. G. E. W. Wolstenholme and Julie Knight), London, p. 4.
Nelson, R. A., and Mayer, M. M. (1949). *J. Exp. Med.*, **89**, 369.
Norris, J. R. (1969). *In* "Spores IV" (Ed. L. L. Campbell), American Society for Microbiology, 45–58.
Oakley, C. L. (1954). *A. Rev. Microbiol.*, **8**, 441.
Oakley, C. L., and Fulthorpe, A. J. (1953). *J. Path. Bact.*, **65**, 49.
Osechinsky, I. V., Mekler, L. B., Petrov, R. V., and Mityushin, V. M. (1969). **16**, 427.
Pappenheimer, A. M., Jr., and Robinson, E. S. (1937). *J. Immunology.*, **32**, 291.
Porterfield, J. S. (1964). *In* "Immunological Methods" (Ed. F. Ackroyd), Oxford, pp. 341–362.
Porterfield, J. S. (1967). *In* "Methods in Virology" (Eds. K. Maramorosch and H. Koprowski), New York and London, vol. I, pp. 525–538.
Prausnitz, C., and Kustner, H. (1921). *Centbl. Bakt.*, I Abt. Orig., **86**, 160.
Preer, R. J., Jr. (1956). *J. Immunology*, **77**, 52.
Preston, N. W. (1963). *Br. med. J.*, **2**, 724.
Preston, N. W. (1965). *Br. med. J.*, **2**, 11.
Rice, C. E. (1948). *J. Immunology*, **60**, 11.
Scolari, L., Picard, J. J., and Heremans, J. F. (1968). *Clin. chim. Acta*, **19**, 25.
Sorkin, E. (1963). *In* "The Immunologically Competent Cell" (Ciba Fdn Symp.) (Ed. G. E. W. Wolstenholme and M. P. Cameron), London, p. 53.
Steele, A. S. V., fand Coombs, R. R. A. (1964). *Int. Archs Allergy appl. Immunology*, **25**, 11.
Thewaini Ali, A. J., and Oakley, C. L. (1967). *J. Path. Bact.*, **93**, 413.
Thomson, R. O., Walker, P. D., and Hardy, R. D. (1966). *Nature, Lond.*, **210**, 760.
Vogel, J., and Shelokov, A. (1957). *Science, N.Y.*, **126**, 358.
Walker, P. D., and Batty, Irene (1964). *J. Appl. Bact.*, **27**, 137.
Walker, P. D., and Batty, Irene (1965). *J. Appl. Bact.*, **28**, 194.
Walker, P. D., Thomson, R. O., and Baillie, Ann (1967). *J. Appl. Bact.*, **30**, 317.

CHAPTER VII

The Localization of Bacterial Antigens by the Use of the Fluorescent and Ferritin Labelled Antibody Techniques

P. D. WALKER, IRENE BATTY AND R. O. THOMSON

The Wellcome Research Laboratories, Langley Court, Beckenham, Kent, England

I.   Fluorescent Labelled Antibodies 219
     A. Introduction 219
     B. The conjugate 222
     C. Staining procedure 228
     D. Examination of preparations stained with fluorescent labelled
         antibodies 230
     E. Results 233
II. Ferritin Labelled Antibodies 237
     A. Introduction—properties of ferritin 237
     B. The preparation of ferritin from horse spleen 238
     C. Conjugation of ferritin to immunoglobulin 239
     D. Uses of the conjugate 241
     E. Results 242
References 246
Appendix 248
Addendum 250

I. FLUORESCENT LABELLED ANTIBODIES

A. Introduction

Fluorescent labelling of antibodies was introduced by Coons *et al.* in 1941. They showed that fluorochromes, i.e., substances that emit light in the visible region of the spectrum when irradiated with ultraviolet light, could be coupled to antibodies without loss of immunological specificity. If such a labelled antibody was reacted with a preparation containing the homologous antigen, combination occurred and the site could be located by its fluorescence when the preparation was illuminated by ultraviolet light. The method was subsequently extended by Coons and Kaplan (1950).

There are two main methods of applying the technique: direct and indirect. A schematic representation of the use of fluorescent labelled antibodies to locate bacterial cell-wall antigens with these two methods is given in

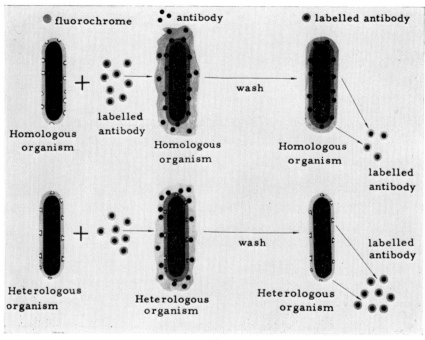

(a)

FIG. 1(a). Schematic representation of direct method of fluorescent staining.

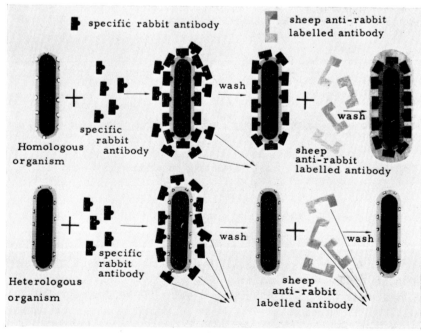

(b)

FIG. 1(b). Schematic representation of indirect method of fluorescent staining.

Fig. 1. In the direct method, the specific labelled antibody is applied to the preparation, it combines with specific receptor sites in the cell walls of the homologous bacteria and this combination is virtually unaffected by washing, when only the excess of labelled antibody is washed away. In the case of a heterologous organism, no combination occurs with the antigen sites in the bacterial cell wall, and during the washing all labelled antibody is removed. If the preparation is then irradiated with ultraviolet light, only the homologous bacteria that have combined with the specifically labelled antibody fluoresce. In the indirect technique, the preparation is first reacted with a specific unlabelled antiserum and again antigen–antibody combination unaffected by washing will occur only with the homologous organism. A fluorescent labelled antibody specific for the animal species from which the first specific globulin came (e.g., sheep anti-rabbit if the original specific antiserum was prepared in rabbits) is then applied to the washed preparation and will combine with any specific antiserum attached to the cell. As specific antiserum is present only on homologous organisms, these will become coated with a further layer of the labelled species specific antibody. No combination will occur with the heterologous organisms, as there was no combination with specific antibody during the first stage. After washing, the preparation is examined with a fluorescent microscope having an ultra-violet light source. Only the homologous organisms to which the fluorescent labelled antiserum has been attached can be seen to fluoresce.

The advantage of the direct technique lies in its simplicity, but in order to apply it each antiserum must be individually labelled with the fluoro-chrome. The indirect technique has the advantage that a single species specific labelled antiserum can be used with various different antisera; the specificity of the staining is determined at the first stage. However, this method involves an additional step with the possibility of increase of non-specific staining due to the double application of serum to the preparation.

The ideal fluorochrome must combine with the antiserum to form a stable conjugate without unduly affecting its specificity or combining power and must emit light of suitable intensity and colour on irradiation (Chadwick et al. 1958; Pearce, 1960). Few fluorochromes fulfil these requirements, and of these, fluorescein isothiocyanate, which fluoresces green, and lissamine rhodamine B200, which fluoresces red, are the most used. The fluorescence obtained with these dyes contrasts well with the naturally occurring blue autofluorescence of tissues. Other fluorochromes, such as fluolite C and lissamine flavine FSS, which fluoresce blue, are unsuitable for many kinds of work (Chadwick et al., 1958) because they lack this contrast. 1-Dimethylaminonaphthalene-5-sulphonic acid (DANS), an acceptable fluorochrome, is little used because, although its fluorescence is very similar to that of fluorescein, it is less intense.

The fluorescent labelled antibody technique has found many applications in bacteriology, virology, pathology, mycology and parasitology, both in the diagnosis of disease, i.e., in the identification of pathogens, and in basic research (Cherry *et al.*, 1960; Gordon, 1962; Lennette and Schmidt, 1964; Zamon, 1965; Cherry and Moody, 1965). Each discipline may require its own special techniques at various stages during preparation, staining and examination of specimens. The authors' experiences with the technique have been confined mainly to its use in the location of bacterial antigens. This has included both diagnostic applications and use as a research tool in the location of antigens during various phases of growth. In the following Sections, the methods of preparation, staining and examination of specimens using fluorescent labelled antibodies will be described in detail. These methods have given consistent and reliable results over the past 5 years. At appropriate points attention will be drawn to possible modifications that may be necessary for application in other fields of work.

B. The conjugate

1. *Preparation of antisera*

The method of preparing antisera suitable for conjugating with fluoro-chromes will vary with the antigen used. For example, if bacterial cell walls are being investigated, the antiserum will be prepared against heat-killed bacterial cells. Alternatively, if it is the heat labile or flagellar antigens that are being studied, a formolized suspension would be a more appropriate antigen. Antisera may also be prepared against spores or soluble antigens. Similar criteria apply to the preparation of antisera against other organisms, for example viruses, where the antiserum may be against soluble or particulate antigens. Rabbits are convenient animals in which to make antisera, although for some antigens the guinea-pig or goat are useful alternatives. The route of administration and dosage will also vary with the type of antigen. For soluble antigens the subcutaneous or intramuscular routes, preferably with the use of an adjuvant, are better than the intravenous route, but the intravenous route is useful for particulate bacterial antigens (Walker, 1963; Batty and Walker, 1963a, b, 1964; Walker and Batty, 1964, 1965). In the case of the heat-stable vegetative cell wall and spore antigens we have found that the agglutination reaction is a good indicator of the potency of the serum, although other workers have found that the estimation of antibodies by conventional serological methods is not necessarily a suitable indicator for fluorescent potentiality (Karakawa *et al.*, 1964; Lewis *et al.*, 1964). Specificity and not absolute titres should be the main object: in general the narrower spectrum type specific antibodies will appear much earlier in the immunization process than the broader reacting group anti-bodies.

2. Conjugation of immunoglobulins

The original method of labelling antibodies with fluorochromes developed by Coons *et al.* (1941) used fluorescein isocyanate. Considerable improvement in stability and simplification of the process resulted from the use of fluorescein isothiocyanate developed by Riggs *et al.* (1958). With the increasing use of this method, refinements both in the conjugation procedures and in the staining techniques have taken place. In this Section simple reliable methods of conjugating fluorescein isothiocyanate (BDH) and the sulphonylchloride of lissamine rhodamine B200 (ICI) to antisera will be described.

(a) *Labelling antisera with fluorescein isothiocyanate.* This method is essentially that of Riggs *et al.* (1958) as modified by Marshall *et al.* (1958) using Rivanol precipitation for the removal of albumin and the α- and β-globulins (Horsjai and Smetna, 1956). This leaves a chromatographically pure immunoglobulin preparation—

   (i) Adjust the pH of the antiserum to 8·5 with 0·1M NaOH.
   (ii) Gradually add a volume of 0·4% Rivanol (2-ethoxy 6,9-diamino-acridine lactate) (Koch-Light Laboratories Ltd, Colnbrook, Bucks) in distilled water equal to $3\frac{1}{2}$ times that of the serum. Stir continuously; a heavy yellow precipitate forms (Fig. 2a, between pp. 240 and 241).
   (iii) Decant off the clear supernatant and if necessary centrifuge the residue to recover any remaining supernatant. Add activated charcoal (LS grade, Farnell, Carbon, Plumstead, London S.E. 18) at a concentration of 1·2 g/100 ml to the supernatant and stir vigorously to remove excess of Rivanol.
   (iv) Filter through a Whatman No. 42 filter paper on a Buchner funnel under vacuum (Fig. 2b). The filtrate should now be colourless or, if any haemolysis has occurred before separation of the serum, slightly red.
   (v) To the clear filtrate add an equal volume of saturated ammonium sulphate (Fig. 2c) while stirring constantly. A white precipitate of the immunoglobulin forms and the mixture is left at 4°C overnight for complete precipitation to occur.
   (vi) Collect the precipitate by centrifugation and discard the supernatant fluid. Dissolve the precipitate in a volume of distilled water approximately equal to one third of the original volume of serum and transfer to Visking cellulose tubing. A sufficient length should be cut to allow for an increase in volume of up to two and a half times on dialysis. Dialyse the globulin against several changes of phosphate buffered saline pH 7·5 (see Appendix for compositions

of buffers, etc.) for 24 h to ensure that no ammonium sulphate remains. The procedure can be speeded up considerably by more frequent changes of buffer and constant stirring with a magnetic stirrer.

(vii) After dialysis adjust the concentration of the serum to 10 mg protein/ml. This figure may be critical (Tokamuru, 1962) if optimum conjugation is to occur (see Section on non-specific staining). The protein concentration can be estimated by the micro-Kjeldahl method for nitrogen or spectrophotometrically by the procedure of Porter (1955). The latter method is simple and convenient, measurements being made of the ultraviolet absorption of appropriate dilutions of the globulin at 260 and 280 nm. For pure protein the ratio of the absorption at these wavelengths should be 0·58 and deviation from this figure indicates the extent to which other absorbing material is present. The concentraton of protein is determined by the formula —

$$\frac{\text{Reading at 280 nm}}{1\cdot3} = \text{Protein concentration, mg/ml.}$$

The accuracy of the values obtained depends to a certain extent on how closely the 260/280 nm absorption ratio approximates to 0·58.

(viii) Add fluorescein isothiocyanate powder to the immunoglobulin at a concentration of 0·05 mg/mg of protein. As the protein concentration has been adjusted to 10 mg/ml this is equivalent to 0·5 mg of fluorescein isothiocyanate/ml. Add 10% (v/v) of 0·5M carbonate-bicarbonate buffer pH 9·0 (see Appendix) to the final reaction mixture. The fluorescein isothiocyanate and globulin are normally stirred together at 4°C overnight by a magnetic stirrer, although successful conjugations have also been achieved by stirring for 2 h at room temperature.

(ix) After the reaction is complete, it is necessary to remove any excess of the fluorochrome and this can be conveniently carried out by either of two methods. In the first method, the reaction mixture is placed in a dialysis bag as previously described and dialysed against phosphate buffered saline pH 7·2 until the dialysate is free from fluorescence when examined under ultraviolet light. This may take up to 14 days even with frequent changes of saline buffer.

In the second method use is made of the technique of gel filtration in which the conjugated globulin is separated from the more slowly running fluorescein isothiocyanate on a column of

Sephadex G-25 (Pharmacia, Uppsala, Sweden) (Fig. 2d). The conjugated globulin passes through the column as the first coloured fraction, while the excess fluorochrome remains behind as the more slowly running band (see Appendix for details of column and elution). Since Sephadex acts as a molecular sieve, separation of all small molecules from the conjugated globulin occurs. It has been our experience that even after equilibrating the column with several changes of buffered saline, separation of the conjugated globulin from buffer may occur. On prolonged storage this may lead to precipitation of the conjugate. To prevent this, the fraction may be given a short dialysis against the buffer, or more simply an additional amount [about 10 % (v/v)] of the buffer may be added to the conjugate fraction from the column.

(b) *Labelling antisera with lissamine rhodamine B*200. This method is essentially that described by Chadwick *et al.* (1958)—

(i) Prepare the immunoglobulin and adjust to 10 mg of protein/ml, as described for conjugation with fluorescein.

(ii) Grind together in a fume cupboard 1 g of lissamine rhodamine B200 and 2 g of P_2O_5 with a pestle and mortar for 5 min. The mixture first becomes sticky and requires frequent scraping from the sides with a spatula, but eventually it becomes dry and more readily workable. Add 10 ml of dry acetone and stir well. Leave for 5 min and filter off any undissolved material. This process gives an acetone solution of the sulphonyl chloride of lissamine rhodamine B200.

(iii) To one volume of the globulin solution add one volume of physiological saline and one volume of carbonate–bicarbonate buffer pH 9·0.

(iv) Carefully add drop by drop 0·1 ml of the sulphonyl chloride for each millilitre of globulin solution, stirring constantly. Continue stirring for 30 min. It is advisable to check the pH of the reaction mixture and if necessary re-adjust to pH 9·0 when all the sulphonyl chloride has been added. If the pH is allowed to fall to any appreciable extent, very poor conjugation will occur.

(v) The excess of fluorochrome can be removed either by dialysis or by fractionation on Sephadex as described previously for the fluorescein conjugate. The final conjugate should be a rich purple. Preparations showing a pink tinge are, in general, poor conjugates.

(c) *Non-specific staining.* The problem of non-specific staining becomes most acute when non-particulate antigens are studied, as for example in the examination of tissues for amorphous virus antigens or for signs of antibody production.

In the study of bacteria and other recognizable particulate entities that can be readily distinguished from surrounding cells or debris in tissue sections and smears, non-specific background staining is often of little consequence. It is therefore usually possible to use untreated labelled globulin for staining. Labelled whole serum is now rarely used.

Non-specific staining may arise from any of several causes—

(a) The presence of shared antigens that will combine with their antibodies present in the conjugate; this can be avoided by prior absorption of the labelled immunoglobulin with the antigen (see, for example, the preparation of type specific streptococcal antisera described by Moody et al., 1958).

(b) The presence of conjugated albumin and/or free fluorochrome in the conjugate; by the proper use of the preparative procedures described, non-specific staining due to these factors should not be encountered.

(c) Non-specific adherence of the conjugated immunoglobulin; this type of staining presents more of a problem and several methods have been suggested to overcome it: (i), absorption with tissue powders; (ii), fractionation of the conjugate on ion-exchange cellulose powders; (iii), the use of counterstains; (iv), dilution of the conjugate; and (v) changing the electrolyte concentration (von Mayersbach, 1966). The first three of these will be considered in greater detail—

1. Liver and other tissue powders are sometimes very effective in removing non-specific staining and have been used extensively in virus work, but the serum may require several absorptions. The main disadvantage of this treatment is that it may well reduce to some extent the intensity of the specific staining. The following absorption procedure described by Heimer (1967) is recommended. Add 1 ml of the conjugate to 100 mg of moistened liver powder (e.g., pig liver powder, Wellcome Reagents Ltd, Beckenham, Kent), mix well and leave for 1 h at room temperature. Centrifuge at 3000 rev/min for 30 min and collect the supernatant for use. The liver powder may be wetted by covering the required quantity with phosphate buffered saline, pH 7·2, centrifuging and discarding the supernatant. For certain purposes it is better to absorb the conjugate with tissue powder from the organ or tissue to be examined.

2. It has been shown that attachment of fluorescein isothiocyanate to protein is not a uniform process, and that within the same fluorescent antibody preparation there may be marked variation in the number of fluorescent groupings per protein molecule (Goldstein et al., 1961;

Curtain, 1961, Peters, *et al.*, 1961). It was also shown that the specific and non-specific staining properties of fluorescent labelled antibodies are related to the number of fluorescent groupings conjugated with antibody. It is thought that the fluorochrome–protein ratio is important in determining the specificity of staining (Goldstein *et al.*, 1961; Brighton 1966). In order to obtain a fraction having bright specific staining with little or no suboptimally coupled specific globulin, Goldstein *et al.* (1962) prepared a series of fluorescent rabbit immuno-globulins by adding different amounts of the dye to provide various new dye–protein ratios. It was found that in these preparations significant fractions of the immunoglobulins were coupled sub-optimally, and inclusion of these in the preparations inhibited the brightness of specific staining, and therefore decreased the sensitivity of the technique. These sub-optimally coupled fractions were eliminated from the preparations by chromatography on DEAE cellulose. The material was applied to a column equilibrated at pH 7·5 with 0·01M sodium phosphate. Uncoupled immunoglobulins and fractions having fluorochrome protein ratios of $0·3 \times 10^3 – 0·6 \times 10^3$ were not absorbed on the column. When these had been washed through, gradient elution to 1M NaCl produced a series of fractions that had increasing fluoro-chrome–protein ratios. Of this series, only those with the brightest ratios ($2·4 \times 10^3$ and above) produced acceptably bright staining. Reference should be made to these papers for further details.

3. Hall and Hanson (1961) described several counterstains suitable for use with fluorescent labelled antisera. These are based on the fluor-escent properties of aluminium chelates of certain *o*-hydroxy-azo dyes. Several 2,2-dihydroxy-azo dyes were screened for their possible use as fluorescent counterstains by dissolving the dyes in 2M NaOH in concentrations of 1×10^{-5} mole of dye/ml, chelating with an excess of aluminium chloride, and adjusting the pH to 5·5 with HCl. Six dyes were found to stain animal tissue in conjunction with immuno-chemical staining; Alizarine Garnet, yellow; Pontachrome Violet, yellow; Pontachrome Blue Black, cherry red; Eriochrome Black, orange; Diamond Red, orange yellow, and Flazo Orange, orange red. Flazo Orange was found particularly useful as a bacterial counterstain. We have had successful results with Superchrome Blue B extra, a dye in the same series as Pontachrome Blue Black. A slight yellowing of the fluorescence due to fluorescein is evident (Fig. 5a, between pp. 240 and 241), but this is found to be acceptable.

Procedure. After staining and washing in the usual manner the prepara-tion is treated with the counterstain; 30–60 sec are sufficient for tissue smears or sections. The counterstain is prepared as follows. Dissolve 1.7×10^{-5} mole of azo dye in 1 ml of NN-dimethylformamide. Add slowly, while shaking, 5 ml of chelating agent, and leave the solution for 30 min. The composition of the chelating agent is 50 ml of NN-dimethylformamide, 20 ml of distilled water, 10 ml of 0.1M $AlCl_3$ and 10 ml of M acetic acid. Adjust to pH 5.2 with M NaOH and make up to a total volume of 100 ml with distilled water.

Successful results with these counterstains have also been reported by Paton (1964). Evans Blue has also been used successfully by White and Kellog (1965) and by Midura, *et al.* (1967). Lissamine rhodamine B200 coupled to a heterologous immune or normal globulin is also an effective counterstain for fluorescein labelled homologous immunoglobulin. This is illustrated in Fig. 8a, b (between pp. 240 and 241), which show the results of staining tissue culture cells infected with canine distemper or canine hepatitis virus with fluorescein labelled distemper antibodies and rhodamine labelled hepatitis antibodies. In Fig. 8(a) the slight non-specific staining by the rhodamine labelled hepatitis serum provides a contrasting background to the specifically stained cytoplasmic inclusion bodies of the developing distemper virus. Conversely, in Fig. 8(b) the slight non-specific staining by fluorescein labelled distemper antiserum provides a good contrast to the specifically stained intranuclear inclusion bodies of the hepatitis virus.

C. Staining procedure

The main features of the staining procedure are illustrated in Fig. 3(a)–(h).

1. *Preparation of Smears (Fig. 3a)*

For the preparation of bacterial smears by far the most satisfactory results are obtained by making smears in saline from young (12 h) colonies from agar plates. The individual worker will quickly gain experience in preparing the right kind of smear. If the smear is too thick, there may be poor penetration of antibody, with the result that only the surface layer, which may be lost on washing, is stained. Alternatively if the smear is too thin difficulty may be found in locating the organisms. If it is necessary to examine liquid cultures, it is advisable to centrifuge a portion of the culture and re-suspend the organisms in a few drops of saline. For examination of tissues for bacteria, either a surface impression or a smear made by rubbing the cut end of a piece of tissue fairly heavily on the slide gives satisfactory results. If a dark-ground condenser is used for examination, slides of suitable thickness should be selected. Otherwise standard microscope slides (e.g., Chance Brothers Ltd, Smethwick, Birmingham 40) are satisfactory.

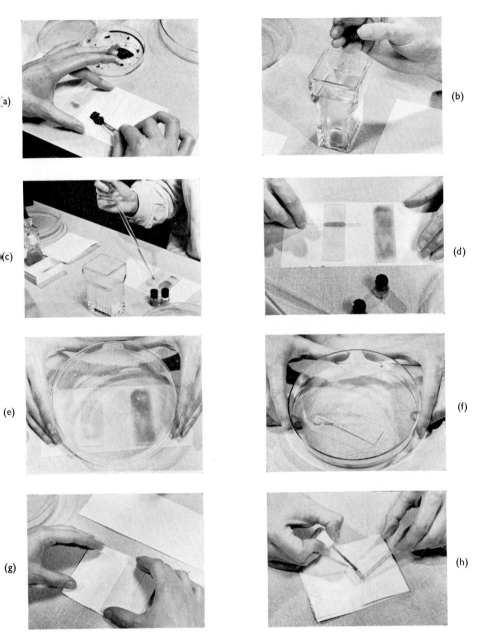

FIG. 3. Staining of preparations with fluorescent labelled immunoglobulins; (a), smearing; (b), fixation; (c), staining; (d), spreading; (e), incubation; (f), washing; (g) and (h), mounting.

2. Fixation (Fig. 3b)

The choice of fixative varies with the antigen under study. Using heat-stable bacterial cell wall antigens and spore antigens we have found that both fixation by gentle heat (60°C) and by acetone for 10 min are very satisfactory; acetone fixation is preferable for examining smears of tissues for the presence of bacteria where heating would be undesirable and in cases where the solvent properties of acetone are useful in removing fats that may interfere with staining. Other fixatives, such as formalin (6% formalin B.P.), are useful, particularly for heat-labile antigens, although strong oxidizing agents, such as Carnoy's solution, should be avoided.

3. Staining (Fig. 3c, d)

Spread the fluorescent labelled antibodies over the fixed smear. To avoid drying during the staining process, a Petri dish placed over damp blotting paper makes a satisfactory moist chamber. A staining time of 30 min is usually adequate, but under certain circumstances it may be appropriate to stain for a longer or shorter time.

4. Washing (Fig. 3e, f)

Drain the slides free of the fluorescent labelled globulin and wash thoroughly by gentle agitation in two or three changes of 0·01M phosphate buffered saline pH 7·5.

5. Mounting (Fig. 3g)

Quickly dry the slides between blotting paper and mount them in 0·5 M carbonate buffered glycerol pH 9·0 (see Appendix) under a coverslip. This pH is recommended as fluorescence is greater at alkaline pH values, and it gives a greater margin of safety to counteract the increase in acidity of glycerol that occurs on aging. Polyvinyl alcohol is another satisfactory mountant, particularly useful for making permanent preparations.

6. Sealing (Fig. 3h)

Collodion in the form of nail varnish or other similar quick drying solution is suitable for sealing the coverslip to the slide if the preparations are to be kept for a long period. It is advisable to store slides in the dark.

D. Examination of preparations stained with fluorescent labelled antibodies

1. Equipment

With the increasing interest in fluorescent microscopy, most of the major manufacturers of miscroscopes produce equipment suitable for use with the fluorescent labelled antibody technique. Although these instruments have the great advantage of compactness and versatility, they are normally multi-purpose and the light reaches the condenser indirectly, with consequent reduction in intensity. Owing to the low intensity of the light emitted by the specimen, it is essential for the purposes of photography that any prism system used for binocular vision should be removable to prevent considerable loss of intensity. Two sources of illumination are available; high-pressure mercury-vapour and quartz–iodine lamps. The merits of these two sources have been discussed by Baerer (1966), but whichever source is used it is important to have the most intense source practicable.

The relatively large expenditure for commercial equipment has prompted many workers to construct their own apparatus. Although this often lacks the advantages of commercial models it is usually good for the particular purpose of fluorescent microscopy, and is both reliable and cheap. Examples of this type of apparatus are described by Young (1961), Batty and Walker (1967) and Heimer (1967). The apparatus illustrated in Fig. 4 has been used

Fig. 4. Apparatus for fluorescent microscopy; see Appendix E for details.

by us for several years, both in the routine examination of specimens and for photography. The apparatus can be assembled fairly easily with equipment readily available (see Appendix) and gives results equal to those of more expensive apparatus although not quite so convenient. In use it is necessary to provide adequate shielding to prevent stray rays from the lamp reaching the operator. Providing this is done, no danger exists in the use of this equipment, as the eye is protected from stray ultraviolet light by the secondary filter. It is possible by mounting the microscope horizontally to focus the ultraviolet beam directly onto the condenser of the microscope, thereby eliminating the loss that occurs by mirror absorption. Alternatively, by using a front-silvered mirror the microscope can be used in the conventional upright position (see Batty and Walker, 1967; Heimer, 1967). However, recently Ledwell et al. (1967) have also described a system in which light is projected directly on to the condenser of the microscope with the latter in the vertical position. A dark-ground condenser is preferable for many types of examination, as fluorescence shows up brightly against the dark background. The advantages and disadvantages of bright-field and dark-field condensers are discussed by Young (1961), who concludes that the dark-field system has decided theoretical and practical advantages.

2. Choice of filters

The range and combinations of filters available, together with data on their transmission, are comprehensively discussed by Young (1961) and Nairn (1964). Most manufacturers supply information on the combinations of filters suitable for their equipment. We agree with Heimer (1967) that the individual worker should experiment with different combinations of filters in order to select the best for a particular system. In order to obtain the brightest fluorescence, blue–violet filters transmitting in the region of 450 nm are recommended. It should be remembered, however, that as the wavelength is progressively increased into the visible region, it becomes necessary to use either a green or orange secondary filter, and the investigator must be thoroughly conversant with the appearance of the image under these conditions. If bacteria are being studied, stained organisms fluoresce brilliantly, although the appearance of unstained cells will vary. Bacteria showing little natural autofluorescence are almost invisible; however, any marked autofluorescence will appear a dull green or orange with this combination of filters (Fig. 5b, between pp. 240 and 241).

Providing the investigator is thoroughly conversant with the appearance of stained and unstained organisms under his system, no difficulty arises. If there is any doubt, ultraviolet filters transmitting in the region of 250–350 nm should be used. At this wavelength the green or red fluorescence is much less intense, but any natural autofluorescence appears blue, as a colourless secondary filter is employed with this primary filter (Fig. 5c). This system has been found particularly useful in the location of Shigella sonnei in faecal smears (Taylor et al., 1964), where difficulties may be encountered in distinguishing weakly fluorescing organisms from the dull-green tissues with blue–violet light.

E. Results

The specificity of fluorescent labelled antibodies when used as stains can be determined by using the controls laid down by Coons and Kaplan (1950), *viz.*—

1. Absence or diminution of staining if unconjugated immune globulin is allowed to react with the preparation before the conjugate is applied.
2. No staining by non-immune labelled globulin.
3. No staining if the conjugate is absorbed with the specific antigen before use.

The advantages of using dark-ground illumination are very evident when culture smears of bacteria are being examined. The organisms fluoresce brightly against a dark background. Labelled antibodies prepared against heat-killed cells combine with the antigen in the cell wall, which fluoresces brightly (Fig. 6a). In the case of motile species, such antisera may also contain small amounts of flagellar antibodies, and this can result in the staining of flagella (Fig. 6b). The specificity of the fluorescent staining technique is illustrated in Fig. 7(a, b). Fig. 7(a) is a photomicrograph of a mixed smear under dark illumination with a tungsten lamp. Large numbers of organisms are seen. A photomicrograph of the same field illuminated by ultraviolet light shows only a small proportion of these organisms, the homologous ones fluorescing (Fig. 7(b)). Specificity is again illustrated by Fig. 8(c) (facing p. 241), which is a photograph of a mixed smear of *Clostridium septicum* and *Clostridium chauvoei* that has been stained with a mixture of the two antibodies conjugated to different fluorochromes. The anti *Cl. septicum* globulin has been conjugated with fluorescein isothiocyanate and the anti *Cl. chauvoei* globulin with lissamine rhodamine B200. Each organism has removed from the mixture its own specific antibody, and as these are conjugated to different fluorochromes the organisms fluoresce green or red. A further use of this technique to follow the location of various antigens during phases of growth is illustrated in Fig. 9. In the case of both *Clostridium sporogenes* and *Bacillus cereus*, specific antisera were prepared against either spores or vegetative cells. By conjugating the antibodies to the spores with fluorescein isothiocyanate and the antibodies to heat killed vegetative cells with lissamine rhodamine B200, it is possible to observe the antigenic changes occurring during germination and sporulation by staining with a mixture of these antisera at various stages. In *Cl. sporogenes* the emerging vegetative cell appears to split or stretch the spore coat, which then dissolves on the surface of the vegetative cells, whereas in *B. cereus* the spore coat remains attached to the end of the emerging vegetative cell even after the cell has divided several times. These differences are clearly illustrated by the colour changes from green (spore antigen sites in the spore coats) to red (vegetative cell wall antigen sites).

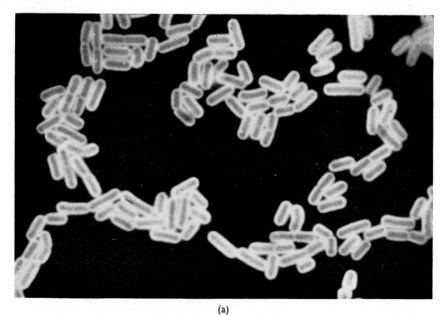

(a)

FIG. 6(a). Culture smear of *Clostridium welchii* Type D, CN 3688, stained with *Cl. welchii* Type D fluorescein labelled immunoglobulin. × 5000.

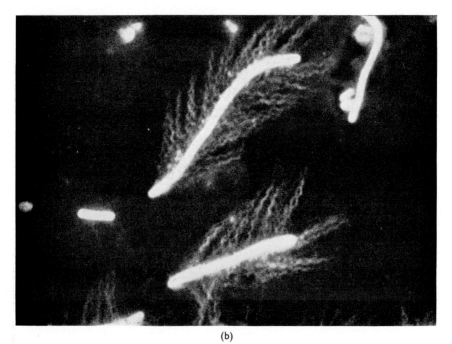

(b)

FIG. 6(b). Culture smear of *Clostridium septicum*, CN 3204, stained with *Cl. septicum* fluorescein labelled immunoglobulin. × 5000.

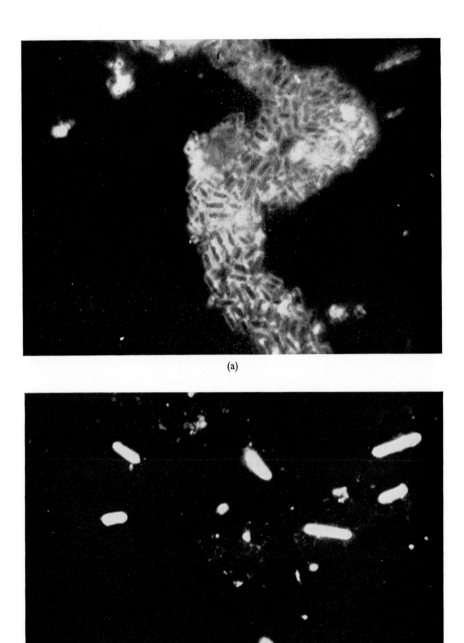

(a)

(b)

Fig. 7(a) and (b). Mixed smear of *Clostridium botulinum* Type E, *Cl. welchii* and *Staphylococcus*, × 5000: (a), dark-ground illumination; (b), same field with fluorescein labelled *Cl. botulinum* Type E immunoglobulins.

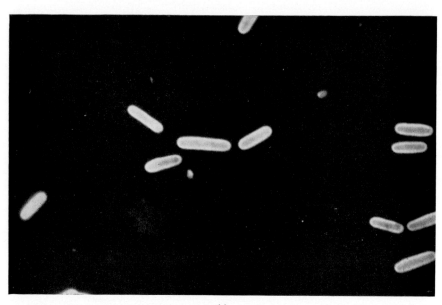

(a)

Fig. 10(a). Smear from oedema fluid of guinea-pig infected with *Clostridium oedematiens* Type B, stained with fluorescein labelled *Cl. oedematiens* Type B immunoglobulins. × 5000.

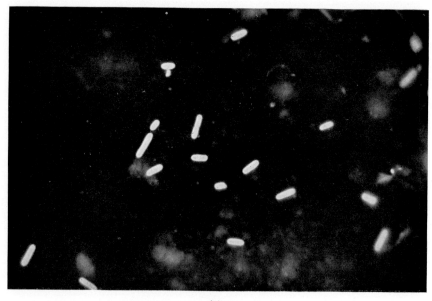

(b)

Fig. 10(b). Impression smear from necrotic lesion in the liver of a sheep, dying with black disease, similarly stained. × 2500.

The appearance of organisms in tissue smears is shown in Fig. 10 (a, b), from which it can be seen that the organisms are sharply differentiated from the background tissues.

These examples illustrate some of the types of information that can be obtained by using fluorescent labelled antibodies. From a diagnostic viewpoint, antibodies fall into two categories; broad spectrum species specific antibodies and narrower spectrum strain specific antibodies, although intermediate categories do occur. The species specific antibody is very useful for identifying the organisms in tissues from pathological conditions or in cultures. The strain specific antibody, although less useful in diagnosis is nevertheless useful when epidemiological studies are being undertaken. It is possible by using such labelled antibodies to follow the distribution and transmission of a particular strain within a population. The application of the technique to the identification of Group A streptococci, enteropathogenic *Escherichia coli*, *Treponema pallidum*, *Neisseria gonorrhoeae*, *Corynebacterium diptheria*, *Bordetella pertussis*, salmonellae, shigellae, brucellas, pasteurellas, bacilli, leptospira and listerias has been covered in the comprehensive review of Cherry and Moody (1965), and for this reason we have chosen to give additional examples from our own experience rather than re-cover this ground.

II. FERRITIN LABELLED ANTIBODIES

A. Introduction—properties of ferritin

The use of ferritin labelled antibodies first described by Singer (1959) has enabled the electron microscopist to identify areas of specific antigen in ultrastructural studies in much the same way as fluorescent labelled antibodies have been used to detect specific antigens at lower magnification with the light microscope. Whereas the latter technique relies for its efficacy on the intense emission of fluorochromes irradiated by certain wavelengths of light, the value of the ferritin labelling technique lies in the high electron scattering power of the ferritin molecule.

Ferritin is an iron-containing protein present in many tissues in the body, but most readily prepared from spleen. It is remarkable in that it may contain as much as 23% Fe by weight, present in the form of a micelle of Fe $(OH)_3$ within the ferritin molecule, the protein part of which has a molecular weight of about 460,000. Under the electron microscope, this micelle shows up clearly as a characteristic electron-opaque spot some 120Å in diameter. The chemistry of ferritin and its relevance in iron metabolism are discussed in definitive reviews by Granick (1946) and Michaelis (1947).

B. The preparation of ferritin from horse spleen

At these laboratories two types of ferritin have been used with equal success in the preparation of labelled antibodies. Both have been derived from horse spleen, the first by the method of Granick (1946) and the second by the method of van Heyningen described by Florey (1967). Both procedures will be described in detail.

It is our experience that the best spleens are those obtained from normal horses between 6 and 12 years old. Spleen from horses under immunization which are bled regularly are unsatisfactory for the production of ferritin.

Whether a spleen is going to be a rich source of ferritin can be determined by teasing a piece with a drop of 10% w/v $CdSO_4$ on a slide and observing the tissue under a microscope; crystals of ferritin should be visible after about 1 min. If no such crystals can be seen after 5 min, the spleen is low in ferritin and should not be used.

1. Preparation of crystalline ferritin (Granick, 1946)

Trim the spleen of fatty connective tissue and homogenize in a Waring blender. Mix with 1–1·5 volumes of de-ionized water. Heat the mixture with constant stirring to 80°C, allow to cool to 50°C and remove the mass of coagulated protein by straining through heavy-duty linen, first under gravity and then by the application of pressure. The dark-brown filtrate should now be filtered through a layer of Hyflo Supercel (Johns, Manville, Lompoc, Calif.) in order to remove residual quantities of suspended denatured protein.

To each litre of clear filtrate add and dissolve 350 g of $(NH_4)_2SO_4$. Collect the precipitate formed by centrifugation and then re-dissolve it by suspending in a small quantity of water; dialyse against water.

Remove insoluble material by centrifugation and add to the clear supernatant 0·25 volumes of 25% (w/v) $CdSO_4.\frac{8}{3}H_2O$. A voluminous precipitate forms immediately. Store the solution for 2–3 days in the cold. Microscopic examination reveals the presence of crystalline ferritin dispersed throughout the amorphous material.

Centrifuge the suspension at low speed to allow differential deposition of the crystalline and the amorphous fractions. Remove and discard the supernatant and skim off and discard most of the amorphous layer.

Wash the residue with water, re-centrifuge and then suspend in a small volume of 2% (w/v) $(NH_4)_2SO_4$ in order to re-dissolve the crystalline ferritin.

Remove the remainder of the amorphous fraction from this solution by centrifugation and add 0·25 volumes of 25% (w/v) $CdSO_4.\frac{8}{3}H_2O$ to the supernatant as before. At this stage crystallization begins immediately and is complete in a few hours. Very little amorphous material is formed.

Treat the suspension of crystals as described above then recrystallize the ferritin twice more.

The final material may be stored in the cold as a wet crystalline mass or as a solution sterilized by filtration through a Seitz pad. It cannot be freeze-dried because of the presence of Cd which renders dried ferritin insoluble. In order to remove residual Cd from the ferritin, the method of Rifkind et al. (1964) has been followed. The ferritin, dissolved in 2% (w/v) $(NH_4)_2SO_4$, is precipitated three times with one volume of saturated $(NH_4)_2SO_4$, the precipitate being amorphous. A concentrated aqueous solution of this final precipitate is dialysed thoroughly against water and then against 0·05M phosphate, pH 7·5. This material may be freeze-dried, although a small fraction of the dried ferritin may still be insoluble. Normally ferritin is kept as a 2·5% (w/v) solution, sterilized by filtration, and the concentration is determined by drying a small volume to constant weight on a filter paper in an incubator.

2. Preparation of Cd free Ferritin (Florey, 1967)

Homogenize one trimmed horse spleen in a Waring blender, suspend in one volume of water and leave overnight at room temperature. Centrifuge the suspension, keep the supernatant and repeat the extraction on the residue.

Combine both supernatants and heat to 80°C. Filter off the heavy precipitate, hot, through a fluted filter paper by gravity.

Precipitate the clear filtrate when cool by the addition of 300 g of $(NH_4)_2SO_4$ to each litre. After 1 h, centrifuge off the amorphous crude ferritin, discard the supernatant and suspend the precipitate in a small volume of 23·4% (w/v) $(NH_4)_2SO_4$.

Centrifuge the suspension and discard the supernatant. Dissolve the residue in a small amount of water, dialyse thoroughly against water then against 0·05M Na phosphate, pH 7·5. Freeze-dry or store as a sterile solution.

C. Conjugation of ferritin to immunoglobulin

Singer and Schick (1961) described the use of m-xylylene di-isocyanate and toluene 2,4-di-isocyanate for coupling ferritin to immunoglobulin. Of the two chemicals, preference was given to the latter since the functional groups of the m-xylylene di-isocyanate have similar reactivities, and the conditions that were required for coupling with this agent promoted the likelihood of electrovalent linkage between the proteins to be coupled. Toluene 2,4-di-isocyanate does not suffer from this disadvantage, the isocyanate group at the ortho position being much less reactive than that at the para position; conditions were described in which this more reactive group could be used in the first stage to attach the molecule to ferritin.

By altering the conditions the isocyanate on the *ortho* position completed the coupling to added immunoglobulin. This procedure resulted in a linkage between the proteins that was completely covalent. Borek and Silverstain (1961) and Rifkind *et al.* (1964) have found *m*-xylylene di-isocyanate to be satisfactory when used to couple ferritin to antibacterial, antiviral and anti-protein immunoglobulin. The bifunctional compound p,p'-difluoro-*m,m'*-dinitrophenylsulphone has been used with considerable success by Taude and Sri Ram (1962).

In this laboratory we have used the procedure described by Singer and Schick (1961) with the coupling agent toluene 2,4-di-isocyanate and consistently good results have been obtained with the ferritin labelled immunoglobulins thus produced. The conditions used are fairly flexible within certain limits. For example, when the amount of immunoglobulin to be coupled is very small more dilute reaction mixtures may be used without apparently interfering with the conjugation.

1. Preparation of immunoglobulins

The preparation of purified immunoglobulin from immune serum has been described earlier. The concentration and purity of the immunoglobulin are determined by the method of Porter (1955).

Before coupling the immunoglobulin solution is concentrated by dialysis against polyethylene glycol ($M = 20,000$) ("Carbowax", Union Carbide Ltd, London W.1). Kohn (1959) thoroughly dialysed against 0·1M borate buffer, pH 9·5, and diluted in this buffer to a concentration of 1·5% (w/v) immunoglobulin.

2. Preparation of ferritin

Centrifuge 6 ml of a 2·5% (w/v) ferritin solution in 0·05M phosphate, pH 7·5, prepared according to methods (1) or (2) above on an ultracentrifuge for 3 h at 130,000 g. Remove most of the colourless supernatant and replace with fresh buffer.

Gradually dissolve the ferritin pellet in 0·05M phosphate, by occasional shaking over a period of several hours. If this stage is hurried by, for example, the too vigorous use of a glass rod to break up the pellet, some denaturation of the ferritin occurs. Adjust the concentration of the dissolved ferritin to 1·5% (w/v) by dilution to 10 ml.

3. Conjugation

Cool 5 ml of 1·5% (w/v) ferritin in 0·05M NaH_2PO_4, pH 7·5, to 0°–2°C and add 0·1 of toluene 2,4-di-isocyanate (K & K Laboratories Inc., Plainview, N.Y.). This reagent is kept frozen and stoppered in the refrigerator,

(a)   (b)

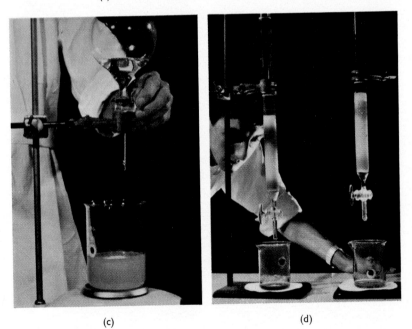

(c)   (d)

FIG. 2. Conjugation of fluorescein to immunoglobulins: (a), rivanol precipitation; (b), filtration after treatment with activated charcoal; (c), $(NH_4)_2SO_4$ precipitation; (d), separation of immunoglobulin on Sephadex G25.

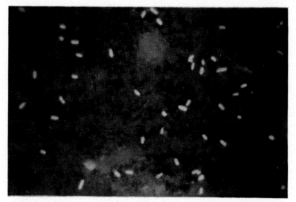

(a)

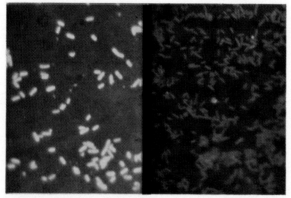

(b)

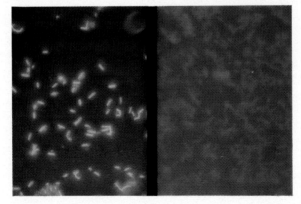

(c)

Fig. 5(a). Tissue smear containing *Cl. septicum* stained with fluoroscein labelled *Cl. septicum* antiserum, counterstained with Superchrome Blue Extra. × 500.

(b). Culture smear of *Cl. botulinum* Type A stained (left) *Cl. botulinum* Type A fluorescein labelled immunoglobulin (right) *Cl. botulinum* Type E fluorescein labelled immunoglobulin; blue–violet illumination, exposure time 10 sec. × 500.

(c). Culture smear of *Cl. botulinum* Type A similarly stained; ultraviolet illumination, exposure time 10 min. × 500.

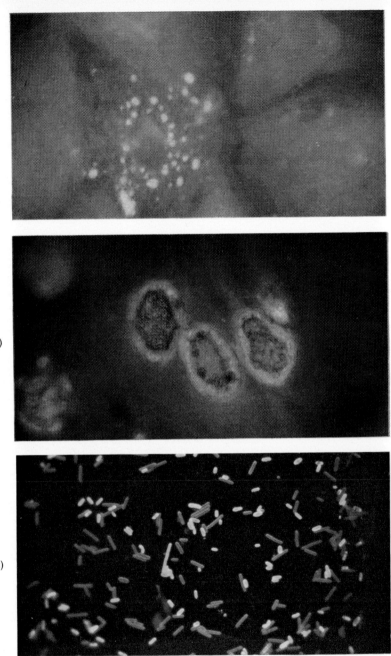

FIG. 8(a). Monolayer of dog kidney tissue culture 4 days after injection with Canine Distemper virus stained with a mixture of fluorescein labelled anti-Canine Distemper virus γ-globulin and rhodamine labelled anti-Infectious Canine Hepatitus virus γ-globulin. Note green fluorescence due to Canine Distemper virus in cytoplasm. × 2000.

(b). Monolayer of dog kidney tissue culture 4 days after infection with Infectious Canine Hepatitus virus stained similarly. Note orange fluorescence of Infectious Canine Hepatitus virus in nuclei. × 2000.

(c). A smear of a mixed *Cl. septicum* and *Cl. chauvoei* culture stained with a mixture of fluorescein isothiacyanate labelled *Cl. septicum* antiserum and lissamine rhodamine B.200 *Cl. chauvoei* labelled antiserum. × 750.

(a)

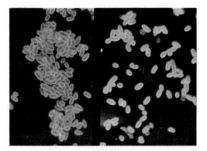

(d)

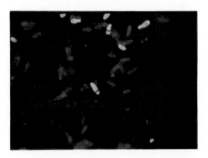

(b)

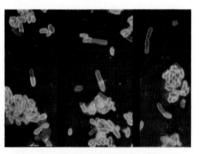

(e)

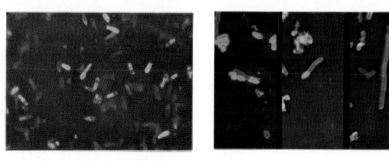

(c)   (f)

FIG. 9. Germinating spores of *Cl. sporogenes* L206 and *B. cereus* stained with fluorescent labelled antiserum; all staining was done as described under (a).

(a) Germinating spores of *Cl. sporogenes* L206 stained with mixture of fluorescein labelled spore antiserum and rhodamine labelled vegetative antiserum.

(b) Germinating spore showing out growth.

(c) Final germination stages.

(d) Ungerminated spore suspension (left) and spore suspension immediately following germination (right) of *B. cereus*.

(e) Germinating spores of *B. cereus* showing out growth.

(f) Final germination stages.

and when required can be thawed sufficiently by warming in the hand. It quickly refreezes when added to the ferritin and should be broken up into a crude suspension with a glass rod.

Stir the mixture vigorously for 25 min with a magnetic stirrer, maintaining the temperature at 0°–2°C. Transfer to a cold centrifuge bottle and spin in a refrigerated centrifuge. Carefully transfer the clear supernatant with a Pasteur pipette to a cold bottle and stand at 0°C for an additional hour to allow any dissolved reagent to react.

To the reaction mixture add an equal volume of 1·5% (w/v) immuno-globulin in 0·1M borate buffer, pH 9·5, warm to 37°C and incubate at this temperature for 1 h. After this period dialyse the solution overnight against 0·1M $(NH_4)_2CO_3$, in order to neutralize any unreacted isocyanate groups. Finally dialyse the mixture against phosphate-buffered saline (see Appendix).

4. *Purification of the conjugate*

The final solution contains ferritin, γ-globulin and the ferritin-γ-globulin conjugate as well as small quantities of other conjugates. Unreacted γ-globulin is removed by sedimenting the ferritin and ferritin-γ-globulin from the mixture by ultracentrifugation for 3 h at 130,000 g and discarding the supernatant containing γ-globulin. Rifkind *et al.* (1964) recommended that this should be repeated twice. Again careful treatment of the sedimented pellet is important. Unreacted ferritin is still present with the conjugate and several methods may be used to eliminate it. In our experience it does not interfere with the staining reaction, and no further purification has been employed. Borek and Silverstein (1961) have used both starch block, and continuous-flow paper electrophoresis to separate the conjugate from ferritin. Wagner (1967) elegantly separated all the components in the reaction mixture by molecular filtration through 4% (w/v) agarose gel. No doubt, the commercially available gels, Sepharose 4B (Pharmacia Limited, Uppsala, Sweden) or Biogel A (Bio-Rad Laboratories, Richmond, Calif.) could be used with equal success.

D. Uses of the conjugate

So far as specificity is concerned, the same criteria apply to ferritin labelled antibodies as apply to fluorescent labelled antibodies.

Ferritin labelled antibodies can be used in two ways, either by pre-embedding staining, in which the antigen and ferritin labelled antibody reaction takes place before the specimen is processed for the electron microscope, or by post-embedding staining in which ultra-thin sections of suitably fixed and embedded organisms are stained with the ferritin labelled antibody. The latter method involves great technical difficulties (Singer and McClean, 1963; Striker *et al.*, 1966; Thomson *et al.*, 1967).

Surface antigens are readily located in the electron microscope with the pre-embedding staining technique. We have found the following procedure very useful in examining the surface antigens of spores and vegetative cells of bacteria (Thomson *et al.*, 1966). Suspensions of spores and vegetative cells or spores at various stages of germination are fixed for 18 h in 0·01M phosphate, pH 7·5 containing 10% (v/v) formalin B.P. After washing, 0·2 ml of the concentrated suspension is shaken with 0·1 ml of the appropriate labelled antibody at 37°C for 30 min, centrifuged and washed in 10% (w/v) phosphate-buffered formol saline. After the final wash the deposit is fixed in 1% (w/v) osmic acid for 1 h, dehydrated in alcohol and embedded in the epoxy resins, Maraglas (Freeman and Spurlock, 1962) or Araldite (Glauert *et. al.*, 1965).

E. Results

Some examples of the results obtained by using this technique are illustrated in Figs. 11, 12 and 13. In Fig. 11(a) ferritin labelled antibody prepared against vegetative cells is located only around the vegetative cell; conversely, ferritin labelled antibody prepared against the spore is located only around spores (Fig. 11(b)). In germinating suspensions of spores, ferritin labelled antibody to the vegetative cell can be seen staining the emerging vegetative cell wall (Fig. 12) (Walker *et al.*, 1966, 1967a, b).

Antibodies or ferritin labelled antibodies do not penetrate freely across cell boundaries of mammalian cells (Loeffler *et al*, 1962; Easton *et al.*, 1962) and in order to locate antigens within the cell by this method some means must be found to facilitate the passage of the antibody into the cell. This has been accomplished by several methods, including fixation, freezing and thawing, and disintegration (see Sternberger, 1966). The amount of damage caused by such a process must be minimal to be compatible with the location of antigens within the cell. Although disintegration destroys the ultrastructure of bacterial spores, it is nevertheless possible to interpret pictures of spore disintegrates stained with vegetative cell labelled antibody. In Fig. 13 (a, b), ferritin labelled antibody to the vegetative cell can be seen along the core or cortical membrane in disintegrated spores and along the developing vegetative cell wall in disintegrated germinating spores.

The difficulties of locating intracellular antigens by using the pre-embedding staining technique have stimulated attempts to find a reliable post-embedding method. This problem is complicated by the non specific attraction of ferritin to the normal embedding media used in the preparation of specimens (Singer and McClean, 1963). The attempt by Singer and McClean to develop an ionic embedding medium for this purpose has been only partially successful. Further work has been directed towards enzyme labelled antibodies (Nakane and Pierce, 1966a, b) in addition to sera

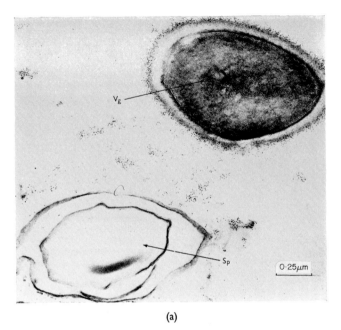

(a)

FIG. 11(a). Section of vegetative cell and spore of *Bacillus cereus* stained with ferritin labelled antibody to the vegetative cell. × 49,000

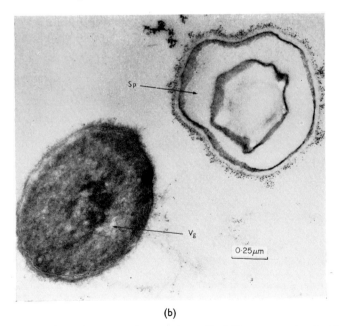

(b)

FIG. 11(b). Section of spore and vegetative cell of *B. cereus* stained with ferritin labelled antibody to the spore. × 49,000

labelled with ferritin. This new and developing field has recently been
reviewed by Sternberger (1966).

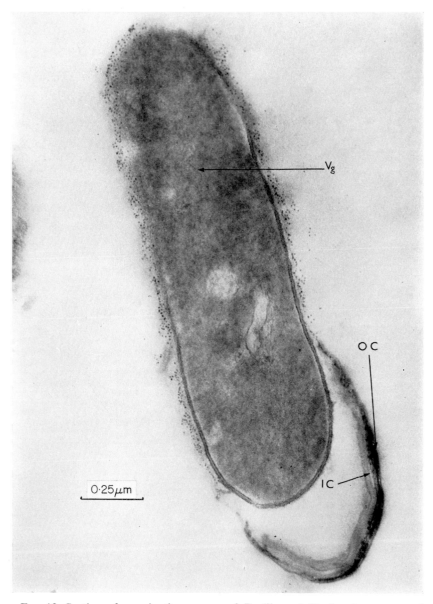

FIG. 12. Section of germinating spores of *Bacillus subtilis* showing outgrowth
stained with ferritin labelled antibody to the vegetative cell. × 91,500

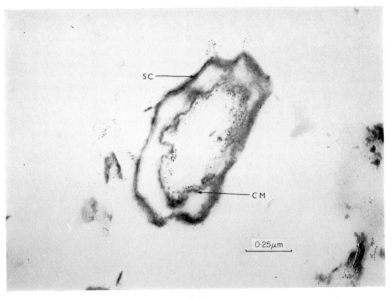

(a)

Fig. 13. (a). Section of disintegrated spores of *B. cereus* stained with ferritin labelled antibody to the vegetative cell. × 102,500

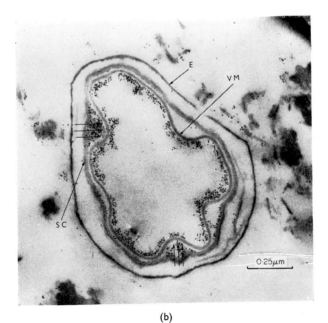

(b)

Fig. 13(b). Section of disintegrated germinating spores of *B. cereus* stained with ferritin labelled antibody to the vegetative cell. × 102,000

246   P. D. WALKER, IRENE BATTY AND R. O. THOMSON

Baerer, R. (1966). *In* "Fluorescent Methods in Histochemistry". Royal Microscopical Society, Sheffield 1966.
Batty, I., and Walker, P. D. (1963a). *J. Path. Bact.*, **85**, 517.
Batty, I., and Walker, P. D. (1963b). *Bull. Off. int. Épizoot.*, **59**, 1499.
Batty, I., and Walker, P. D. (1964). *J. Path. Bact.*, **88**, 327.
Batty, I., and Walker, P. D. (1967). *In* "International Symposium on Immunological Methods of Biological Standardization," Volume, 4, p. 73. Karger, Basle & New York.
Borek, F., and Silverstein, A. M. (1961). *J. Immun.*, **87**, 555.
Brighton, W. D. (1966). *J. clin. Path.*, **19**, 456.
Brighton, W. D., Taylor, C. E. D., Tomlinson, A. H., and Wilkinson, A. E. (1966). Mon. Bull. Minist. Hlth, **26**, 179.
Chadwick, C. S., McEntergert, M. G., and Nairn, N. C. (1958). *Immunology*, **1**, 315.
Cherry, W. B., and Moody, M. D. (1965). *Bact. Rev.*, **29**, 222.
Cherry, W. B., Goldman, M., and Carske, T. R. (1960). "Fluorescent Antibody Technique in the Diagnosis of Communicable Diseases". U.S. Dept. of Health, Education and Welfare.
Coons, A. H., and Kaplan, M. H. (1950). *J. exp. Med.*, **91**, 1.
Coons, A. H., Creech, H. J., and Jones, R. N. (1941). *Proc. Soc. exp. Biol. Med.*, **47**, 200.
Curtain, C. C. (1961). *J. Histochem. Cytochem.*, **9**, 484.
Easton, J. M., Goldberg, B., and Green, H. (1962). *J. exp. Med.*, **115**, 275.
Florey, Lord. (1967). *Proc. R. Soc.*, **B166**, 375.
Freeman, J. A., and Spurlock, B. O. (1962). *J. Cell. Biol.*, **13**, 437.
Glauert, A. M., Rogerts, G. E., and Glauert, R. H. (1965). *Nature, Lond.*, **178**, 803.
Goldstein, G., Slizys, I. S. and Chase, M. W. (1961). *J. exp. Med.*, **114**, 89.
Goldstein, G., Spalding, B. H., and Hunt, W. B. Jr. (1962). *Proc. Soc. exp. Biol. Med.*, **111**, 416.
Gordon, M. A. (1962). "Fungi and Fungous Diseases," Chapter XV, p. 207. Charles & Thomas, Illinois.
Granick, S. (1946). *Chem. Rev.*, **38**, 379.
Hall, G. T., and Hanson, P. A. (1961). *Zentbl. Bakt. ParasitKde, Abt. I. Orig.*, **184**, 548.
Heimer, G. V. (1967). *In* "Progress in Microbiological Techniques", p. 15. Butterworths, London.
Horsjai, J., and Smetna, R. (1956). *Acta. med. scand.*, **155**, 68.
Karakawa, W. W., Borman, E. K., and McFarland, C. R. (1964). *J. Bact.*, **87**, 1377.
Kohn, J. (1959). *Nature, Lond.*, **183**, 1055.
Ledwell, D. M., Taylor, C. D. E., Clark, S. P., and Heimer, G. V. (1967). *J. clin. Path.*, **20**, 214.
Lennette, E. H., and Schmidt, N. J. (1964). "Diagnostic Procedures for Viral and Rickettsial Diseases", 3rd Edn. America Public Health Association Inc., New York.
Lewis, J. J., Jones, W. L., Brooks, J. B., and Cherry, W. B. (1964). *Appl. Microbiol.*, **12**, 343.
Loeffler, H., Henle, G., and Henle, W. (1962). *J. Immun.*, **88**, 763.
Marshall, J. D., Evaland, W. C., and Smith, C. W. (1958). *Proc. Soc. exp. Biol. Med.*, **98**, 898.

Michaelis, L. (1947). *Adv. Protein Chem.*, **3**, 53.
Midura, T. F., Inouye, Y., and Bodily, H. L. (1967). Publ. Hlth Rep., *Wash.*, **82**, 275.
Moody, M. D., Ellis, E. C., and Updyke, E. L. (1958). *J. Bact.*, **75**, 553.
Nairn, R. C. (1964). "Fluorescent Protein Tracing", Vol. 2. Livingstone, London.
Nakane, P. K., and Pierce, G. B. Jr. (1966a). *J. Histochem. Cytochem.*, **14**, 929.
Nakane, P. K., and Pierce, G. B. Jr. (1966b). *J. Cell Biol.*, **33**, 307.
Paton, A. M. (1964). *J. appl. Bact.*, **27**, 237–243.
Pearce, A. G. E. (1960). "Histochemistry, Theoretical and Applied", p. 137. Churchill, London.
Peters, J. H., McDevitt, H. O., Pollard, L. W., Harter, J. G., and Coons, A. H. (1961). *Fedn Proc. Fedn Am. Socs exp. Biol.*, **20**, 17.
Porter, R. R. (1955). *Biochem, J.*, **59**, 405.
Rifkind, R. A., HSU, K. C., and Morgan, C. (1964). *J. Histochem. Cytochem.*, **12**, 131.
Riggs, J. L., Sewald, R. J., Burkhalter, J. H., Downs, C. M., and Metcalf, J. J. (1958). *Am. J. Path.*, **34**, 1081.
Singer, S. J. (1959). *Nature, Lond.*, **183**, 1523.
Singer, S. J., and McClean, J. D. (1963). *Lab. Invest.*, **12**, 1002.
Singer, S. J., and Schick, A. F. (1961). *J. biophys. biochem. Cytol.*, **9**, 519.
Sternberger, L. A. (1966). *J. Histochem. Cytochem.*, **15**, 139.
Striker, G. E., Donah, E. J., Petrali, J. P., and Sternberger, L. A. (1966). *Expl molec. Pathol. Suppl.*, **3**, 52.
Taude, S. S., and Sri Ram, J. (1962). *Archs. Biochem. Biophys.*, **97**, 430.
Taylor, C. D. E., Heimer, G. V., Lea, D. J., and Tomlinson, A. J. H. (1964). *J. clin. Path.*, **17**, 225.
Thomson, R. O., Walker, P. D., and Hardy, R. (1966). *Nature, Lond.*, **210**, 760.
Thomson, R. O., Walker, P. D., Baillie, A., and Batty, I. (1967). *Nature, Lond.*, **215**, 393.
Tokamuru, T. (1962). *J. Immun.*, **89**, 2.
Von Mayersbach, H. (1966). In "Fluorescent Methods in Histochemistry". Royal Microscopical Society, Sheffield, 1966.
Wagner, M. (1967). *Naturewissenschaften*, **15/16**, 444.
Walker, P. D. (1963). *J. Path. Bact.*, **85**, 41.
Walker, P. D., and Batty, I. (1964). *J. appl. Bact.*, **27**, 137–139.
Walker, P. D., and Batty, I. (1965). *J. appl. Bact.*, **28**, 194.
Walker, P. D., Baillie, A., Thomson, R. O., and Batty, I. (1966). *J. appl. Bact.*, **20**, 512.
Walker, P. D., Thomson, R. O., and Baillie, A. (1967a). *J. appl. Bact.*, **30**, 317–320.
Walker, P. D., Thomson, R. O., and Baillie, A. (1967b). *J. appl. Bact.*, **30**, 44–449.
White, L. A., and Kellog, D. S. Jr. (1965). *Appl. Microbiol.*, **13**, 171.
Young, M. R. (1961). *Q. Jl. microsc. Sci.*, **102**, 419.
Zamon, V. (1965). *Trans. R. Soc. trop. Med. Hyg.*, **59**, 80.

APPENDIX

A. Preparation of carbonate buffer, 0·5 M, pH 9·0

Solution A

Na₂CO₃, 5·3 g   Distilled water to 100·0 ml

Solution B

NaHCO₃, 4·2 g   Distilled water to 100·0 ml.

Theoretically, a pH of 9·0 should result from mixing 4·4 ml of solution A with 100 ml of solution B. In practice, it is sometimes necessary to add as much as 17 ml of solution A to 100 ml of solution B. The pH should be checked on a meter.

B. Preparation of phosphate buffer, 0·01M, pH 7·2

Solution A

Na₂HPO₄, 1·4 g   Distilled water to 100·0 ml

Solution B

NaH₂PO₄.H₂O, 1·4 g   Distilled water to 100·0 ml

Add 84·1 ml of solution A to 15·9 ml of solution B; add distilled water to make 1 litre. For buffered saline, add 8·5 g of NaCl before adding water.

C. Preparation of mounting fluid

Add one part of phosphate-buffered saline to 9 parts of reagent-grade glycerol. The refractive index of this mixture is approximately 1·46. Since glycerol tends to become acid, only a small quantity of mounting fluid should be prepared at a time, and the pH of this should be adjusted to 7·5. The authors prefer to add one part of the carbonate buffer to 9 parts of glycerol. This furnishes a margin of safety, and may enhance fluorescence.

D. Preparation of Sephadex column

Use G25 medium-grade Sephadex; suspend 25 g in 250 ml of physiological saline, mix well, allow to settle and pour off the suspension of fine particles. Repeat this three times in all. Re-suspend the deposit in 250 ml of 0·01M phosphate saline, pH 7·5, and pour into glass column, ca. 3 cm dia, with a sintered-glass disc. Allow buffered saline to drip out until a column ca. 15 cm in height has formed. This will cope with 25 ml of fluorochrome labelled γ-globulin. Close the tap. Place a circular absorbent cotton wool disc on the surface of the column. Pipette fluorochrome-conjugated γ-globulin so that an even layer forms, open the tap, and when the top of the fluorochrome-conjugated γ-globulin is level with the top of the cotton wool fill up with buffered saline. Collect the first fluorescent fraction. Overnight dialysis against the phosphate buffer is then sufficient to render the labelled antisera ready for use. A slightly longer column is advisable for the separation of lissamine rhodamine B200.

The fluorescent labelled antiserum can be stored unpreserved at 4°C.

E. Equipment for routine diagnostic fluorescent microscopy

1. *Lamp*

Box-type mercury compact lamp. Osram ME/D, 250 W, 3-pin base. Available from Young & Wildsmith Ltd, 35 Little Russel St, New Oxford St, London W.C.1.

2. *Choke*

Z. 1879. Available from General Electric Co. Ltd, Magnet House, Kingsway, London W.C.2.

3. *Capacitor*

60 μF, F. 8513. Available from General Electric Co. Ltd, Magnet House, Kingsway, London W.C.2.

4. *Microscope*

Straight tube, monocular.

5. *Objective*

× 45, fluorite.

6. *Condenser*

Substage, adjustable, dark ground.

7. *Microscope mirror*

"Beck", surface aluminized (N.B., this is needed in place of the standard substage mirror when the microscope is used in the vertical position.)

8. *Filters for critical visual examination.*

(a) *Primary*. Kodak ultraviolet-transmitting glass filter (Wratten No. 18A, 2 in. square). 12% $CuSO_4$ in glass filter cell, 50 × 20 mm internal dimensions, with sealed top and stopper.

(b) *Secondary*. Kodak ultraviolet filter (Wratten 2B gelatin, 2 in. square). Available from C. J. Whilens Ltd, Ilford Optical Works, Forest Road, Barkingside, Essex.

9. *Filters for photography or scanning*

(a) *Primary*. 2·5% ammoniacal $CuSO_4$ in glass filter cell (as above).
(b) *Secondary*. Kodak (Wratten No. 9 gelatin).

10. *Immersion oil*

Fluorfree oil, 200 ml. Available from G. T. Gurr Ltd, 136–140 New King's Road, London S.W.6.

11. *Condensing lens*

Nelson type 2, in metal mount only, 2–5 in. focal length. Available from W. Watson & Sons Ltd, 25 West End Lane, Barnet, Herts.

12. *Optical bench*

The use of an optical bench is an aid to stability, particularly when many examinations are being made, but it is possible to support the various pieces of equipment by more primitive means.

13. *Saddles*

(a) *One to support lamp.*

(b) *One metal table.* The 3-pin socket for the lamp will be fixed to this and slotted into saddle (a).

(c) *One to support Nelson lens.*

(d) *One to support filter assembly.*

(e) *One to support microscope.* If photographs are needed, the tube of the microscope must be horizontal, and an additional saddle will be needed to support it. Saddles are available from Precision Tool & Instrument Co. Ltd, 353 Bensham Land, Thornton Heath, Surrey.

14. *Levelling screws*

Four, for the optical bench.

ADDENDUM

Since this article was submitted for publication at the beginning of 1968, considerable progress has been made in the field of enzyme labelling of antibodies and it is appropriate to consider this in more detail. With this technique enzymes such as horseradish peroxidase and acid and alkaline phosphatase are coupled to antibodies which are then allowed to react with the antigen. The antigen antibody complex is then made visible by incubating with an appropriate substrate for the particular enzyme coupled to the antibody (Avrameas and Uriel, 1966; Nakane and Pierce, 1966). The relatively small molecular weight of the enzyme labelled antibody complex compared to that of the ferritin labelled antibody allows better penetration of tissues and hence its use to locate tissue and virus antigens within cells (e.g., Leduc, Wicker, Avrameas, and Bernhard, 1969; Wicker and Avrameas, 1969; Leduc, Scott, and Avrameas, 1969). In the context of the location of bacterial antigens this technique is only an advantage if it can be used to stain appropriately fixed and embedded sections. Leduc, Scott, and Avrameas (1969), using peroxidase or alkaline phosphatase labelled antibody, were unable to demonstrate any staining on glycol methacrylate embedded tissue sections, although a faint positive reaction was found on ultrathin frozen sections with alkaline phosphatase labelled antibody. On the other hand, Kawarai and Nakane (1970) were able to localize tissue antigens directly on ultrathin sections using peroxidase labelled antibody. Using this method luteinizing hormones, growth hormone and prolactin were localized on ultrathin sections of methacrylate embedded anterior pituitary gland of the rat. Before staining part of the embedding medium was removed with water saturated solutions of xylene or benzene.

Preliminary experiments in our laboratories using acid phosphatase labelled antibody to the vegetative cell on ultrathin sections of *Bacillus cereus* embedded in pre-polymerized methacrylate have been partially successful. However, a number of unknown factors affect the reproducibility of this method and these are currently under investigation.

A. Method of coupling

Several methods using different coupling agents have been followed for the conjugation of wheat germ acid phosphatase (Koch-Light Laboratories Colnbrook, Bucks) to antibody, for example, those of Nakane and Pierce (1967) (p,p'-difluoro-m,m'-dinitrophenyl sulphone or l-ethyl-3 (3-dimethyl aminopropyl) carbodiimide) and Avrameas (1969) (glutaraldehyde). Although all were equally effective the latter is preferred because of its simplicity and will be described in detail.

Prepare antisera to young organisms as previously described and isolate γ-globulin therefrom. To 5 mg γ-globulin in 0·5 ml 0·1M phosphate buffer pH 6·8 add 20 mg acid phosphatase. Make to 1 ml with buffer then slowly add with stirring 0·5 ml of 0·1% (v/v) glutaraldehyde in water. Allow to stand for 2 h at room temperature then dialyse the mixture for 48 h against several changes of phosphate buffer. Use immediately.

B. Method of staining

(a) *Pre-embedding staining.* Fix suspensions of spores and vegetative cells as previously described (p. 242) and after washing shake 0·2 ml of the concentrated suspension with 0·1 ml of the acid phosphatase labelled antibody at 37°C for 30 min. Wash the organisms three times in sterile saline and resuspend in the incubation mixture consisting of:

Solution A.—Dissolve 120 mg sodium β-glycerophosphate in 30 ml 0·05M cacodylate buffer containing 6·5% (w/v) sucrose pH 5·1.

Solution B.—Dissolve 60 mg lead acetate in 20 ml cacodylate sucrose buffer.

Add Solution B dropwise to Solution A. Allow to stand at room temperature overnight and then for one hour at 37°C. Filter before use.

Incubate the reaction mixture at 37°C for $\frac{1}{2}$ h. Collect the organisms, wash in saline. After a further final wash in saline resuspend in 20 ml saline containing 1% ammonium sulphide; the organisms turn black (Pearse, 1968).

Dehydrate in alcohol and embed in Maraglas as previously described.

(b) *Post-embedding staining.* Cut appropriate sections of the organism embedded in pre-polymerized methacrylate and collect on 200 mesh formvar coated grids. Invert the grids on the surface of a drop of the labelled antibody at room temperature for 30 min. Wash thoroughly in saline and transfer to the surface of a drop of the incubation mixture.

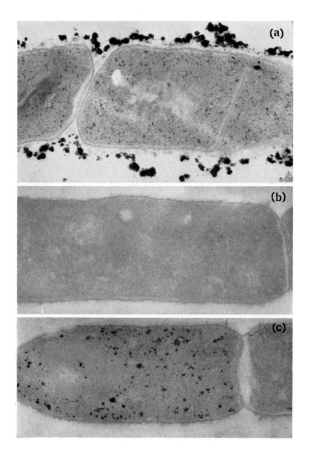

FIG. 14.

(a). Section of vegetative cell of *B. cereus* stained with acid phosphatase labelled antibody to the vegetative cell. × 34,000; note deposits of lead sulphide surrounding the cell and also within the cell.

(b). Similar section stained with heterologous acid phosphatase labelled antibody. × 20,700; note absence of deposits of lead sulphide both in and surrounding the bacterial cell.

(c). Similar section to Fig. 14b. × 20,700; note deposits of lead sulphide within the cell but not surrounding the cell.

Following a further wash in saline treat the grids in one of two ways. Transfer one grid on to the surface of a drop of ammonium sulphide for 10 sec, quickly wash in distilled water and dry. Wash the second grid quickly in distilled water and dry.

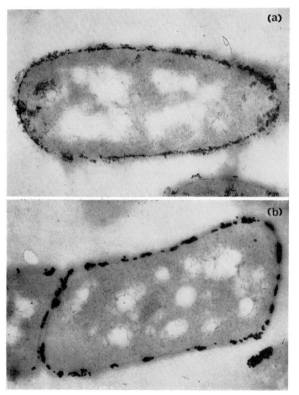

FIG. 15

(a). Ultrathin section of *B. cereus* fixed in glutaraldehyde and embedded in pre-polymerized methacrylate stained with acid phosphatase labelled vegetative cell antibody. × 20,700; note deposits of lead phosphate on the vegetative cell wall in the sections not treated with ammonium sulphide.

(b). Similar section treated in addition with ammonium sulphide. × 38,300 note deposits of lead sulphide on the vegetative cell wall and cross walls.

Results

(a) *Pre-embedding staining.* Deposits of lead sulphide can be seen surrounding the vegetative cells stained with homologous enzyme labelled antibody (Fig. 14a). No deposits are seen with heterologous enzyme labelled antibody (Fig. 14b). Deposits seen in the bacterial cell itself are presumably due to endogenous acid phosphatase activity following aldehyde fixation (Fig. 14c).

(b) *Post-embedding staining*. In the untreated section deposits of lead phosphate can be seen on the vegetative cell wall and cross walls (Fig. 15a). Sections post treated with ammonium sulphide show much more opaque deposits and some aggregation has occurred (Fig. 15b). Specific staining of the cell wall is still apparent, however.

No deposits are observed with heterologous enzyme labelled antibody.

Discussion

It is quite clear that enzyme labelled antibodies can be used to locate bacterial antigens either by the pre-embedding staining technique or post-embedding staining technique. Unfortunately, post-embedding staining of sections is not entirely reproducible from day to day. Using identical solutions and techniques good staining will occur on one occasion followed by several days of failure to demonstrate any staining whatsoever. One suspects that any one of a number of factors, e.g., surface tension, which are operative during staining, may influence the final result although Kawarai and Nakane (1970) apparently did not have this difficulty. Bearing these limitations in mind the technique of enzyme labelled antibodies appears to offer a promising solution to the location of intracellular antigens.

REFERENCES

Avrameas, S. (1969). *Immunochemistry*, **6**, 43–52.
Avrameas, S., and Uriel, J. (1966). *C.r. Seanc. Acad. Sci., Paris*, **262**, 2543.
Kawarai, Y., and Nakane, P. K. (1970). *J. Histochem. Cytochem.*, **18**, 3, 161–166.
Leduc, E. H., Scott, G. B., and Avrameas, S. (1969). *J. Histochem. Cytochem.*, **17**, 211.
Leduc, E. H., Wicker, R., Avrameas, S., and Bernhard, W. (1969). *J. gen. Virol*, **4**, 609–614.
Nakane, P. K., and Pierce, G. B., Jr. (1966). *J. Histochem. Cytochem.*, **14**, 929.
Nakane, P. K., and Pierce, G. B., Jr. (1967). *J. Cell. Biol.*, **33** (2), 307–318.
Pearce, A. G. E. (1968). "Histochemistry, Theoretical and Applied", Vol. 1. Churchill, London.
Wicker, R., and Avrameas, S. (1969). *J. gen. Virol.*, **4**, 465–471.

CHAPTER VIII

Toxin-Antitoxin Assay

IRENE BATTY

Wellcome Research Laboratories, Beckenham, Kent

I.	Terminology	257
II.	Selecting a Laboratory Standard Antitoxin	261
	A. Avidity	261
	B. Other conditions governing toxin/antitoxin combination .	262
	C. Danysz phenomenon	263
III.	Indicators	263
	A. Whole animals	263
	B. Skins	263
	C. Red cells	264
	D. Lecithovitellin	264
	E. Toxoid sensitized sheep red cells	264
	F. Tissue cultures	265
	G. The Ramon flocculation test	265
IV.	Flocculation Tests	268
	A. Materials required	268
	B. To test a toxin	268
	C. To test an antitoxin	270
V.	Total Combining Power Tests	275
VI.	Accuracy and Reproducibility	277
	Appendix I	277
	Appendix II	278
	Appendix III	278
	References	279

It has been known since the late 1880's that certain bacterial species produce toxins—very poisonous substances of high molecular weight—(Roux and Yersin (1888); Kitasato (1891)) which when introduced into susceptible animals give rise to characteristic lesions and may even cause death. Almost simultaneously both Von Behring and Kitasato (1890) showed that the serum of animals which received repeated injections of non-lethal doses of tetanus or diphtheria toxin acquired the property of specifically neutralizing these toxins and preventing their harmful effects. It very soon became clear to these early workers that if the effects of toxins and antitoxins were to be studied in anything more than the most super-ficial way, it was necessary to quantify their interactions and by 1897 Ehrlich had reported his classical work on the standardization of diphtheria toxin.

The first stage in any attempt at quantification is the definition of the units by which the activity is to be measured. Initially, the potency of toxins was measured in terms of the least volume of culture filtrate which, when injected by a stated route, would kill an animal of designated species and weight within a certain time; that is, it was measured in terms of the minimum lethal dose (M.L.D.). This parameter measures the toxicity of a toxin, and although independent of ability to react with antitoxin, it has the grave disadvantage that such measurements are particularly sensitive to changes in the indicator system. For example, the apparent toxicity of a toxin has been found to vary with the diet of the indicator animal, and even with the season of the year, as this may affect the susceptibility of the animal. The reproducibility of such tests is therefore poor either in the same laboratory at different times, or between laboratories at the same time. Added to this poor reproducibility was the early discovery that toxin tended to lose toxicity, either on storage or following treatment with a variety of chemical reagents, (toxoiding) without any dimunution in ability to com-bine with antitoxin. Therefore, although Ehrlich originally defined a unit of (diphtheria) antitoxin as the smallest amount of antiserum which would neutralize 100 M.L.D. of toxin, a unit of antitoxin is now defined as that amount of an antiserum which has the same ability to combine with a volume of toxin (some of which may be toxoided) as had the arbitrary unit of antiserum laid down by the original worker, i.e., as a certain weight of the dried standard preparation of antitoxin. For example one international unit of diphtheria antitoxin has the same neutralizing activity as 0·0628 mg of the International Standard.

It should be the usual practice of workers discovering a new toxin to prepare its antitoxin, and to allocate to this an arbitrary value, which is then used in all preliminary work until the World Health Organization is persuaded of the need for a reference preparation against which other workers can standardize similar antitoxins. Although it is not possible to make a reproducible direct measurement of biological activity, it is possible to make accurate comparative measurements of such activity. This is particularly important where antitoxins or vaccines are being used for prophylactic and therapeutic purposes. In this context, as these materials are mainly completely innocuous to man, it is usual to have minimum requirements rather than absolute standards. It is in this area, of vaccine and antitoxin production and use, that problems of assaying toxins and antitoxins most commonly occur, and it is from workers in this field that most of the accepted methods of assay have stemmed.

I. TERMINOLOGY

The following terms occur throughout the literature on toxin–antitoxin assay and have by usage acquired the meanings given.

Toxin. A pharmacologically or biologically active antigenic product of a bacterial cell, for example *Clostridium perfringens* α toxin, which is a lethal, necrotic, haemolytic, lecithinase.

Antitoxin. That part of the serum protein, usually gamma globulin, produced by the body in response to stimulation by a toxin or toxoid. It has the property of combining with and neutralizing the specific pharmacological or biological activity of the toxin, as does tetanus antitoxin.

Toxoid. Toxin retaining its ability to combine with, and stimulate the production of, antitoxin, but rendered harmless by chemical or physical treatment. Formalin is probably the most commonly used chemical for this purpose. It was first used to prepare diphtheria toxoid.

Minimum effective dose (M.E.D.). The smallest amount of a toxin that can be reliably detected. In specifying this property of a toxin it is necessary to state which specific indicating effect is being titrated. It is usual to adapt this term to the system; for example, M.H.D.—minimum

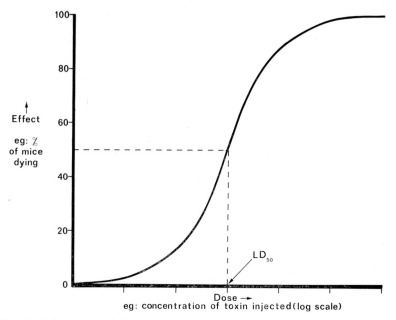

FIG. 1. A hypothetical dose-response curve—effect plotted against dose.

5a–13

haemolytic dose, M.L.D.—minimum lethal dose, etc. If a titration is performed to establish the M.E.D. by measuring the effect of a series of doses in the biological system, and if the size of the effect—be it size of a necrotic lesion, degree of haemolysis, or percentage of injected animals dying—is plotted against the logarithm of the dose, a sigmoid dose-response curve is obtained. From an examination of such a curve it is clear that a determination of the minimum effective dose is very imprecise, and requires the expenditure of many animals or much time in making the large number of tests necessary to nullify the effect of biological variability. It is therefore increasingly the accepted practice to establish the dose-response curve (Fig. 1) and to read off from this the point at which a 50% effect is produced, for example, 50% of the animals die or 50% of the red cells are haemolysed. This property of the toxin is known as the ED_{50}; or again specifying the indicating system, LD_{50} for the lethal dose, HD_{50} for the haemolytic dose, and so on.

Table I lists some of the common pathogenic bacteria, their toxins and the indicating effects by which they can be both detected and assayed.

TABLE I

Some of the exo-toxins produced by bacteria together with their indicating effects

Organism	Toxins	Indicator effects
Corynebacterium diphtheria	Diphtheria toxin	Lethal erythrogenic
Clostridium perfringens types A, B, C, D and E	α	Lethal necrotic haemolytic lecithinase
	β	Lethal necrotic
	ε	Lethal necrotic
	ι	Lethal necrotic
	θ	Lethal haemolytic
	κ	Lethal necrotic collagenase
	λ	Proteinase
	μ	Hyaluronidase
	γ	Desoxyribonuclease
Cl. botulinum	A,B,C,D,E,F	Lethal haemagglutinating
Cl. tetani	Tetanospasmin	Lethal neurotoxic
	Tetanolysin	Oxygen labile haemolysin
Cl. septicum	α	Lethal necrotic haemolytic
Staphylococcus aureus	α	Lethal necrotic haemolytic
	β	Lethal erythrogenic haemolytic
	Leucocidin	Destroys leucocytes
	Enterotoxin	Causes kittens to vomit
Streptococcus pyogenes	Streptolysin O	Oxygen labile haemolysin
	Streptolysin S	Oxygen stable haemolysin
	Streptokinase	Destroys fibrin clot

Test Dose. The amount of toxin or toxic filtrate which when mixed with the stated number of units or fractions of a unit of the standard antitoxin, under suitable conditions, produces a standard effect on the indicator system (S.I.E.). Again the abbreviation adopted for the test dose is related to the indicator system used: for example, L+ is the test dose determined using the lethal effect of a toxin, Lr is that determined using the necrotic effect, and Lh that derived from a haemolytic system.

Endpoint. In calibrating an antitoxin using its power to reduce the activity of a toxin to a given level, or in establishing the test dose of a toxin by its combining power for antitoxin, the end point of the titration is reached when the standard indicating effect (S.I.E.) is achieved. Commonly this is when the activity of the toxin is reduced to its ED_{50}— although practical considerations sometimes dictate that the null end-point is used (see Section on avidity).

Units. These are analogous to the international and national physical units of length and mass, in that they are defined by reference to the activity of a given weight, or more rarely volume, of an arbitrarily designated standard preparation. Although the weight of the standard preparation to which unit activity is assigned was originally related to a particular degree of biological activity, it was not necessarily related to the therapeutic use of the serum or the potency at which a vaccine was issued. It happens therefore that some units are small for one purpose, so that a therapeutic dose may contain many thousand units (in diphtheria a minimum of 5000–30,000 is recommended), but large by other criteria; and a reasonable level of circulating antitoxin in an immune animal may be only a fraction of a unit. The protective level for a horse immune to tetanus toxin is usually agreed to be around 1/100 I.U. A unit of tetanus antitoxin will neutralize ca. 100,000 lethal doses of tetanus toxin whilst a unit of *Cl. perfringens* α antitoxin will only neutralize ca. 50 lethal doses of α toxin. A unit of diphtheria antitoxin will neutralize ca. 8000 skin reacting doses of diphtheria toxin whilst a unit of *Clostridium septicum* α antitoxin will only neutralize ca. 30 skin reacting doses of septicum toxin. It is clear from the foregoing that there is no relationship between the unit of one specific antitoxin and the unit of a second, different, antitoxin. A unit of *Cl. perfringens* antitoxin may be composed of much more immuno-globulin than say a unit of diphtheria antitoxin.

Unit equivalents. The reciprocal of the test dose equivalent to 1 unit of antitoxin, and a term of convenience used to describe the concentration of toxins or toxoids.

Level of test. The number of units equivalent in the test dose with which it has been found convenient to perform the titrations. If this level is too high materials are wasted—if it is too low there will not be sufficient indicating doses present to ensure a clear reproducible end point. Thus, tetanus antitoxin is titrated at the 0·1 unit level (L + /10), whilst septicum antitoxin is titrated at the 4 unit level (4L +). In titrating antitoxins in sera of doubtful avidity (q.v.), such as those produced following two badly spaced injections of a poor antigen, or at an early stage in the response to a large dose of good antigen, or when comparing sera of differing avidity as in following the antitoxin levels during the course of hyperimmunization, it is important to perform all tests at the same test level (cf. Section on choice of laboratory standard).

Selection of test toxins for assaying sera. In the early days of toxin/antitoxin assay, the most important attribute for a test toxin was stability, but now that methods have been found for successfully freeze drying (lyophilizing) toxins it is reasonable to assume that each vial from the same lot has the same potency, and that most toxins are stable for the period of the test. Exceptions to this statement are the oxygen labile haemolysins, e.g. streptolysin O, and with such toxins tests should be arranged so that the minimum time elapses from the reconstitution of the dried reduced toxin to the addition of the serum. Although the purity of a test toxin has little significance for its efficiency, it is obviously important in selecting a test toxin to choose one which, at least under the conditions of the test, is monospecific. It would not affect the results of the titration if the culture filtrate used as a test toxin for tetanus antitoxin contained the haemolysin, tetanolysin, if the indicator system was the effect of subcutaneous injection into mice and the end point was extreme paralysis on the fourth day, but it would be unwise to use a *Cl. perfringens* type C culture filtrate for titrating δ antitoxin by the haemolytic test when the filtrate would contain not merely δ but a second haemolysin α and it would not be possible to distinguish which haemolysin was being neutralized nor which antitoxin was being assayed. Obtaining a monospecific culture filtrate without employing difficult and sophisticated methods of purification is sometimes difficult and in some cases it may be necessary to add antitoxin to a culture filtrate to neutralize one component before the filtrate can be used as a test toxin for another.

For most purposes the most important attribute of an efficient test toxin is its toxicity; the greater the number of ED_{50} in the test dose, the more accurately will it be possible to titrate the antitoxin, i.e., the smaller the proportionate differences that can be detected. For example, with a diphtheria toxin capable of giving a good reaction when the test dose at one unit is diluted 8000 times and 0·2 ml is injected intradermally into a guinea-pig skin, it is possible, despite the biological variability of the indicator—guinea-pigs skin—to test with an accuracy of $\pm 5\%$; whereas when titrating at low levels with a *Cl. perfringens* β toxin which is capable of giving a good necrotic reaction only down to a dilution of the test dose (Lr)/8, it is not possible to test with an accuracy of greater than $\pm 50\%$.

Affinity. This is for a toxin roughly what avidity is for an antitoxin; it depends on the goodness of fit between determinant groups on the toxin molecule and the combining sites on the antibody molecule.

II. SELECTING A LABORATORY STANDARD ANTITOXIN

When setting up a test procedure for the first time it is clearly necessary to avail oneself of the appropriate International Standard Preparation; but these are not intended for everyday use and the selection of a suitable laboratory standard will play a large part in the validity or otherwise of the subsequent results.

A. Avidity

It is important that the serum selected whether crude or purified should be avid; this means it must react irreversibly and completely with toxin at low concentrations. A serum which is non-avid will not only give an indistinct endpoint, but will have an apparent value which will vary with the concentration of the test mixture. With an avid serum and a good *Cl. perfringens* α test toxin, when testing at 10% differences one would expect to find that a 10% increment of antitoxin would reduce the indicating effect of the toxin from 100% haemolysis to $10-15\%$ haemolysis or from 60% haemolysis to no detectable haemolysis whereas with a non-avid antitoxin it might take from 5–7 such 10% increments to obtain the same effects.

The easiest method of demonstrating the avidity of the serum is by the method of dilution as outlined by Glenny and his colleagues (1932). In this method the ratio between the amount of the antitoxin required to neutralize one Lr dose of toxin in a total volume of 2 ml of mixture and that required to neutralize the same dose of toxin in a total volume of 200 ml of mixture is calculated. In each case 0·2 ml of the mixture (diphtheria toxin and antitoxin) is injected intradermally into the skin of a guinea-pig. The closer the ratio approximates to unity the more avid the antitoxin and the more suitable the serum is for use as a laboratory standard. Although Glenny's work was at the time confined to diphtheria antitoxin the method has since been applied successfully to most of the common anti-exotoxins.

Care is however needed in fixing the level at which the first mixture is made when dealing with toxin of low toxicity, that is with few ED_{50} per unit combining power, and it is sometimes necessary to compare the amount of antitoxin required to neutralize say 40 Lr or L+ doses in 2 ml with that required to neutralize the same amount of toxin in 200 ml. This ratio is known as the *dilution ratio* of the serum.

The reality and importance of this dissociation was well demonstrated by Glenny and his colleagues (1932). They showed that a mixture of toxin with non-avid antitoxin, which was apparently harmless when injected intravenously into a rabbit in a dose of 10 ml was lethal when injected in a dose of 0·1–0·5 ml, the greater dilution of the smaller dose in the blood of the animal leading to dissociation thus releasing a lethal dose of toxin from combination.

It is now regarded as good practice amongst those concerned with antitoxin titration to include two control sera (laboratory standards) with each batch of sera to be tested. One would be as already stated, a serum selected for its avidity and for its affinity to the test toxin. It would give clear cut reproducible results under the test conditions and any deviation from the designated value should indicate errors of technique only. The second serum would be one which, although not necessarily avid, has similar properties both physical and biological to the sera under test. For example if the sera under test were rabbit sera where the antitoxin present was the result of the response to two injections as is commonly the case in antigenicity tests, then the second standard should be a rabbit serum produced by a similar regime with a titre comparable to those under test. The first standard would probably be a pepsin-refined antitoxin made by pooling sera from a number of horses bled at a late stage of hyperimmunization.

B. Other conditions governing toxin/antitoxin combination

In all neutralization tests, the conditions under which the tests are performed, whilst not necessarily critical in the sense that if they are not adhered to the combination will not take place, are important in the sense that variations in any one of the conditions, e.g., time of combination, type of buffer, pH or temperature will upset the reproducibility of the test. It is therefore important in setting up a new test procedure first to ascertain the optimal conditions of time, temperature etc. for the combination and whilst practical considerations may make it necessary to work at less than optimal conditions it is essential for good results to use as nearly as possible identical conditions each time the test is performed. For example in a test involving the α lecithinase of *Cl. perfringens* it is important to use a buffer which ensures an adequate level of calcium as this enzyme is calcium dependent (Oakley and Warrack (1941)). In all tests it is always advisable to put toxin into the reaction tube and then to add the antitoxin because of the phenomenon described by Danysz as early as 1902 and designated by his name.

C. Danysz phenomenon

Danysz (1902) observed that the toxicity of a mixture of a constant amount of toxin and antitoxin varies with the method by which they are mixed, for example one unit of antitoxin added to 0·1 ml of toxin gives a neutral mixture but if the toxin is added to the antitoxin in two lots of 0·05 ml at an interval of say 10 min then the result will be a toxic mixture as the first 0·05 ml will have used up most of the unit of antitoxin by a process of multiple combinations leaving insufficient antitoxin free to neutralize the second addition of toxin.

III. INDICATORS

A. Whole animals

Mice are the animals most commonly used where death of the animal is the only readily demonstrable effect of the toxin. Logically it is desirable to use specific pathogen free mice of a known strain but practically, unless money is unlimited, it is more important to use healthy mice of a robust constitution—less expensive and less likely to die from non-specific causes e.g., temperature variations, intercurrent infections etc. White mice are preferred for intravenous injection and the lateral vein of the tail is the preferred site of injection. For intraperitoneal and subcutaneous injection colour is unimportant except insofar as it is easier to render white mice individually identifiable by the use of dabs of alcohol based dyes. It must be remembered that mice are not susceptible to all toxins equally; for example they are resistant to diphtheria toxin. It is usual to use the intra-venous route for *Cl. perfringens* α, β, ε and ι *Cl. septicum* α and *Staphylococcus* α toxins and the subcutaneous route of injection for tests involving tetanus and *Cl. oedematiens* (*novyi*) α toxin.

B. Skins

Both rabbits and guinea-pigs are commonly used for intradermal tests. In making a choice between the two it should be remembered that the extremely sensitive rabbit skin can prove too sensitive for a toxin when the degree of erythema is the indicator and that whilst albino guinea-pigs are ideal for testing mixtures of diphtheria toxin and antitoxin, albino rabbits are useless for skin tests because they tend to have rapidly growing soft short hair which it is almost impossible to clip. (It is not possible to use a depilatory paste (see Appendix I) on rabbits—the lethal dose is too close to the depilating dose.) It is possible to inject, and read the results of the injection of about 20 mixtures of diphtheria toxin/antitoxin on a 400–450 gm guinea-pig, but only 10 mixtures of *Cl. perfringens* type B containing a necrotic toxin β with a hyaluronidase and only 6 mixtures of *Cl. perfringens* type D, ϵ, a necrotic toxin, where the LR_{50} is very close to the LD_{50}. In rabbits it is possible to test 48 mixtures on a 1 kg rabbit but here it is advisable to mark the rabbits with rows of Indian ink circles (use a metal bottle cap and an ink pad) into the centres of which the injections are made. A record should be made at the time of injection giving the exact position on the rabbit (see Fig. 2, 3 and 4, between pp. 280 and 281).

C. Red cells

The choice of cell will depend first on their susceptibility to the toxin, e.g., horse cells are relatively resistant to *Cl. perfringens* α toxin but particularly susceptible to its θ toxin, secondly on availability and ease of bleeding— in many laboratories it may be difficult to obtain fresh cells other than rabbit or human—and lastly on the normal homogeneity and stability of the cells; in this regard sheep are preferable to rabbits. Efficient washing without too fast or prolonged spinning and careful calibration of red cell content are important in ensuring reproducible results; (2000 rpm for 6 min—washing with isotonic saline (0·85% NaCl) and spinning, repeated a minimum of three times) (Fig. 5, between pp. 280 and 281).

D. Lecithovitellin

A method of preparing the substrate for the lecithinase test is given in the appendix but it should be remembered that its life is relatively short and if used more than one month after preparation the crispness of the result will suffer (Fig. 6, between pp. 280 and 281).

E. Toxoid sensitized sheep red cells

Tetanus toxins, toxoids and antitoxins can be titrated fairly accurately by a haemagglutination inhibition method in which washed sheep red cells are sensitized with purified tetanus toxoid by the method of Stavitsky (1954) and preserved with buffered formalin (Fulthorpe, 1957, 1958). Mixtures of toxin, or toxoid, and antitoxin over a suitable range are prepared and allowed to stand for 1 h at room temperature; the sensitized cells are added, the tests are left overnight at room temperature, and read by observing the pattern of cells at the bottom of the tube. A smooth carpet of cells with a faint ring of unagglutinated cells round the periphery is found to give the most consistent endpoint. This is a convenient and economical method for titrating tetanus antigens and antitoxins where the alternative *in vivo* test takes four days, but it is not a method which can be applied indiscriminately to all toxin/antitoxin assays. With diphtheria (Fulthorpe, 1962) anti-*Cl. novyi* (*oedematiens*) it fails completely although *Cl. perfringens* type D ε and *Cl. septicum* α have been titrated routinely by this method.

F. Tissue cultures

A continuously maintained cell line HEp2 has been shown to be a suitable alternative substrate to the mouse for titrating *Cl. novyi* (*oedematiens*) α toxins and antitoxins. Interestingly, the mouse is four times as sensitive as a tissue culture cell to the action of this toxin, but in both death takes place slowly unless high concentrations of toxin are used. For greater sensitivity it is best to use freshly seeded tubes containing a measured number of HEp2 cells. Over a long series the agreement between the results of *in vivo* tests in mice and *in vitro* tests using tissue cultures showed a correlation coefficient of 0·99. As with sensitized red cells, tissue cultures cannot be used as an indicator for all systems—they fail with *Cl. perfringens* ε toxin, and tetanus toxin, but are useful for diphtheria (Lennox and Kaplan, 1957; Pappenheimer and Brown, 1968) *Cl. septicum* α, and possible but insensitive in *Cl. perfringens* α tests. A 50% endpoint is used and tests can be performed with 10% differences between tubes but good aseptic technique must be used.

G. The Ramon flocculation test

This is the only commonly used toxin/antitoxin neutralization test which is self indicating, and as originally used was basically a constant antigen-optimal ratio (OR) precipitation test in which the time element was emphasized, the tube which showed the first visible signs of a floccular precipitate was the one which contained the most nearly neutral mixture, and in which there were most nearly equivalent amounts of toxin or toxoid and antitoxin. It can also be used as a constant antibody OR test and it is a fortunate coincidence that in titrating exo-toxin or anti-exotoxins it has been found that unlike most of the other optimal ratio precipitation tests the OR constant antigen is usually very close to the OR constant antibody. In selecting laboratory standard reagents for these tests the criteria are somewhat different from those already described. Selection of the laboratory standard antitoxin is ideally made by flocculating a large number of different batches of avid antitoxin, of known titre in some other test system, with a representative selection of toxins or toxoids, and noting the first tube to flocculate, and the times of flocculation of all mixtures. The ratio between the antitoxin value in the flocculation test and in a neutralization test *in vivo* is calculated and the distribution of these ratios, known as the serum ratios, is plotted. A curve such as that shown as a solid line in Fig. 7 will result if the serum used to determine the test dose of the flocculating toxin was a representative serum. If it was not a representative serum either of the two dotted lines may result. The best sera for this

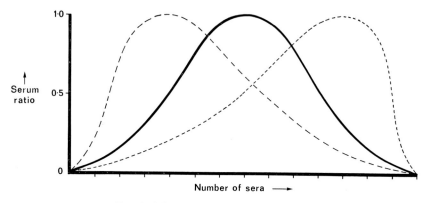

FIG. 7. Three hypothetical plots of serum ratios.

purpose are those with serum ratio most nearly approaching the average for the group and ideally ones having a ratio of 1·0. The time dose response curve is then plotted and Fig. 8 shows (A) that for a good and (B) that for a poor standard serum. A good flocculating standard serum flocculates rapidly in the tube containing the equivalent (neutral) mixture and relatively much more slowly in the tube containing any antigen or antibody excess. It also gives only one zone of flocculation, and the value calculated on the basis of the first tube to flocculate corresponds closely to the *in vivo* value. A good flocculating test toxin or toxoid is one that has similar properties to a good flocculating standard serum.

It is not necessary when titrating antitoxins by this method to use toxin, and for reasons of stability and safety it is common practice to use for this test a toxoid, calibrated against the standard antitoxin chosen as shown. The advantages of flocculation tests are speed and the fact that they are self indicating (one variable less); the disadvantages lie first in the large amounts of material required, for whilst it is possible to demonstrate the presence of as little as 0·001 units of *Cl. perfringens* α antitoxin and to titrate with between 10–20 % accuracy sera containing as little as 0·01 units using red cells as indicator, there will be no detectable flocculation between *Cl. perfringens* α toxin and antitoxin at levels below two units, and second in the difficulty of ensuring that the flocculation is specific. The method of testing blends of toxin of known titre with equal volumes of toxin of un-

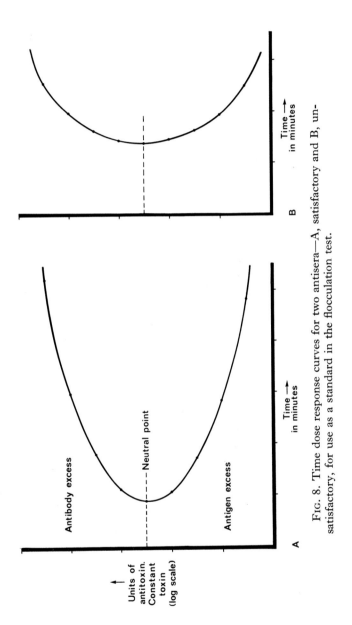

Fig. 8. Time dose response curves for two antisera—A, satisfactory and B, unsatisfactory, for use as a standard in the flocculation test.

known but expected roughly equivalent titre is of some help in this regard. It should be noted that not all species of sera are suitable for this test. Horse serum is very good but sheep and cattle sera give results very far removed from their *in vivo* values. With some toxins, e.g., *Cl. perfringens* ε toxin only a pepsin refined serum will give results which are related to the *in vivo* values (Batty and Glenny, 1947).

In the following Section reasonably detailed test procedures are set down. These are not the only methods for performing these tests but are methods that have been successfully used by technicians in laboratories for many years with many types of toxin and antitoxin and with sera from many animal species.

IV. FLOCCULATION TESTS

A. Materials required

Tubes (ca. 8 mm × 80 mm).

Rack to hold tubes.

Constant temperature water bath set at 50°C with sufficient water to cover the bottom third of the mixtures in the tubes so that stirring by virtue of the convection currents will occur.

Stop watch.

Buffered saline as diluent if necessary.

Laboratory standard flocculating serum.

Flocculating test toxin or toxoid.

Toxin, toxoid or antitoxin to be titrated.

1 ml contain and deliver pipettes.

Blotting (filter) paper.

Non-absorbent paper.

B. To test a toxin

Tubes

	1	2	3	4	5	
First Row						
1. Place	0·2	0·3	0·45	0·7	1·0 ml	of unknown toxin or toxoid in tubes of first row.
Second Row						
2. Place	0·4	0·45	0·5	0·55	0·60 ml	of test toxin or toxoid in tubes of second row.

Both Rows

3. Add 1·0 ml of standard serum containing 50 units of antitoxin to each tube.

Fig. 9. A useful magnifying apparatus for reading flocculation tests.

4. Mix by inversion using the non-absorbent paper to cover tubes.
5. Place rack in 50°C bath, note time.

Experience will tell the suitable time intervals for examining the racks; initially it is convenient to use a glass or perspex-sided water bath so that the tubes are visible at all times. A viewer as shown in Fig. 9 helps the observer to detect the earliest stages of flocculation.

Note first and second tubes to flocculate.

In the second row tube 3 should flocculate first. In the first row of the test if

Tube 1 flocculates first the value is ca. 250

$$
\begin{array}{cc}
2 & 167 \\
3 & 110 \\
4 & 70 \\
5 & 50
\end{array}
$$

The test would then be repeated with the mixture in the first tube to flocculate placed in the tube in the middle of the range e.g., if tube 2 flocculated first

$$
\begin{array}{cccccc}
 & 1 & 2 & 3 & 4 & 5 \\
\text{Place} & 0\cdot24 & 0\cdot27 & 0\cdot3 & 0\cdot33 & 0\cdot36 \text{ ml toxin in tubes 1--5.}
\end{array}
$$

Add 50 units of antitoxin to each tube. Make tubes up to constant volume with buffer.

If the fourth tube now flocculates first followed by the third tube the value of the unknown toxin is equivalent to ca. 140 units.

C. To test an antitoxin

For example, the laboratory standard of 110 units and an unknown serum

Tubes		1	2	3	4	5
Row 1	(a) Place the test dose of the toxin equivalent to 50 units in in each tube					
	(b) Add	0·36	0·4	0·45	0·5	0·55 ml of standard serum.
Row 2	(a) As for row 1					
	(b) Add	0·3	0·45	0·7	1·0	1·5 ml of unknown serum.

Mix by inversion covering tubes with non-absorbent paper.

In Row 1 tube No. 3 should be the first to flocculate.

In the repeat test on Row 2 the volume of serum added will be so arranged as to place the first flocculating mixture in the middle of the range. For example, if tube No. 3 flocculated first proceed as follows:

1st Repeat	(a) As for row 1					
Row 2	(b) Add	0·6	0·65	0·7	0·8	0·9 ml of unknown serum.

If tube 4 now flocculates first followed by tube three proceed as follows:

2nd Repeat (a) As for row 1
Row 2   (b) Add 0·65   0·7   0·8   0·9   1·0 ml

Tube 3 should now flocculate first and may be followed by tube 4 in which case the value of the unknown serum would be 62·5 units.

It is usual when titrating antitoxin to make the first test with serum additions in adjacent tubes differing by 50% to get an indication of possible value, this is followed by two tests with 10% differences between tubes the results of which should agree.

The method of test has been set out in full because it embodies the system which is followed throughout. With the other methods of toxin/antitoxin assay differences are due to the extra stage necessitated by the use of an indicator system and the various times and temperatures of incubation of the mixtures depending on the toxin/antitoxins which are under test and the indicator systems used.

For example, in the haemolytic test for *Cl. perfringens* α toxin the mixtures are allowed to stand at room temperature for 30 min before the indicator, ca. 6% suspension of sheep red cells in buffered saline is added. The test is then incubated at 37°C for 1 h and allowed to stand overnight in the cold before it is read.

For the guinea-pig intracutaneous test for diphtheria, *Cl. perfringens* α, β and ε toxins the mixtures are held at room temperature for 1 h before being injected into the guinea-pig. Preliminary readings of the erythematous or necrotic reactions are made at 24 h and the final reading is taken at 48 h. The final reading will tend to agree closely with the preliminary reading for diphtheria and *Cl. perfringens* ε but the two readings may be very different for *Cl. perfringens* β—a toxin where the necrotic lesion develops slowly over a period of time. Nonspecific erythematous reactions occurring soon after injection will usually have disappeared after 48 h.

In the test for tetanus toxin using the subcutaneous route in mice, although the mice are examined twice daily so that those showing marked paralysis may be killed, the final reading is not taken until the fourth day. The endpoint for this assay is accepted as killed on the third day with marked symptoms of tetanus or dead on the fourth day after showing symptoms of paralysis on the third day.

Examples

1. *Lecithinase test*

   To find the α value of a *Cl. perfringens* antiserum

Materials—Lambeth tubes. (ca. 8 mm × 80 mm).
Racks.
Water bath at 37°C.
1 ml Contain and deliver pipettes.
Diluent calcium gelatin saline (CagSal) (see Appendix).
Broth saline.
Lecithovitellin (see Appendix).
Test toxin.
Laboratory standard antitoxin.
Unknown serum.

Dilute the laboratory standard antitoxin in CagSal so that the dilution contains 1·2 units of antitoxin per ml. If there is no information as to the concentration of the unknown serum dilute 1 ml by the addition of 4 ml, i.e. 1/5 and test for 5, 10, 25, 50, 100 units as follows.

Reconstitute test toxin in stated vloume of 50% broth 50% saline. Place the test dose of toxin in each of two rows of 5 tubes.

Units tested for	1·2	1·1	1·0	0·9	0·8
Row 1 Add	1·0 ml	0·91 ml	0·83 ml	0·75 ml	0·67 ml Std. serum

Units tested for	5	10	20	50	100
Row 2 Add	1·0 ml	0·5 ml	0·25 ml	0·1 ml	0·05 Unknown serum diluted 1/5

Mix by inversion. Stand 30 min at room temperature—add 0·5 ml lecithovitellin, mix again by inversion—place in water bath for 1 h, leave overnight at room temperature, read.

Expected result

Row 1	Clear	Clear	Trace of cloudiness	Opaque	Coagulum formed at surface
	—	—	T	O	C
Row 2	Clear	Clear	Clear	Coagulum	Coagulum
	—	—	—	C	C

The unknown serum therefore has a value between 20 and 50 units per ml.

Suggested repeat. Dilute unknown 1 + 19 i.e., 1/20.

Place the test dose of toxin in each tube:
Row 1 As before

Units tested for	20	27	36	50
Row 2 Add Serum 1/20	1·0 ml	0·74 ml	0·54 ml	0·4 ml

Possible result

Clear   Clear   Opaque   Coagulum
                          formed
—   —   O   C

Suggested repeat Dilute unknown 1/27

Row 1 As before.

Units tested for	27	30	33	36
Row 2 Add Serum 1/27	1·0 ml	0·9 ml	0·82 ml	0·75 ml

Possible result

Clear   Trace of   Opaque   Opaque
        cloudiness
—   O   O

in which case the unknown serum had a value of 30 units per ml: N.B. when making dilutions of serum always remember to draw the serum up to the zero, not beyond, wipe the outside of the pipette carefully and wash up and down at least ten times in the diluent—serum is very readily adsorbed on to glass.

2. *Capillary method. Haemolytic test* (Fig. 10, between pp. 280 and 281)
   This method which can be modified for any indicator system is useful to test either small volumes of serum or low levels of antitoxin. The example given shows its use in the titration of anti-staphylococcal α values of normal human sera.

Materials—Capillary glass Pasteur pipettes so made that ca. 3 in. will contain 0·1 ml.
       Beaker containing boiling distilled water.
       Beaker containing small volume cold distilled water.
       Dreyer's tubes.
       Filter paper.
       Racks to hold tubes.
       Test toxin.
       Standard serum.
       Water bath at 37°C.
       Rabbit or sheep red cells.
Diluent   50% broth 50% saline (BS).

Prepare the following toxin dilutions:

Test dose in 1 ml BS	TD/1
0·5 ml TD/1 + 0·5 ml BS	TD/2
0·5 ml TD/1 + 2 ml BS	TD/5

$$0.5 \text{ ml TD}/5 + 0.5 \text{ ml BS} \quad \text{TD}/10$$

and so on to cover the range over which the serum is to be tested.

The capillary pipette is marked with grease pencil to contain about 0·1 ml and 0·1 ml of each toxin dilution required for a test of the serum is transferred to the appropriate Dreyer's tube and blown in with a rubber teat. The toxin dilutions are put in the tubes, the weakest first, proceeding to the strongest.

The pipette is then rapidly washed out by sucking boiling water into it above the grease pencil mark and expelling the liquid into the cold beaker—repeating four times. The pipette is then dried and cooled by blowing air in and out on to filter paper and then using the same pipette add ca. 0·1 ml of the serum under test to each tube containing toxin. The pipette is then discarded and a fresh one used for the next serum.

When the rack is completed [two controls, (a) toxin dilutions + an equal volume of saline, (b) toxin dilutions + an equal volume of the laboratory standard diluted to contain a concentration in the middle of the range of the tests, will have been included] the tubes are mixed either by rotation or by inversion covered by non-absorbent paper. Either method requires care as only small volumes are present.

Allow the mixtures to stand for 30 min then add 0·1 ml of a 6% suspension of sheep or rabbit red cells in saline to each tube, mix again and place in the 37°C water bath for 2 h. Leave in the refrigerator overnight read percentage haemolysis either by comparison with a row of tubes containing amounts of dilutions of 10% lysed red cells differing from 0–100% or using some form of absorptiometer. The end point for this test is 50% haemolysis (Fig. 10).

Intracutaneous tests in guinea-pigs

Used to titrate the β toxin in filtrates of *Cl. perfringens* types B and C.

Materials—Deposit glasses.
           Racks.
           Contain and deliver pipettes.
           Diluent—BBS borate buffer saline pH 8·2.
           Absorbent and non-absorbent paper.
           Small syringes—preferably all glass with sharp size 214-SWG
             needles.
           Depilated guinea-pigs.
           Standard β antitoxin diluted with ϵ antitoxin of sufficient
             concentration to over neutralize any likely concentration of ϵ
             toxin present in type β filtrates so as to contain 2 β units in
             0·5 ml of serum dilution.

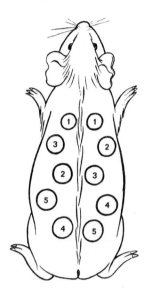

FIG. 11. Pattern used when injecting *Cl. perfringens* (*welchii*) β toxin antitoxin mixtures designed to prevent masking of one area of spreading necrosis by another.

Place varying amounts of the unknown filtrate in a row of tubes, add 0·5 ml of the standard β+ε antitoxin to each tube, mix by inverting covered by non-absorbent paper, stand at room temperature for 1 h, inject 0·2 ml of each mixture intradermally into the depilated guinea-pig skin according to the above pattern which was arrived at as a result of experience (Figs. 11 and 3). Type B filtrates contain hyaluronidase and the necrotic reactions tend to spread downwards and backwards. This arrangement decreases the chances that a large reaction might mask a small reaction.

The guinea-pig is examined at 24 and 48 h.

V. TOTAL COMBINING POWER TESTS

It was for many years the practice to assay toxoids by the flocculation test; it was not possible to test them by an *in vivo* or *in vitro* test requiring an indicator system as by definition toxoids have lost their pharmacological or biological activities whilst retaining the ability to combine with anti-toxin. For many purposes this was satisfactory with the proviso that the flocculating serum was known not to give more than one zone of floccula-tion with either the toxin or toxoid and that this value agreed reasonably with the *in vivo* value of the toxin. It was discovered by several workers (Pope, 1953) that pure toxins were very difficult to flocculate and also (Fulthorpe, 1958) that it was possible for a toxoid to apparently retain its

TABLE II
Statistical assessment of the bio-assay of *Cl. perfringens* β and ϵ antitoxins

Material under test	Indicator system route and animal	Number of tests	Geometric mean value units/ml.	Standard deviation (log units)
Cl. perfringens	Intradermal rabbit	463	174	$\pm 0 \cdot 0278$
ϵ antitoxin	Intravenous mouse	31	172	$\pm 0 \cdot 0282$
β antitoxin	Intradermal guinea-pig	313	1150	$\pm 0 \cdot 0363$
	Intravenous	238	1170	$\pm 0 \cdot 0268$

TABLE III
P.95 limits of error as percentages of the mean observed potency

Material and test system	Number of tests*		
	1	2	3
Cl. perfringens ϵ antitoxin			
i.d. rabbit	$88 \cdot 2$–$113 \cdot 3$	$91 \cdot 5$–$109 \cdot 2$	$93 \cdot 0$–$107 \cdot 4$
i.v. mice	$88 \cdot 0$–$113 \cdot 6$	$91 \cdot 4$–$109 \cdot 4$	$92 \cdot 9$–$107 \cdot 6$
Cl. perfringens β antitoxin			
i.d. guinea-pig	$84 \cdot 9$–$117 \cdot 8$	$89 \cdot 1$–$112 \cdot 3$	$91 \cdot 0$–$109 \cdot 9$
i.v. mice	$88 \cdot 6$–$112 \cdot 8$	$91 \cdot 8$–$108 \cdot 9$	$93 \cdot 3$–$107 \cdot 3$

* One "test" means 4 mixtures of toxin/antitoxin at 10% intervals injected into 1 rabbit or 1 guinea-pig or 2 mice with each mixture.

TABLE IV
P.99 limits of error as percentages of the mean observed potency

Material and test system	Number of tests*		
	1	2	3
Cl. perfringens ϵ			
i.d. rabbit	$84 \cdot 8$–$118 \cdot 0$	$89 \cdot 0$–$112 \cdot 4$	$90 \cdot 9$–$110 \cdot 0$
i.v. mice	$84 \cdot 6$–$118 \cdot 2$	$88 \cdot 8$–$112 \cdot 6$	$90 \cdot 8$–$110 \cdot 1$
Cl. perfringens β			
i.d. guinea-pig	$80 \cdot 6$–$124 \cdot 0$	$85 \cdot 9$–$116 \cdot 4$	$88 \cdot 3$–$113 \cdot 2$
i.v. mice	$85 \cdot 3$–$117 \cdot 2$	$89 \cdot 4$–$111 \cdot 9$	$91 \cdot 2$–$109 \cdot 6$

* One "test" means 4 mixtures of toxin/antitoxin at 10% intervals injected into 1 rabbit or 1 guinea-pig or 2 mice with each mixture.

flocculation value with one serum whilst the value apparently dropped when tested with a serum of different quality or to retain its flocculating capacity whilst losing its ability to stimulate the production of antibody. This discrepancy between flocculating capacity and immunogenicity stimulated workers to try to find other methods of testing toxoids for their ability to combine with antitoxin.

In the total combining power test varying volumes of the toxoid and a fixed amount of antitoxin are mixed, combination is allowed to take place and then a known amount of the test toxin is added to each mixture. The mixtures stand for a further length of time then the indicator is added (red cells etc.) or the mixtures are injected into mice and by this means it is possible to calculate the combining power of the toxoid by as it were back titrating the antitoxin remaining uncombined for its power to neutralize the added test toxins.

VI. ACCURACY AND REPRODUCIBILITY

A statistical assessment of the results of various combining power tests for *Cl. perfringens* β toxin and its antitoxin is given in Tables II, III and IV. These tests were performed over a period of 2 years by an assortment of laboratory technicians using the techniques already described. The results show that these tests are eminently reproducible and for a biological system yield a high degree of accuracy for the expenditure of relatively few animals and laboratory time. They also demonstrate that in such test systems using avid antitoxins and test toxins which fulfil the requirements of affinity and toxicity already set out, results obtained are, as they should be, independent of the indicator systems.

APPENDIX I

Depilating Guinea-pigs

(a) Clip the hair from the flanks, leaving a line of hair along the backbone.
(b) Apply a depilatory paste to the clipped areas. The paste can be made by mixing:

Barium sulphide	35 g
Flour	35 g
Talc	35 g
Powdered soap	5 g

The mixture is best kept in air-tight tins. Sufficient water is added to make a thick cream, which is brushed on to the guinea-pig and left for the minimum time for efficient depilation, ca. 4 min. All traces of the paste are carefully washed away and the guinea-pig gently patted dry.

APPENDIX II

Borate Buffer Saline (BBS)

Dissolve 30 g of a mixture of:

Borax	59 g
Boric acid	84 ml
Sodium chloride	99 g

in 2 litres of 0·3% (w/v) sodium chloride solution. This buffer is very stable.

Calcium Gelatin Saline (Cagsal)

Add 138·75 ml of 1% calcium chloride solution.
100 ml 5% gelatin saline.
22·5 g NaCl.
Make up to 2·5 litres with distilled water.
Add 0·2 g thiomersalate.
2·5 g phenol.
Autoclave at 10 lb for 10 min.
Keep in cold.

Lecithovitellin

The preparation of lecithovitellin is as follows:
The yolks from 12 eggs are washed in 0·85% saline to remove the albumin. 1400 ml of 0·85% saline is added. The material is clarified through paper and Seitz filtered to sterilize.

APPENDIX III

HYPERIMMUNIZATION IN HORSES

Horses are the animals most commonly injected for the preparation of antisera for therapeutic or prophylactic use.

The antisera produced can be grouped as follows:

Antitoxic sera: tetanus, dipthheria, botulinum, gas gangrene (*Cl. septicum, Cl. oedematiens, Cl. perfringens (welchii)* Type A), lamb dysentery (*Cl. perfringens* type B), pulpy kidney disease (*Cl. perfringens* type D), staphylococcal, streptococcal and dysentery (shiga).

Antibacterial sera: anthrax, leptospira, pneumococcal, meningococcal and coli.

Antiviral sera: dog distemper.

Many of these antisera have been rendered obsolete by improvements in chemotherapy. There are situations where antitoxic sera are still important therapeutic agents, for example, in lamb dysentery, *Cl. perfringens* type B infection in lambs from non-immune mothers where the disease strikes before there is time for the production of homologous antibody as a response to active immunization. A similar situation exists in other cases of intoxication when no active immunity has been stimulated, for example, in botulism, tetanus and enterotoxaemia due to *Cl. perfringens* type D (pulpy kidney).

It is the usual practice in establishments where horses are immunized, first to select healthy horses of middle age so that there is the possibility of some naturally acquired immunity. The horses are test bled and the sera tested for the presence of possible relevant antibodies such as those to diphtheria, *Cl. perfringens* α toxin,

streptolysin O etc. It has been found (Barr and Glenny, 1945–46) that hyperimmunization may result in the production of poor quality non-avid sera, if the antibody
producing mechanism is forced by too frequent injections of too much antigen
before the animal is equipped to deal with such large amounts of the antigen, that
is before a good basal immunity is established (Barr 1950–51). Good basal immunity
has been shown to exist when animals have reasonable levels of circulating antitoxin as a result of natural infection regardless of whether the infection has been
accompanied by clinical symptoms. This is the reason for testing incoming horses.
A horse with say, circulating diphtheria antitoxin would be injected intramuscularly
with repeated and gradually increasing doses of diphtheria toxoid two or three
times a week. From time to time small samples of blood will be taken and the serum
titrated for antitoxin. When a satisfactory titre is achieved larger quantities of
blood, circa 8 litres, will be taken and processed. After a resting period the horse
will be subjected to a shorter recall course of immunization of possibly five doses
of diphtheria toxin accompanied by an adjuvant such as potash alum. At the con-
clusion of such a course the animal will again be bled a total of roughly 24 litres
over a period of 10 days. This procedure of alternating immunization periods,
bleeds and rest periods will be continued until either the fitness of the horse
shows signs of deterioration or the titres achieved begin to fall below a useful level.

In the case of antibodies such as tetanus which are never naturally found in the
horse (the lethal dose is smaller than the immunogenic dose) it is necessary to
prepare the horses for hyperimmunization. In some countries it is possible to rely
on being able to select horses from an artificially actively immune population; in
others it is the practice to give the horse two doses of toxoid with adjuvant, spaced
by at least a four-week interval and then allow the horse to rest for at least 3 months
and with advantage for even longer periods up to 1 year whilst the basal immunity
becomes firmly established. The hyperimmunization then begins with first the
repeated increasing doses initially of toxoid, later of toxin, both without adjuvant–
followed by shorter courses of toxin with adjuvant as described for diphtheria.
With some antigens which are less toxic for the horse and when natural antibody is
present as can happen with *Cl. welchii* (*perfringens*) type A, it is possible to dis-
pense with the use of toxoid and begin the immunization course with small doses
of toxin.

REFERENCES

Barr, M., and Glenny, A. T. (1945–46). *J. Hyg. Camb.*, **44**, 135.
Barr, M. (1950). *Brit. J. exp. Path.*, **31**, 615.
Barr, M. (1951). *J. Path. Bact.*, **63**, 557.
Batty, I., and Glenny, A. T. (1947). *Brit. J. exp. Path.*, **28**, 110.
Behring, E. A. Von, and Kitasato, S. (1890). *Dtsh. med. Wschr.*, **16**, 1113.
Danysz, J. (1902). *Ann. Inst. Pasteur*, **116**, 331.
Ehrlich, P. (1897). *Klin. Jb.*, **6**, 299.
Fulthorpe, A. J. (1957). *J. Hyg. Camb.*, **55**, 382.
Fulthorpe, A. J. (1958). *Immunol.*, **I**, 365.
Fulthorpe, A. J. (1962). *Immunol.*, **5**, 30.
Glenny, A. T., Barr, M., Ross, H. E., and Stevens, M. F. (1932). *J. path. Bact.*,
    35, 495.
Kitasato, S. (1891). Experimentelle Untersuchungen uber das Tetanusgift.
    Zeitsch. f. Hyg., **10**, 267.
Lennox, E. S., and Kaplan, A. S. (1957). *Proc. Soc. exp. Biol. Med.*, **95**, 700.

Oakley, C. L., and Warrack, G. H. (1941). *J. Path. Bact.*, **53** (3), 335.
Pappenheimer, A. M., and Brown, R. (1968). *J. exp. Med.*, **127**, 1073.
Pope, C. G., and Stevens, M. F. (1953). *Brit. J. exp. Path.*, **34**, 241.
Roux, E., and Yersin, A. (1888). Contribution a l'etude de la diphtherie. *Ann. Inst. Past.*, **2**, 629.
Stavitsky, A. B. (1954). *J. Immunol.*, **72** (5), 360.

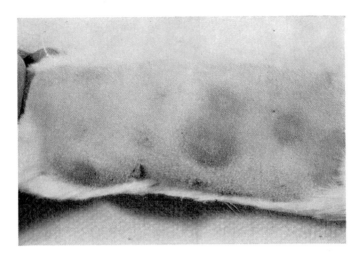

Fig. 2. Guinea-pig 48 h after the intradermal injection of a series of ranges of mixtures of diphtheria toxin and antitoxin.

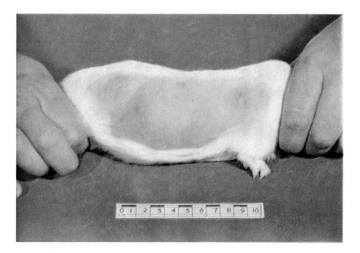

Fig. 3. Guinea-pig 48 h after the intradermal injection of a range of *Cl. perfringens (welchii)* β toxin and antitoxin mixtures.

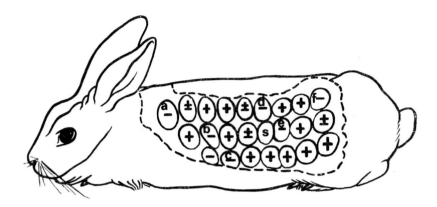

Fig. 4. Rabbit 72 h after the intradermal injection of a series of ranges of mixtures of Trypsin-activated *Cl. perfringens* (*welchii*) type D ϵ toxin and antitoxin. The plan shows the readings allocated to the reactions in each range of a titration. The rows should be read from top to bottom.

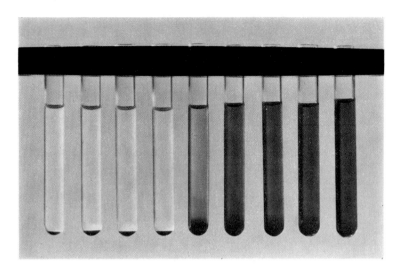

FIG. 5. A titration of *Cl. perfringens* (*welchii*) α antitoxin using sheep red cells as indicator. The concentration of antitoxin differs by 5% per tube.

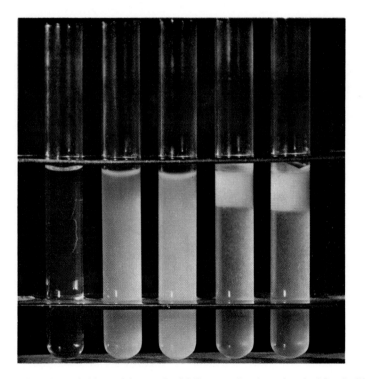

FIG. 6. Titration of *Cl. perfringens* (*welchii*) α antitoxin using lecithovitellin as indicator. The concentration of antitoxin differs by 10% per tube.

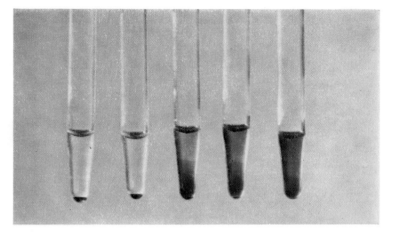

Fig. 10. Capillary method used to titrate small volumes of human serum for *Cl. perfringens* (*welchii*) α antitoxin.

CHAPTER IX

Techniques for Handling Animals

W. H. Kingham

Shell Research Limited, Milstead Laboratory of Chemical Enzymology,
Broad Oak Road, Sittingbourne, Kent

I.	Introduction	282
II.	General Considerations	283
	A. Species	283
	B. Working conditions	283
	C. Injection routes and techniques	285
	D. Depilation	288
III.	Mice	288
	A. Varieties	288
	B. Handling	289
	C. Injections	289
	D. Anaesthesia	290
	E. Withdrawal of body fluids	290
IV.	Rats	291
	A. Varieties	291
	B. Handling	291
	C. Injections	291
	D. Anaesthesia	292
	E. Withdrawal of blood	292
V.	Guinea-pigs	292
	A. Varieties	292
	B. Handling	293
	C. Injections	293
	D. Withdrawal of blood	294
VI.	Rabbits	294
	A. Varieties	294
	B. Handling	295
	C. Injections	295
	D. Withdrawal of blood	296
VII.	Euthanasia	297
	A. Mice	297
	B. Rats	297
	C. Guinea-pigs	297
	D. Rabbits	297
VIII.	Appendix	297

I. INTRODUCTION

The microbiologist and the biochemist often find it necessary to use experimental animals in the course of their work. Sometimes an experimental programme will involve the extensive use of animals for infectivity experiments or toxin assays, and detailed autopsies which may be required of animals which die during an experiment, or are killed at the end of it. Such experimental work will normally be carried out in laboratories where facilities are good and experienced personnel are at hand to assist with the work. Often, however, the research worker will make frequent use of small numbers of animals. For instance, many workers need occasionally to prepare antisera (cf. Oakley, this volume, p. 173), or to inoculate small numbers of animals with micro-organisms suspected of pathogenicity. The object of this Chapter is to describe methods for handling the commoner experimental animals; for the injection of materials by the more common routes; and for the withdrawal of body fluids for subsequent examination.

There are many variations in technique used in performing these more routine animal procedures, some of which can cause considerable distress to the experimental animal. Nervousness arising from inexperience and lack of practice on the part of the operator is a major factor in causing distress to the animal. Wherever possible a demonstration of techniques by an experienced worker should be sought by a newcomer to the field before any attempt is made to work with an animal himself. The present Chapter describes methods which, in my experience, cause the least discomfort to the experimental animal, whilst allowing the experimental procedure to be carried out quickly and effectively. Where more unusual techniques are to be used or autopsies performed, the reader is referred to the Universities Federation of Animal Welfare Handbook, (E. & S. Livingstone), Institute of Animal Technicians Manual (Crosby Lockwood & Son) and National Institute of Health—Methods by W. Gray, all of which, in addition to describing the handling of a wide range of experimental animals, also contain valuable information concerning the design and operation of animal houses, the breeding and rearing of animals and the diseases to which experimental animals are subject.

Important Note. In certain countries, such as the United Kingdom, the performance of many of the procedures to be described can be carried out only by persons in possession of a valid animal licence. It is essential that anyone embarking on experiments involving animal techniques should familiarize himself with the requirements of the country in which he is working, and ensure that he is satisfying them.

II. GENERAL CONSIDERATIONS

A. Species

The animals most commonly used in the laboratory are mice, rats, guinea-pigs and rabbits. There are several varieties of each of these species in common use in laboratories, and the advantages and disadvantages of the different varieties for various experimental procedures will be described briefly.

B. Working conditions

Experimental procedures should be carried out in a room separate from that in which animals are reared. Rats and mice can conveniently be carried in their cage to an adjoining room. This is less easy with guinea-pigs and rabbits, and these animals are usually removed from their cages and carried to the working area. Laboratory animals are often remarkably experienced at resisting removal from their cages. Rabbits can be particularly stubborn, and a prolonged battle to remove a recalcitrant animal from its quarters can result in a badly frightened and excited subject. Methods of removing animals from their cages will be described in the subsequent sections; a little time spent practising the technique, before animals are to be handled for experimental purposes, is time well spent.

A suitable working area is one sufficiently narrow to enable two people to sit on opposite sides whilst working comfortably together on the surface. A table measuring 3×2 ft. is ideal for most purposes. The surface should be easily cleaned, but not slippery. Animals rapidly become frightened if unable to get a firm footing on the table surface. Many workers prefer to use a wooden surface with heavy polyurethane varnish, or a laminar plastic finish, covered with a towel or a piece of cloth. Removable ridged rubber coverings can also be used.

The experimental room should be maintained at a high standard of cleanliness, and should be designed in such a way that it is as free as possible from dust-retaining surfaces. Instruments, when not in use, should be housed in glass-fronted cupboards, and there should be adequate supplies of sterile instruments, cotton wool, and the various solutions required for the experimental procedures. Table I lists the equipment required for an operating room in which small animal routine procedures are performed.

Adequate facilities must be available for the disposal of soiled swabs and cloth and for the safe removal of potentially contagious material. Soiled glassware should be discarded into a suitable disinfectant/cleaning solution (1% hypochlorite solution). Used, disposable syringes and animal carcases should be discarded into a plastic-lined paper bag, suitable for sealing and disposal by incineration. The room should be provided with adequate

TABLE I

Basic requirements for a small animal house experimental room

A small, easily cleaned table approximately 3 × 2 ft or an animal operating table. Non-slip top, i.e., ridged rubber.

Instruments

An instrument steriliser.

Scissors: Sharp/sharp points) ⎫ One pair of each type of the following sizes:
Sharp/blunt points ⎰ $3\frac{1}{2}$ in., 5 in., 6 in.

These must be properly maintained

Spencer Wells artery forceps

Curved ends : Two pairs of each size: $3\frac{1}{2}$ in., 5 in., 6 in.
Straight ends : Two pairs of 5 in.

One small size retractor, quadrant type

Forceps: : Blunt ends   : Two pairs 5 in.
       Straight fine points : Two pairs of each size: $4\frac{1}{2}$ in., 5 in., 6 in.
       Trewes toothed ends: Two pairs of each size: $4\frac{1}{2}$ in., 5 in., 6 in.
       Needle forceps   : One pair

Swann Morton or Gillette Scalpel handles : 3 large size, 3 small size.

Swann Morton or Gillette blades: Two packets of each number—10, 11 and other numbers requested by individual operators.

Mixed selection of ligature needles, curved and straight.

Sterile equipment

Surgical rubber gloves.
Gowns and caps.
Linen towels.
Gauze.
Ligatures—dissolvable and silk.
Petri dishes (disposable type, "Sterilin", supplied by Sterilin Limited, The Crescent, Richmond, Surrey, England).
Sterile pipettes : 1, 2, 5 and 10 ml capacity.
(Disposable type by Falcon Products, U.S.A. or sterilized glass pipettes).
Sterile McCartney bottles : wide neck universals.
Sterile disposable syringes : 1, 2, 5, 10, 20 and 30 ml capacity.

	Gauge	Length in.
Needles disposable, sizes : No. 1	21 ×	$1\frac{1}{2}$
No. 12/15	23 ×	$1\frac{1}{4}$
No. 18	25 ×	$\frac{5}{8}$
Serum No. 2	19 ×	2

Syringes and needles: "Steriseal"—suppliers: Arnold Horwell, Cricklewood, London.

Suggested American suppliers:

"Monoject" (Roehr Products Co., Inc., Waterbury, Conn.) Suppliers American Hospital Supplies Corporation, Evanstown, Illinois.

"Tomac" (American Hospital Supplies Corporation, Evanstown, Illinois).

TABLE I *continued*

Miscellaneous equipment

A 9 in. desiccator with grid.

Electric hair clippers.

A 2½ in diameter × 4 in. length metal tube with a metal gauze base.

Ether, Sodium pentabarbitone, chloroform and surgical spirit, all of which should be stored in a locked metal cabinet.

Cotton wool, absorbent.

Liquid plastic preparation called "New skin"—suppliers: Harwoods Laboratories, St. Helen's, Lancashire.

---

hand-washing facilities and a dispenser of germicidal soap should be available.

At the end of each day, or at the end of each working period the working surfaces should be cleaned thoroughly by scrubbing with hot, soapy water and finally treating with a disinfectant solution such as hypochlorite, 70% ethyl alcohol or proprietary disinfectant.

C. Injection routes and techniques

Injections may be carried out by a number of different routes, the experimental requirement largely dictating the choice. The use of pre-sterilized disposable syringes avoids the time-consuming procedures of servicing and sterilizing glass syringes and sharpening needles which used to be a feature of animal work. The choice of the correct size of needle for the various injection routes in different animals is a matter of considerable importance and Table II summarizes the appropriate needle size for use.

TABLE II

Needle sizes

Disposable needles can be obtained in a variety of sizes, but the following four sizes are useful for the more general types of work:

Steriseal	No. 1	(21 swg × 1½ in.)
Steriseal	No. 12/15	(23 swg × 1¼ in.)
Steriseal	No. 18	(25 swg × ⅝ in.)
Steriseal Serum	No. 2	(19 swg × 2 in.)

	Intra-venous	Intra-peritoneal	Intra-muscular	Intra-dermal	Intra-cardiac	Sub-cutaneous
Mice	No. 18	No. 18	No. 12/15	No. 18	No. 12/15	No. 12/15
Rats	No. 18	No. 18	No. 12/15	No. 18	No. 12/15	No. 12/15
Rabbits	No. 18	No. 1 or 12/15	No. 1 or 12/15	No. 12/15	No. 1	No. 12/15
Guinea-pig		No. 12/15	No. 12/15	No. 12/15	No. 1	No. 12/15

If bleeding a rabbit by cardiac puncture use needle Serum No. 2.

TABLE III

Maximum recommended injection volumes

	Intra-venous (ml)	Intra-peritoneal (ml)	Intra-muscular (ml)	Intra-dermal (ml)	Intra-cardiac (ml)	Sub-cutaneous (ml)
Rat	0·5	5	0·5	0·1	0·2	0·2
Rabbit	5	10	2	0·1	5	0·5
Guinea-pig	*	5	1	0·1	1	0·2
Mouse	0·2	2	0·2	0·1	0·5	0·1

* Not a suitable subject.

These figures are intended only as a guide and they should be interpreted with caution and common sense. Experience has shown that the injection of large volumes of material into conscious animals can be both arduous and difficult; in principle, the volume of the inoculum should be kept to a minimum. It is essential when carrying out injection procedures to avoid injecting volumes of air. Syringes should be loaded carefully. The needle is attached aseptically to the syringe and the inoculum drawn from its receptacle through the needle into the syringe. Care must be observed to prevent the tip of the needle becoming damaged by contact with the container. When the syringe contains sufficient inoculum, it should be inverted needle tip uppermost and a small piece of sterile gauze placed over the point. By gently depressing the plunger all the air can be expelled, the gauze hood can now be removed and the syringe is loaded and ready for use.

Equally important is a knowledge of the volumes of liquid which may be introduced by the different routes without causing undue discomfort to the experimental animal. Table III summarizes the volumes which may be introduced by various routes into the common laboratory animals.

1. *Intravenous injection*

The details for selecting suitable veins are given under the sections on the various animals. The vein should be filled with blood in order to render it readily visible. This can be achieved in a variety of ways, such as by blocking off the vein and stimulating the blood supply to the area either by use of warm water or by applying a mild rubifacient such as xylol. The use of an irritant is not normally recommended and, if used, it must be removed by washing with alcohol at the end of the injection. The hair should be removed over the injection site and this is usually done by the use of electric hair clippers followed by shaving with a sharp razor.

The needle is inserted into the vein with the bevel facing upwards and preferably in the direction of the blood flow along the vein. Whether the tip of the needle has actually entered the vein may be difficult to decide and withdrawing the plunger of the syringe a little before starting injection to ascertain whether blood is drawn up through the needle will confirm the correct location of the needle tip. Not infrequently the needle will pass into the sheath surrounding the vein when the needle should be withdrawn and the insertion repeated. A considerable difference of opinion exists as to whether the swabbing of the skin surface before injection serves any useful purpose. If the surface is to be swabbed, 70% ethyl (absolute) alcohol is the preferred agent. This pre-treatment is really only necessary if pathogenic bacteria are to be injected, it then serves to minimize the contamination of the skin of the animal and so helps to eliminate undesirable spreading of the pathogen.

2. Intramuscular injection

Here the needle is inserted at right angles to the body surface. The needle must penetrate into the muscle block. Again the area should be shaved or clipped before injection and the injection followed by swabbing of the skin.

3. Intradermal injection

The object here is to introduce a little material into the thickness of the dermis. Usually a small portion of skin is pinched between the fingers, the needle is inserted at a very oblique angle into the skin and a finger of the other hand used to locate the point of the needle and ascertain that it is almost emerging again from the skin fold. An alternative method is to hold the syringe, needle bevel uppermost, parallel to the skin and gently push the point under the dermis until the lumen is covered, then make the injection.

4. Intraperitoneal injection

The injection is made through the skin and body wall into the cavity of the lower abdomen. An area of skin and abdominal wall is lifted between the finger and thumb and the needle pushed through the two layers with the bevel facing downwards. It is usual to move the syringe slightly from side to side when it is in position to ensure that the needle is not fouling any part of the intestine before making the injection.

5. Subcutaneous injection

The technique here is similar to that used for intraperitoneal injection, except that only the skin is taken up in the initial fold and the needle passes through the skin layer, with the bevel downwards.

6. Scarification

This method allows material to penetrate through the skin. An area of skin is shaved, possibly a depilatory may be used (see (D) below). When the surface of the skin has been properly prepared the required amount of liquid is deposited on the surface of the skin and a sharp needle used to scratch and stab gently through the liquid. This treatment is usually continued until a small amount of blood comes from the skin into the liquid droplet. Swabbing with alcohol or ether is not recommended as it may destroy organisms or protein being placed on the skin.

In addition to these normal routes of injection there are many more specialized procedures, e.g. corneal implants, intracardiac injections intracranial injections. The beginner should not attempt to carry out these routines without having the techniques carefully demonstrated by an experienced worker.

D. Depilation

In some cases the hair must be removed over a substantial area of skin before an injection can be made. This is the case particularly where skin toxin reactions are being examined. The surface of the skin is first shaved, or clipped, using electric hair clippers, and a depilation paste consisting of one part barium sulphide, one part zinc oxide, two parts soluble starch, mixed to a thick paste in water, is spread liberally over the area and after a few minutes is carefully scraped off. The residual hair will be removed with the paste, and it is usual to apply lanolin cream to the treated area immediately afterwards. Depilation is usually carried out some hours before the injection procedure.

III. MICE

A. Varieties

There are many strains of mice available for laboratory purposes. Many of these are inbred to produce animals for specific research such as work on hormone or enzyme deficiencies, genetic work or studies on animals which are similar to humans in their reaction to disease.

For more general research purposes there are many laboratory strains bred by particular establishments and the experimenter will often find himself using these. Of the more generally available strains the Swiss White is popular and is often used for microbiological work. If the reader has specific requirements he is advised to consult the Catalogue of Uniform Strains of Mice published by the Laboratory Animal Centre, Carshalton, Surrey, England or the Handbook on Genetically Standardized Jax Mice. The Jackson Laboratory, Bar Harbour, Maine, Edited by Earl L. Green, 1968.

B. Handling

Up to 10 mice are often housed together and the cage can be conveniently transferred to the experimental room. Young mice can leap 12–14 in. vertically, and may escape if the cage top is completely removed. They are easily handled, transfer from one cage to another or from a cage to a working surface being achieved by means of the tail.

To remove a mouse from a cage lift the top of the cage sufficiently with one hand to allow the other hand to be slipped into the cage, and then grasp the tail of the selected animal firmly between the finger and the thumb. The mouse can now be withdrawn and the lid replaced. Place the animal's feet on the table top, or on the cage top. Hold the tail in one hand whilst running the thumb and forefinger of the other hand up the back to the neck. Pushing the skin gently, take hold of the loose skin at the back of the neck firmly so that the animal cannot twise round and bite; do not hold the mouse so tightly as to strangle it. Turn the animal over and, still holding the tail, gently stretch the mouse over the clenched fingers. The tail may finally be hooked under the little finger so as to leave the other hand free if desired. Presented in this way it is comparatively easy for the animal to be given an intraperitoneal injection.

An experienced operator can hold a mouse with one hand and inject with the other. However, for the inexperienced it is desirable to have a colleague hold the animal for injection.

C. Injections

1. *Intraperitoneal, subcutaneous and intradermal injections*

The fingers of the left hand are used to raise a ridge of skin (including the body wall for an intraperitoneal injection) along the centre line of the lower abdomen. The injection is then made into this fold along the mid-line of the body.

2. *Intramuscular injection*

The operator gently holds the rear leg with the left hand, draws it out straight and inserts the needle into the muscle of the inner thigh.

3. *Intravenous injection*

Intravenous injection is given into a vein in the tail of the animal. Place the mouse in a gauze tube of about 1 in. diam., the end of which is blocked by a cork to prevent exit. The tail is then threaded through a hole in the cork used to block off the open end. The tail veins can be dilated by dipping the tip of the tail into warm water. Injection should be made at the tip of the tail, so that if the first attempts fail, the injection can be repeated

5a–15

by working towards the body. The tail is held across the first finger and is trapped between the first finger and the thumb. One of the problems encountered in performing an intravenous injection into the mouse is that the vein becomes occluded by the pressure from the point of a needle and empties of blood, so becoming very difficult to see. When the needle is thought to be in position the syringe piston should be gently withdrawn, whereupon the entry of a little blood into the syringe indicates that the needle is in the vein.

D. Anaesthesia

Mice are conveniently anaesthetized in a small desiccator, the lower part of which contains a pad of cotton wool soaked in ether. The animal is watched carefully, and it will be seen that the breathing gradually becomes slower and deeper; finally the animal collapses. At this stage, the animal may be bled from the tail, and will recover from the anaesthetic in about 5–10 min. Anaesthesia must be continued to a deeper stage, until respiration actually ceases, for exsanguination or for the removal of peritoneal fluid. For more prolonged anaesthesia an intraperitoneal injection of sodium pentobarbitone is recommended.

E. Withdrawal of body fluids

1. *Blood*

Small volumes of blood (about 0·2 ml) can be obtained by removing the tip of the tail with a sharp scalpel while the mouse is under light anaesthesia. Bleeding can be encouraged by lightly drawing the fingers along the tail from the base towards the tip. Another method said to be less traumatic, but which is not generally accepted by the United Kingdom Licencing Authority, is to bleed from the venous plexus, situated in the orbit behind the eyeball. As this is a rather delicate operation it should only be attempted after tuition. The animal is held with the thumb and forefinger encircling the neck. Slight pressure will cause veins to dilate, a finely drawn Pasteur pipette is inserted in the middle angle of the eye, with a drilling motion, between the eyeball and the orbit, through the conjunctiva towards the optic nerve for a depth of about 5 mm. As soon as the plexus is reached blood will gush into the pipette. When sufficient sample is obtained pressure on the neck is released and bleeding stops immediately. (Method prescribed by Noller H. G. Hliwo. 1955 S. 770. Published in The Blood Morphology of Lab. Animals. S. Schermer. Pub. F. A. Davis Company, U.S.A.)

Two to three ml of blood may be obtained by killing the animal and bleeding from the axilla. The animal is deeply anaesthetized and then pinned out on its back on a dissecting board. The skin over the sternum is lifted with toothed forceps and a vertical cut made with a sharp scalpel or fine scissors down to the chest wall. The flap of skin is lifted away from the chest wall and the connective tissue cleared to the arm-pit on one side. The axillary vein is then severed and the blood allowed to accumulate in the pocket formed by the skin flap and the chest wall. The blood can then be pipetted from this pocket.

2. Peritoneal fluid

The animal is killed by exposure to ether, then is pinned out on its back on a dissecting board. The surface of the abdomen is swabbed with alcohol and the skin lifted in toothed forceps. A small incision, about a quarter-of-an-inch long, is made with a sharp scalpel or fine scissors in the mid-line of the abdomen passing through the skin and body wall. A fine Pasteur pipette is then passed through into the peritoneum and the fluid withdrawn. The pipette is moved from side to side during the withdrawal of the fluid in order to keep the end free from parts of the intestine.

IV. RATS

A. Varieties

Three strains are commonly available; the Hooded Lister and two albinos, the Wistar and the Sprague Dawley. In addition to these a number of inbred strains are bred by particular laboratories and some success has attended attempts to produce strains with specific hormone, enzyme and organ deficiencies. No catalogue of strains appears to be available at the present time.

B. Handling

Like the mouse, the rat is normally kept in a small cage with a removable grilled top, and it is usual to carry the cage containing the animal to the operating room. Rats are stronger and more agile than mice, and they need to be handled with caution. They may be picked up by the tail in a similar manner to the mouse, but it is more usual to handle them by placing the palm of the hand on the back near the head with the thumb and the fore-finger under the throat in front of the forelegs and close to the chin. In this position the rat is under control and cannot bite.

C. Injections

1. Intraperitoneal, subcutaneous and intradermal injections

The animal is held as described above and turned over so that its back lies in the palm of the hand and the tail is held with the other hand. In this position it can be presented for injection in the same way as previously described for mice.

2. *Intravenous injection*

This injection is carried out using the tail veins in an apparatus similar to that previously described for mice, but of larger dimensions. Alternatively, since this is a somewhat painful procedure for the animal, some workers prefer to lightly anaesthetize the rat first (see D below).

3. *Intracardiac injection* (this technique is preferably demonstrated by an experienced operator).

The animal is anaesthetized and laid on its back. The needle is inserted into the thorax in a horizontal direction at about the third intercostal space, 1–2 cm below the sternum. The exact position of the heart is determined by palpation. When the needle is in position, the plunger of the syringe is withdrawn slightly to check that blood flows through the needle, indicating that it is indeed in the heart.

D. Anaesthesia

Rats are anaesthetized by giving an intraperitoneal injection of 0·08 ml of sodium pentabarbitone per 200 g body weight using a short, fine needle and allowing about 10 min for the drug to take effect. Anaesthesia is then completed using ether as described previously for mice.

E. Withdrawal of blood

Small amounts of blood may be obtained from the tail vein. The animal is placed under light anaesthesia and laid on its side on the bench top, with its tail towards the edge of the bench. A vein is incised towards the tip of the tail longitudinally, using a sharp, pointed scalpel. The tip portion is placed into a test tube and the tail massaged firmly from the base towards the tip.

Larger quantities of blood may be obtained by axillary bleeding as described for mice, or by withdrawal direct from the heart. In the latter case, the bleeding needle is inserted as for a cardiac injection and 2–3 ml of blood are withdrawn slowly into the syringe. With a little experience this quantity of blood can be removed from the animal without jeopardizing its survival.

V. GUINEA-PIGS

A. Varieties

The three main varieties of guinea-pigs are English, Abyssinian and Peruvian. The English and Abyssinian are both short-haired, the Abyssinian differs by having tufts of longer hair (rosettes) arranged irregularly over its body, giving it a chunky appearance. The Peruvian has very long hair and is seldom used for experimental purposes. All can be obtained in a variety of colours, black, white, cream, brown etc. They also occur with combinations of these colours. The English, probably the most common, is the most used for laboratory experiments and varies in weight from about 250 g at 6 weeks to 500 g at 12 weeks. Because of the obvious advantage for skin reaction experiments, the English pink eyed white is used extensively. Coloured pigs, however, should not be rejected completely. Provided they are healthy animals from healthy stock, they are quite adequate for most experiments.

B. Handling

Guinea-pigs are normally housed in pens or in cages. They are agile and sometimes difficult to capture but seldom bite. The animal can be conveniently picked up by grasping from behind, around the thorax or by slipping the hand underneath. The animal will kick vigorously in order to escape and must be held firmly, it should be put down on the table top as soon as possible and a firm hold retained.

To present the animal for injection, grasp firmly with the right hand over the back with the thumb under the chin. The forelegs may be trapped between the thumb and thorax or allowed to lie freely. The animal is then turned over on its back and the posterior restrained with the left hand, placing the thumb over the legs.

C. Injections

1. Intravenous injection

The guinea-pig is not a good subject for intravenous injection. Using a large pig and a fine needle the marginal vein of the ear can be injected. Another method seldom practised is to anaesthetize and cut down to the saphenous vein about $\frac{3}{8}$ in. above the hock; this however should be tackled only after tuition.

2. Intraperitoneal injection

With the animal held as described above, intraperitoneal injections can be carried out as for a mouse or rat.

3. Intradermal and subcutaneous injections

These procedures are conveniently carried out on the abdomen, the flank or the back of the animal. Scarification is usually carried out on the back. The guinea-pig is often the preferred animal for observing skin reactions, and for this type of experiment careful depilation is necessary.

4. Intramuscular injection

The animal is turned on its back with the head towards the handler who, with the animal resting in the palm of the right hand, secures the front legs with the thumb. The handler uses the left hand to hold the rear legs, with the thumb across the lower abdomen in front of the knees and fingers across the spine. Gentle pressure with the thumb will straighten the legs and the animal is securely held. In this position it is easy for the operator to take hold of a hind leg and draw it gently towards him exposing the inner aspect of the thigh for injection.

5. Intracardiac injection

Anaesthetize the guinea-pig with sodium pentabarbitone and ether as described for the rat, and an intracardiac injection is carried out in the same way as for the rat, i.e. between the second and third intercostal space, 1–2 cm below the sternum. Under ether anaesthesia the guinea-pig has the habit of suddenly curling up and scratching the ears and head. Anaesthesia is maintained by placing a pad of cotton wool moistened with ether in a tin which has holes punched through the bottom. The tin is held periodically over the face of the animal. The animal's respiration must be closely observed during this procedure, to ensure that anaesthesia is not becoming too deep.

D. Withdrawal of blood

5–10 ml of blood can be withdrawn by the intracardiac route without jeopardizing the animal's survival. Larger quantities of blood may be obtained by killing the animal, and bleeding from the axilla as described for the rat, or by opening the thorax as soon as respiration has ceased (using either ether or chloroform anaesthesia) and exposing the heart. The heart is incised and the blood allowed to drain into a container. The yield of blood may be increased by manual pressure on the body of the animal during this process. Exsanguination by severing the jugular vein is not recommended. It is normally done while the animal is anaesthetized, and the pumping of the heart causes the blood to spurt in a manner difficult to control. If this method is attempted, a wide-necked bottle or beaker is required to catch and retain the blood. If sterile blood is required an alternative method is to exsanguinate under terminal anaesthesia by withdrawing blood from the heart with a sterile syringe and needle.

VI. RABBITS

A. Varieties

More than 30 breeds of rabbits, weighing from 2–7 kilos are listed in the Universities Federation of Animal Welfare Handbook and the reader is referred to this book for details. If serological experiments are being carried out, and large quantities of blood are required, the larger animals should be considered. Lop-eared rabbits are very suitable for this purpose, having a body weight of up to 7 kilos, and their large ears and large ear veins allow them to be bled with considerable ease. An alternative is the New Zealand white, a large rabbit with a weight of 5–6 kilos with the added advantage of a white skin which is useful for experiments involving skin reactions. This breed is quite popular for laboratory diagnostic and research purposes and is usually easily obtainable. Obviously the breeds of rabbits used and maintained in the animal house will be directly related to the type of work carried out by a particular establishment.

B. Handling

The rabbit is normally a docile creature and seldom bites. Any damage inflicted on the operator is usually by scratching with the hind legs. The animal is picked up by grasping the loose skin at the scruff, and supporting the weight with the other hand, allowing the animal to lie along the forearm. In this position the animal will tend to tuck its head under the arm, and can be carried in this way without struggling. The rabbit reacts favourably to petting and gentle talk, and readily reflects any nervousness on the part of the handler. More than the other experimental animals discussed in this Chapter, the rabbit will tend to panic if it finds that it cannot get a good grip on the surface on which it is being handled. A great deal of difficulty can be avoided by the provision of a suitable piece of cloth or other rough surface material. Like most animals the rabbit becomes much more peaceful when in the dark and simply covering the head will often induce an excited animal to become more docile.

Rabbits are housed in a variety of types of cage, usually singly, but occasionally in pairs. Often the cages are not very large, and the wire mesh in the base gives the animal an excellent purchase to resist removal from its quarters. It is important to remove an experimental animal from its cage with the minimum of fuss, since a prolonged struggle will result in the rabbit being nervous even before the start of experimental work. Persuade the rabbit to take up a position with its head facing the operator's left and then grasp the scruff firmly in the right hand. Slip the left hand under the hind quarters and use it to roll the rabbit sideways so that it can no longer obtain a grip on the floor of the cage. In this position it is usually easy to lift the animal clear of the floor and out of the cage.

B. Injections

1. *Intramuscular injection*

Intramuscular injections are usually given into the hind legs or the scruff. Injection of material likely to cause ulcers or granuloma into the scruff will make subsequent handling painful, and the thigh or calf muscles are to be preferred. The practice of injecting into the foot pad is to be deplored since this is a most sensitive site.

Place the animal on the bench, with one hand on the neck close behind the ears. Hook the fingers of the other hand under one hind foot with the thumb over the spine, and roll the rabbit onto its side, thus presenting the other hind leg for injection. The fur should be clipped or shaved and the injection is normally made into the thigh or calf muscles.

2. Intravenous injection

Wrap the rabbit in a piece of blanket or towel with the head and ears protruding. Shave the ear, over the marginal vein, using a sharp scalpel blade, apply firm pressure with the thumb and forefinger at the base of the ear and allow the vein to dilate. Inject, using a small gauge needle with the bevel down, pointing towards the base of the ear. The needle can easily slip into the outer sheath of the vein and a little blood should be gently withdrawn before injecting to ensure that the needle is in the vein lumen. The first injection should be made as near as possible to the tip of the ear, then, should occlusion of the vein occur, subsequent attempts can be made towards the base of the ear.

N.B. A box for holding rabbit has been constructed which leaves only the head and ears exposed. This box is useful if experiments have to be performed single handed. However, as rabbits usually kick quite violently if hurt, being trapped in a box can result in a dislocated spine.

3. Intraperitoneal injection

Roll the rabbit onto its back and place one hand across the lower abdomen close to the hind legs. The other hand is placed across the thorax, close to the forelegs. The animal should be firmly held, but not tight enough to cause distress. Intraperitoneal and intradermal injections can now be given in the abdominal region.

D. Withdrawal of blood

The animal is wrapped and positioned as for an intravenous injection, and the ear shaved over the marginal vein. A paperclip is placed across the base of the marginal vein and the tip of the ear is wiped with a cotton wool swab, moistened with xylol. This causes the vein to dilate within a few minutes. Alternatively, the tip of the ear may be massaged with warm water. If xylol is used, it is important to remember that this is a mild irritant and that it should be removed by swabbing the ear with alcohol at the end of the operation. There are several types of needle available for inducing bleeding, these are usually of the type that will produce a small slit in the vein wall, a triangular leather sewing needle is excellent for the purpose. The small, sterile needles called Sterilets (Arnold Horwell, London) are also suitable. The needle is inserted into the vein in such a way as to make a small, longitudinal cut and withdrawn. The ear is held in such a position that the emerging drops of blood drip straight from the ear into a suitable container. A screwcapped glass bottle is ideal for this purpose. If difficulty is experienced because the blood drops tend to run down the edge of the ear, this can be prevented by smearing the margin of the ear lightly with a little vaseline. 20–30 ml of blood can easily be obtained by superficial venesection in this way. Often, after the initial needle puncture, the vein will contract and blood flow is very slow, but this will last only for a minute or so when bleeding usually commences vigorously. When sufficient blood has been obtained bleeding may be staunched by covering the wound with dry cotton wool, and holding firmly while the paperclip is removed and xylol washed from the tip of the ear with alcohol. If sterile blood is required, it should be taken direct from the vein into a syringe and needle, or into a Vacu-taner (vacuum phials complete with needle, commercially available in a range of volumes from Becton, Dickinson Ltd., Wembley, Middlesex, England, and Becton Dickinson and Company, Rutherford, New Jersey, 07070, U.S.A.).

Larger volumes of blood may be withdrawn by direct cardiac bleeding on an anaesthetized animal, but this technique is not to be recommended for routine use and should certainly be learned by demonstration. Alternatively the rabbit may be killed and exsanguinated by bleeding from the heart. The animal is anaesthetized with sodium pentabarbitone (60 mg per kg body weight), then etherized until respiration has ceased. The thorax is opened and the sternum removed exposing the heart. The heart is incised and the blood allowed to drip into a wide-necked receptacle. Manual pressure on the animal's body will increase the volume of blood obtained. 75–150 ml of blood can be obtained by this method.

VII. EUTHANASIA

Experimental animals may have to be killed. The method may be governed by the nature of the experiment. Where experimental considerations dictate a particular method the technique must be demonstrated and taught by an experienced practitioner. Attempts to do otherwise cause a great deal of distress to the animal, and indeed to the operator. The object of this Section is to describe methods of killing the common experimental animals which in my experience, are rapid and humane. These methods are recommended wherever possible.

A. Mice

The simplest method is to continue chloroform anaesthetization until the animal is dead. In all the methods using chloroform or other inhaled anaesthetics, it should be remembered that direct contact with the anaesthetic can be extremely painful to an animal and must be avoided. A chloroform soaked cotton wool pad is placed in the bottom half of a desiccator and the vessel allowed to stand for a few minutes with the lid slightly open so that the vapour of the chloroform passes through the grill into the top half and displaces most of the air. The mouse is then placed in the top part of the desiccator and the lid replaced. Three minutes exposure will usually suffice to kill the animal.

Some workers prefer to despatch mice by breaking their necks. The technique, when quickly and expertly performed, is certainly rapid and highly efficient. The animal is held by the tail while standing on all four legs. A pencil is placed firmly behind its ears and pressed down sharply at the same time as the hind quarters are lifted vertically and the tail slightly stretched. The neck will break immediately and the animal is dead.

B. Rats

Rats may also be killed by the use of chloroform in a similar way to mice, but perhaps a more satisfactory way is to inject 0·5 ml of sodium pentabaritone intraperitoneally. Death will follow in a matter of 10 min.

C. Guinea-pigs

A guinea-pig may be killed by chloroform or by the use of 0·5 ml of sodium pentabarbitone injected intraperitoneally.

D. Rabbits

The use of chloroform or ether is not recommended for killing rabbits since the animals tend to react violently to the anaesthetic vapours and may on occasion scream in a distressing manner. The simplest way of killing rabbits is to give an intravenous injection of 1–2 ml sodium pentabarbitone when death will follow rapidly.

A method preferred by some workers is to inject 2 ml of a saturated solution of magnesium sulphate intravenously to an anaesthetized animal. Death is virtually instantaneous.

VIII. APPENDIX

A note on the preparation of serum from blood

To prepare serum from blood the blood is allowed to stand in a closed vessel, such as a screw-capped glass bottle, at 37°C until coagulation is complete. The clot is then eased away from the wall of the vessel—an orange stick is an ideal instrument for doing this—and the bottle reincubated at 37°C. In some 4 h the clot will have contracted and the bottle is then placed at 4°C overnight. During this time contraction of the clot continues and any red cells contaminating the serum settle at the bottom of the bottle. The serum is then carefully pipetted off and centrifuged at 2,000 r.p.m. for 5 min. The supernatant serum is freed from sedimented cells, and should be a straw-coloured, clear liquid. There may occasionally be slight coloration due to haemolysis.

The method of storage will be governed to a certain extent by the subsequent use that is to be made of the serum. For many purposes storing in a deep freeze at −20°C is the most acceptable method. If this method is used, it is advisable to split the initial volume of serum into a number of small containers before freezing. For experimental purposes the container can then be thawed and the entire contents used. Repeated freezing and thawing cycles can do a great deal of damage to some of the properties of serum.

Various preservatives can be added to serum for storage at refrigerator temperatures. Common ones are thiomersalate at a concentration of 1/10,000 and sodium azide at a concentration of 0·3% (w/v).

CHAPTER X

The Determination of the Molecular Weight of DNA Per Bacterial Nucleoid

J. De Ley

Laboratory for Microbiology, Faculty of Sciences, State University,
Gent, Belgium

I.	Chemical Method	303
	A. Preparation of the cell suspension	303
	B. Cell count per unit volume	303
	C. Average number of nucleoids/cell	304
	D. Chemical determination of DNA	304
	E. Calculation	305
	F. Example	306
II.	Measurement of the Length after Electron Microscopy . . .	306
	A. Principle	306
	B. Lysis by osmotic shock	306
	C. Lysis by detergent	306
	D. Electron microscopy	307
	E. Calculation of the molecular weight	307
III.	Measurement of Length after Autoradiography	308
	A. Preparation of labelled bacteria	308
	B. Lysis of bacteria	308
	C. Autoradiography	308
IV.	Other Methods	308
References	309

It is useful to know the molecular weight of DNA per bacterial nucleoid because the information gives an insight into the genetic potential of the organism. Dividing the molecular weight of the bacterial DNA by the molecular weight of an average bound nucleotide pair as the sodium salt (663) gives the total number of nucleotide pairs. Since an average of some 1500 nucleotide pairs make up one cistron, the maximum total number of cistrons/nucleoid can be estimated. Part of the bacterial DNA may not be expressed phenotypically, because of, for example, unused genes or nonsense

J. DE LEY

DNA, although this unexpressed part appears to be rather small (De Ley et al., 1966).

TABLE I
The size of the bacterial genome

The number of cistrons is calculated, assuming that there are about 1500 nucleotide pairs/cistron. Data for a yeast and two moulds are given for comparison.

Organism	Molecular weight in daltons	Nucleotide pairs per nucleoid	Estimated number of cistrons	Reference
Mycoplasma gallisepticum	ca. $0 \cdot 2 \times 10^9$	ca. $0 \cdot 3 \times 10^6$	200	Morowitz et al (1962)
Mycoplasma hominis	$0 \cdot 51 \times 10^9$	$0 \cdot 77 \times 10^6$	570	Bode & Morowitz (1967)
Haemophilus influenzae	$0 \cdot 72 \times 10^9$	$1 \cdot 1 \times 10^6$	730	Berns & Thomas (1965)
Dialister pneumosintes	$0 \cdot 8 \times 10^9$	$1 \cdot 2 \times 10^6$	800	Chen & Cleverdon (1962)
Aerobacter aerogenes	$1 \cdot 2 \times 10^9$	$1 \cdot 8 \times 10^6$	1,200	Caldwell & Hinshelwood (1950)
Pseudomonas campestris var. pelargonii	$(2 \cdot 1 \pm 0 \cdot 3) \times 10^9$	$3 \cdot 2 \times 10^6$	2,100	Park & De Ley (1967)
Pseudomonas fluorescens	$(2 \cdot 5 \pm 0 \cdot 7) \times 10^9$	$3 \cdot 8 \times 10^6$	2,500	Park & De Ley (1967)
Pseudomonas putida	$(2 \cdot 7 \pm 0 \cdot 3) \times 10^9$	$4 \cdot 0 \times 10^6$	2,700	Park & De Ley (1967)
Bacillus subtilis	$1 \cdot 3 \times 10^9$	$2 \cdot 0 \times 10^6$	1,300	Dennis & Wake (1966)
Bacillus subtilis	$2 \cdot 4 \times 10^9$	$3 \cdot 6 \times 10^6$	2,400	De Ley & Park (to be published)
Bacillus subtilis	2 to 4 $\times 10^9$	$3 \cdot 0$ to $6 \cdot 0 \times 10^6$	2,000– 4,000	Massie & Zimm (1965)
Escherichia coli	$2 \cdot 8 \times 10^9$	$4 \cdot 2 \times 10^6$	2,800	Cairns (1963b)
Escherichia coli	$(3 \cdot 1 \pm 0 \cdot 2) \times 10^9$	$4 \cdot 7 \times 10^6$	3,100	Park & De Ley (1967)
Saccharomyces cerevisiae	$14 \cdot 5 \times 10^9$	22×10^6	14,000	Esser & Kuenen (1965)
Neurospora crassa	28×10^9	43×10^6	29,000	Esser & Kuenen (1965)
Aspergillus nidulans	27×10^9	41×10^6	27,000	Esser & Kuenen (1965)

A few data on the molecular weight of bacterial DNA per nucleoid are available (Table I). They show that the values can vary at least some 10-fold. For the sake of comparison it may be mentioned that the molecular weight of double-stranded DNA from most phages ranges from 4×10^6–150×10^6 daltons. Exact knowledge of the relationship amongst, and the classification of, bacteria will be provided by a combination of numerical analysis, DNA base composition, DNA hybridization and the molecular weight of the bacterial chromosomal DNA. The latter values have to be known before the possible number of shared similar cistrons can be estimated.

I. CHEMICAL METHOD

In essence the amount of DNA is determined chemically in a certain volume of a bacterial suspension, containing a known number of bacterial cells (determined microscopically). In separate preparations, cells are stained to reveal nucleoids, and the average number of nucleoids/cell is determined. The molecular weight of DNA per nucleoid can then easily be calculated. It is advisable to repeat these determinations at different stages of the growth curve; early, middle and late log, and early stationary phase.

A. Preparation of the cell suspension

A suitable medium is selected in which the bacteria grow well, with—if possible—the production of little polysaccharide. The growth curve is established, for instance by measuring turbidity changes. Thirty minutes before harvesting the cells, chloramphenicol (50 μg/ml of medium) is added. The cells are harvested by centrifugation for 20 min at 10,000 g, washed by repeated suspension in SSC buffer and centrifuged again. SSC buffer contains 0·15 M NaCl and 0·015 M trisodium citrate at pH 7·0.

Finally a very dense suspension of about 10^{11}cells/ml buffer is prepared as stock.

B. Cell count per unit volume

After thorough mixing of the cell suspension, a suitable dilution is prepared in the same buffer, which is added into a counting chamber, such as manufactured by Hausser & Son, Philadelphia, Penn. Cells must be well separated. At least 50 squares are counted, and the procedure is repeated at least three times. Suspensions should not be prepared in distilled water, because some cells may be disrupted. One source of error that is hard to avoid is the distinction between one long cell and two adhering cells that are on the verge of dividing. Clumps of cells should be avoided. They can be prevented by very brief (30 sec) treatments with a sonic oscillator or by adding dilute solutions of detergents, e.g., 0·001–0·01% sodium lauryl sulphate.

Some types of motile organisms may be difficult to count. Their motility may be arrested by adding a trace of formaldehyde, $HgCl_2$, brief heating to 80°C or simply waiting a few minutes before counting until all oxygen in the suspension is used up. Knowing the average number of cells/square in the Petroff–Hausser chamber, which has a volume of 5×10^{-8} ml, it is easy to calculate the number/ml stock suspension.

C. Average number of nucleoids/cell

2×10^{10}–3×10^{10} bacterial cells from the stock suspension are centrifuged onto 1·5% agar. A thin surface slice is cut out of the tube and placed on glass beads over a solution of 2% OsO_4 for 3 min to fix the bacteria. The method of Smith (1950) (modified) is used to stain the nucleoids. The fixed bacteria on the agar are smeared on a coverslide and hydrolysed in M HCl for 20 min at 60°C to break down RNA, rinsed several times with distilled water and stained with Giemsa dye at 37°C for various lengths of time (5–15 min). The optimal time must be established for every strain. Likewise, the commercially available Giemsa stock solution has to be diluted 50 or 100 times according to the strain, for optimal pictures. The average numbers of nucleoids/cell is determined by examining 200–500 randomly selected cells.

The main source of error and uncertainty is that it is sometimes difficult to decide whether a cell has two nucleoids or one long and irregular one. Likewise, a single long cell is sometimes difficult to distinguish from two cells on the verge of division.

D. Chemical determination of DNA

An aliquot, containing some 2×10^{10}–7×10^{10} cells, from the stock suspension is digested in 0·5 M NaOH for 20–24 h at 37°C. After adjusting the digest to 0·6 M $HClO_4$ with conc. $HClO_4$ and cooling for 30 min in an ice bath, 2 ml of cold acetone are added. After about 2 h, the DNA-containing precipitate is centrifuged and washed twice with 1 ml of cold 0·6 M $HClO_4$. The DNA is hydrolyzed with 1·5 ml of 0·6 M $HClO_4$ for 15 min at 90°C. The protein residue is removed by centrifugation and an aliquot of the acid-soluble supernatant is treated with diphenylamine reagent for the estimation of DNA according to Burton (1956).

E. Calculation

DNA content in grams per nucleoid =

$$\frac{\text{Weight (g) of DNA (sodium salt)/ml of stock bacterial suspension}}{\text{Number of cells/ml of stock bacterial suspension} \times \text{average number of nucleoids/cell}}$$

Molecular weight of DNA (sodium salt), daltons/nucleoid = $0 \cdot 60 \times 10^{24} \times$ DNA content (g)/nucleoid.

In cases where it has been examined it has been established by direct and indirect methods that bacterial DNA forms a closed structure, a so-called "circular" chromosome. This has been demonstrated in *Escherichia coli* (Cairns, 1963b), *Salmonella typhimurium* (Sanderson and Demerec, 1964), *Streptomyces coelicolor* (Hopwood, 1965), *Mycoplasma hominis* (Bode and Morowitz, 1967) and in several phages. There are good reasons to believe that circular chromosomes will be found in all bacteria and actinomycetes.

Episomes, of all kinds, make up only a few per cent of the total DNA content of the bacterial cell. They too can contain genetic information. This amount of extrachromosomal DNA falls within the limits of experimental error of the method. The amount of DNA per nucleoid, expressed in daltons, can thus, with a high degree of confidence, be considered as a good estimate of the molecular weight of the DNA of the circular bacterial chromosome. Since the molecular weight of a nucleotide pair (as the sodium salt bound in DNA) is 663, the number of nucleotide pairs per nucleoid = $\frac{\text{molecular weight of DNA}}{663}$. An estimate of the maximal number of cistrons is obtained by dividing the latter result by 1500.

TABLE II

Determination of the molecular weight of DNA/nucleoid with *E. coli*

	Logarithmic phase			Stationary phase
	early	middle	late	late
Time, h	2	3	4·5	16
Turbidity, Klett units	63	178	320	450
μg DNA/cell $\times 10^9$	9·56	9·17	9·70	7·35
Average number of nucleoids/cell	2·00	1·93	1·89	1·35
Molecular weight, daltons $\times 10^9$	2·88	2·86	3·09	3·28

The error in these determinations is usually $\pm 0 \cdot 2 \times 10^9 - 0 \cdot 3 \times 10^9$ daltons, or about $\pm 10\%$

F. Example

An example is provided in Table II (Park and De Ley, 1967).

II. MEASUREMENT OF LENGTH AFTER
ELECTRON MICROSCOPY

A. Principle

The principle of the method of displaying cellular DNA on a protein film was worked out by Kleinschmidt and Zahn (1959) and Kleinschmidt *et al.* (1962); a modification was applied by MacHattie and Thomas (1964). Cells are lysed by osmotic shock in a high salt solution. In the presence of a protein, the contents are spread out gently on the surface of water. Samples of the resulting film are shadowed rotationally and examined in the electron microscope.

In a modification, Bode and Morowitz (1967) described a method in which the cells are lysed by detergent before the formation of the film.

The following recipe is mainly taken from the latter paper, which should be consulted for further details, as well as Lang *et al.* (1967).

B. Lysis by osmotic shock

Log-phase cells are filtered through a membrane filter of suitable pore size to remove clumps, and centrifuged. They are washed by suspension in 0.2 M $NaCl + 0.02$ M $NaHCO_3$ and re-centrifuged.

About 1.5×10^9 cells are suspended/ml of the following solution: 5 M ammonium acetate; 0.01 M-EDTA (disodium salt); 0.015 M NaCl; 0.02% cytochrome c in doubly distilled water. To produce the film the following accessories are prepared—

(a) A stainless-steel knife and ramp (see Bode & Morowitz, 1967); both are cleaned with acetone and flamed.

(b) a paraffin-covered Petri dish, filled with doubly distilled water (hypophase), in a dust-free compartment.

The water is swept clean with the knife.

The ramp is introduced into the water; 0.05 ml of the lysed cell suspension is allowed to run down gently and spread over the entire water surface.

C. Lysis by detergent

About 7×10^9 cells are suspended/ml of 0.2 M $NaCl + 0.02$ M $NaHCO_3$. A 0.3 ml portion of this is added to 0.6 ml of a solution containing 60% (w/v) sucrose, 0.1 M EDTA, 0.1 M Tris pH 8 in a dialysis bag. Either 0.5 mg/ ml lipase (Worthington Ltd, Freehold, New Jersey, U.S.A.) or 2 mg/ml pronase (a protease available from Calbiochem, Lucerne) is added. Dialyse against 120 ml of 40% sucrose, 0·1 M EDTA, 0·1 M Tris pH 8·0, 1% Duponol for 15 h at 37°C to lyse the cells. This high osmotic solution is now slowly removed by dripping in SSC buffer over a period of 9 h, while keeping the volume constant with an overflow. A 0·03 ml portion of the lysed suspension is added to 0·02 ml of 10 M ammonium acetate+ 0·05% cytochrome c. on the ramp and left for 10 min to mix by diffusion.

The handling of DNA solutions with pipettes requires great care; only broad-tipped ones and very slow suction or addition should be used to avoid breakage by shearing of the very long and extremely labile DNA molecules. The ramp is introduced into the Petri dish, prepared as above, and the liquid is allowed to run down gently to spread over the water surface.

D. Electron microscopy

Photographic plates of suitable, isolated closed DNA molecules are enlarged by projection on white paper. The DNA is traced and its length measured with a suitable map measure. For further details see Bode and Morowitz (1967).

E. Calculation of the molecular weight

Knowing the exact magnification factor (ca 7500) of the electron microscope, e.g., by using calibrated polystyrene beads, and the subsequent enlargement (about 12), the real length of the DNA molecule can be calculated.

Both Lang et al. (1967) and Inman (1967) have shown that the length of the DNA molecule depends largely on the ionic strength of the hypophase. Increasing salt concentrations in the medium onto which the DNA solutions are spread, decreases the length. When spread onto distilled water, the length depends on the region of the protein film sampled. From samples, taken within 2 cm of the point at which the protein solution runs onto the water hypophase, it can be calculated from the results of Bode and Morowitz (1967) that the average molecular weight is $2·05 \times 10^6$ dalton/μm of DNA. The error is $\pm 10\%$. With an average molecular weight of 663 for a bound nucleotide pair, it can be calculated that the distance between nucleotide pairs in these conditions is 3·24Å, a structure much closer to the paracrystalline B form than to the crystalline.

The total molecular weight is thus—

$$\frac{\text{Measured length on enlargement } (\mu\text{m})}{\text{Total magnification factor}} \times 2·05 \times 10^6 \text{ daltons}$$

J. DE LEY

III. MEASUREMENT OF LENGTH AFTER AUTORADIOGRAPHY

The method given here is essentially that of Cairns (1962, 1963a) for thymine-requiring strains.

A. Preparation of labelled bacteria

Bacteria are grown in a suitable medium, containing 2 μg of tritiated thymidine (9 C/mmole)/ml, for a few generations, to allow sufficient labelling of the DNA.

B. Lysis of bacteria

The bacteria are diluted to about 10^4 cells/ml in the lysis medium (1·5 M sucrose, 0·05 M NaCl, 0·01 M EDTA, 0·01 M KCN). This mixture is placed in a cylindrical chamber 2 cm diameter, 2·5 mm deep, faced on one side with glass and on the other with a VM Millipore filter, 50 nm pore size It is dialyzed against 1% Duponol C in the same medium for 2 h at 37°C. The medium is then replaced by 0·05 M NaCl, 0·005 M EDTA by dialysis against repeated changes of the latter solution for 18–24 h. DNA is collected on the dialysis membrane by draining on filter paper.

C. Autoradiography

The Millipore filters are stuck onto microscope slides with a cement, overlaid with Kodak AR 10 stripping film, and exposed for 2 months at 4°C over $CaSO_4$ in a CO_2 atmosphere. The developed photograph (Kodak D19b for 20 min at 16°C) is suitably enlarged and the length of the DNA molecules are measured with a map measure. The molecular weight is calculated as described for the electron-microscopic method.

It may be noted that the molecular weight of *E. coli* DNA, determined by the above method by Cairns (1963b), agrees very well with the result of the chemical method, obtained in our laboratory (Park and De Ley, 1967).

IV. OTHER METHODS

Several methods for determining DNA in isolated nuclei of higher organisms have been reviewed and discussed by Vendrely (1955). They are

1. Estimation of deoxyribose.
2. Estimation of phosphorus or purine nitrogen.
3. Histophotometry in visible light.
4. Photometry in ultraviolet light.

Except for the first of these methods, which was applied above, the others have not been adapted for use with bacteria.

REFERENCES

Berns, K. I., and Thomas, C. A. (1965). *J. molec. Biol.*, **11**, 476–490.
Bode, H. R., and Morowitz, H. J. (1967). *J. molec. Biol.*, **23**, 191–199.
Burton, K. (1956). *Biochem. J.*, **62**, 315–323.
Cairns, J. (1962). *J. molec. Biol.*, **4**, 407–409.
Cairns, J. (1963a). *J. molec. Biol.*, **6**, 208–213.
Cairns, J. (1963b). *Cold Spring Harb. Symp. quant. Biol.*, **28**, 43–46.
Caldwell, P. J., and Hinshelwood, C. (1950). *J. chem. Soc.*, 1415–1418.
Chen, C., and Cleverdon, R. C. (1962). *Life Sci.*, **8**, 401–403.
De Ley, J., Park, I. W., Tijtgat, R., and Van Ermengem, J. (1966). *J. gen. Microbiol.*, **42**, 43–56.
Dennis, E. S., and Wake, R. G. (1966). *J. molec. Biol.*, **15**, 435–439.
Esser, K., and Kuenen, R. (1965). "Genetik der Pilze". Springer-Verlag, Berlin.
Hopwood, D. (1965). *J. molec. Biol.*, **12**, 514–516.
Inman, R. B. (1967). *J. molec. Biol.*, **25**, 209–216.
Kleinschmidt, A. K., and Zahn, R. K. (1959). *Z. Naturf.*, **14b**, 770–779.
Kleinschmidt, A. K., Lang, D., Jacherts, D., and Zahn, R. K. (1962). *Biochem. biophys. Acta*, **61**, 857–864.
Lang, D., Bujard, H., Wolff, B., and Russell, D. (1967). *J. molec. Biol.*, **23**, 163–181.
MacHattie, L. A., and Thomas, C. A. (1964) *Science, N.Y.*, **144**, 1142–1144.
Massie, H. R., and Zimm, B. H. (1965). *Proc. natn. Acad. Sci. U.S.A.*, **54**, 1636–1641.
Morowitz, H. J., Tourtelotte, M. E., Guild, W. R., Castro, E., Woese, C., and Cleverdon, R. C. (1962). *J. molec., Biol.*, **4**, 93–103.
Park, I. W., and De Ley, J. (1967). *Antonie van Leeuwenhoek*, **33**, 1–16.
Sanderson, K. E., and Demerec, M. (1964). *Microbiol. Genet. Bull.*, **20**, 11.
Smith, A. G. (1950). *J. Bact.*, **59**, 575–587.
Vendrely R. (1955). *In* "The Nucleic Acids" (Ed. E. Chargaff and J. Davidson), Vol 2, pp. 155–180. Academic Press, London.

CHAPTER XI

Hybridization of DNA

J. DE LEY

Laboratory for Microbiology, Faculty of Sciences, State University,
Gent, Belgium

I. Introduction 311

II. Preparation of the Organisms 311

III. Preparation of Labelled Reference Organisms 313

IV. Preparation of Pure DNA 314

V. The DNA–Agar Method 315
   A. Preparation of stock DNA solutions 318
   B. Preparation of DNA–agar 318
   C. The competition hybridization method 319
   D. The direct hybridization method 322
   E. Collecting fractions 322
   F. Calculations 323
   G. Comparison between the direct and the competition methods . 324

VI. The Membrane-filter Technique 324
   A. The albumin-coated filter method 324
   B. Removing non-hybridized DNA at low ionic strength and high
      pH 325
VII. The Ultracentrifugal CsCl Gradient Method 326

VIII. Introductory Reading on DNA Hybridization with Different
   Organisms 328

References 328

I. INTRODUCTION

In the last decade, bacterial taxonomy has been considerably enriched by the results of several new approaches: numerical analysis, DNA base composition and DNA–RNA and DNA–DNA hybridization. In particular the latter method opens up vistas that were thought to be beyond reach even 10 years ago: it allows one to measure quantitatively and directly, on a chemical basis, the fraction of nucleotide sequences, which is common between a set of organisms. It is possible to determine directly by simple means similarities between genomes, and to estimate the number of cistrons shared. This is not only of importance for determining the relationship between bacteria, thus helping to improve their classification, but it is also a major breakthrough towards an experimental study of the phylogeny and evolution of bacteria. We shall describe and discuss the different methods that have been used and proposed to determine genetic similarity by DNA–DNA hybridization. These methods are not yet sufficiently refined to distinguish between perfect (guanidine–cystosine and adenine–thymidine) and imperfect (other combinations and open loops) pairing of DNA strands from different organisms, but this is now being studied intensively. The results of DNA–DNA hybridizations (also called binding or re-association) are therefore to be regarded as indicative of DNA similarities rather than as perfect homologies.

II. PREPARATION OF THE ORGANISMS

Because the composition and the size of the bacterial genome is independent of the medium on which the bacteria have been grown (in the absence of mutagens), a growth medium can be selected, suitable for each strain, on which the organisms grow quickly and extensively. With some slime-producing bacteria, polysaccharides are often difficult to separate from DNA during the purification of the latter. In some cases, as with *Agrobacterium* and *Rhizobium*, this difficulty can be partially obviated by selecting suitable media (Herberlein *et al.*, 1967); in other cases, as for example with the azotobacters, no suitable medium has yet been developed and much more polysaccharide than cell matter is produced during growth. It is very important to remove as much polysaccharide as possible after growth, by repeatedly suspending the cells in a large volume of buffer, followed by centrifugation and separating the polysaccharide layer (usually on top) with a spatula.

In order to avoid contamination, we prefer to grow the organisms on solid media in Roux flasks. If a fortuitous contamination occurs it can usually be spotted right away; it remains localized on the agar. In liquid media an occasional contamination can spread unseen. Needless to say, every bacteriological care has to be taken that one is working with a pure culture.

It is advantageous to arrest growth before the stationary phase. Older cells are sometimes difficult to disrupt. An example is *Chromobacterium*; young cells break open quite readily with detergents and the yield of DNA is excellent, whereas old cells are very resistant to breakage and practically no DNA can be isolated. Common heterotrophs (Enterobacteriaceae, *Pseudomonas*, achromobacters, etc.) are usually grown for some 20 h. Slow growers (*Xanthomonas*, azotobacters, etc.) usually require 2–4 days.

We find it advantageous to dry the thoroughly washed cells by lyophiliza-tion overnight. In many cases the cells will break open much more readily and the yield of DNA is better. Lyophilization of the cells does not noticeably affect the molecular weight of the final DNA preparation; it should be stressed that solutions of DNA cannot be lyophilized without seriously decreasing the size of the molecules.

It is important to have all the cells from the different strains before making the DNA. This is easily accomplished because the cells, after growth and lyophilization, can be stored in a freezer until needed, without appreciable loss of DNA. One should plan ahead the number of organisms to be used and the number of hybridizations to be made. Common heterotrophs usually yield about 1 g of living cells (250 mg of dry cells) per Roux flask. Since DNA constitutes about 3–4% of the dry weight of many bacteria, some 8–10 mg of DNA may be expected from one Roux flask. In practice, depending on the type of organisms used about 1–6 mg of pure DNA is collected from one Roux flask. To stay on the safe side it is best to remember that one Roux flask will yield 1 mg of pure DNA. For the direct DNA–agar hybridization method, frequently 10–20 Roux flasks per strain suffice. When the competition hybridization method is used with strains that yield little DNA, such as *Rhizobium japonicum*, up to several hundred Roux flasks may be required. When using the competition method, 5 mg of DNA is needed for one set of experiments. This is usually enough to do the experi-ment at three competition levels (500, 1500 and 2500 μg) with each reference labelled DNA and to allow for some loss in shearing and absorbance measurement. More DNA will be needed from the reference strain in the competition method: about 10 mg of ordinary DNA for every 10 experiments to be carried out with the different competing DNAs. This is enough to allow for making DNA–agar and for running the necessary control curves.

III. PREPARATION OF LABELLED REFERENCE ORGANISMS

DNA is usually labelled with either ^{32}P or [14]C. Tritiated DNA is some-times used in the membrane filter method. The former isotope is much cheaper and allows higher specific activities; its disadvantage is the short halflife of 2 weeks, necessitating the frequent inclusion of controls 2-^{14}C uracil is used for incorporating [14]C. In spite of the disadvantages (the rather high cost of the starting material and the relatively low specific activity: ca. 600 counts/min μg of DNA) we prefer the use of the ^{14}C label because of its stability.

Before the labelling experiment proper, the bacteria are grown in a suitable liquid medium in a flask with sidearm to determine the shape of the growth curve. The latter is followed in a Klett colorimeter (Klett Mfg. Co., New York, U.S.A.) or any other suitable device.

In a typical labelling experiment a strain of, for example, *Pseudomonas fluorescens* is inoculated in a few broad-bottomed 1 litre flasks each containing 200 ml of medium—

Medium

Proteose peptone	10 g
Yeast extract	1 g
NaCl	5 g
Glucose	10 g
Distilled water	1 litre

pH 7·2

The organisms are grown at 30°C on a reciprocal shaker. Growth is followed turbidimetrically in a Klett colorimeter. As soon as growth becomes detectable, a solution containing 100–200 μC of 2–^{14}C uracil, sterilized cold by pressure filtration through a bacterial or membrane filter, is added per 200 ml of medium, and the organisms are allowed to grow almost to the end of the log phase. They are harvested and washed 3–4 times as usual. DNA is then prepared from the bacterial paste by the method described in Section IV. The specific activity of the labelled DNA thus prepared ranges from 600 to 1200 counts/min (in a Geiger counter)/μg DNA. The yield from 200 ml of labelled liquid medium is usually about 1–1·5 mg of pure DNA. One should plan on 50 μg of ^{14}C-DNA for every hybridization.

IV. PREPARATION OF PURE DNA

There exist several methods for preparing pure DNA. Probably the most popular is the one described by Marmur (1961), which is used by several laboratories, frequently with slight modification. It is also our method of choice and has been successfully used with a great variety of bacteria. Some strains do not break open on treatment with detergent (Duponol), but the combined action of lysozyme, followed by detergent, may disrupt these strains. In other cases, even this combination is ineffective. The ultimate DNA yield can often be improved by two or three cycles of freezing (immersion of stainless-steel tubes, containing the bacteria, in a mixture of dry ice and ethanol for 15 min) and thawing (10 min in a water bath at 20°–30°C). In some extreme cases even this may not disrupt the cells, and one has to resort to more drastic treatment, such as breakage with ultrasonic devices, use of the Hughes press or mechanical vibration. The latter methods are, however, not recommended, since the DNA molecules are sheared and their trapping later in agar becomes much more difficult. For the detailed steps of Marmur's (1961) method, we refer to the original paper, which should be read by anyone intending work in this field. Our modifications are—

1. Lyophilization of the cells before disruption.
2. Adapted procedures of cell disruption.
3. At the perchlorate stage one has to shake vigorously until there is only foam; this improves the yield of DNA and does not impair its molecular weight. In all subsequent steps treatment must be very gentle, avoid shaking and pipetting or use only broad-tipped pipettes.
4. A series of 3–6 deproteinizations is carried out one after the other; precipitation with ethanol is only effected after these steps.
5. After the RNAase treatment, 40 μg pronase (DNAase free; Calbiochem, 3625 Medford St., Los Angeles, California 90063, U.S.A.; also Lucerne, Switzerland)/ml of liquid is added and incubated overnight at 37°C.

When using four different cell types, with about 7·5 g of dry cells of each, one person can prepare pure DNA in about 1·5–2 weeks. If more strains are required, it is best to obtain the assistance of additional persons at this stage, so that the DNA is not too old when hybridizations start.

Other methods of DNA isolation and purification are reviewed by Kirby (1964). A new modification has recently been published (Kirby et al., 1967).

All DNA preparations should be as fresh and as pure as possible. Concentrated DNA solutions (1 mg or more of DNA/ml of buffer) may be stored in SSC buffer with a drop of chloroform for several months at 4°C. (SSC buffer is 0·15 M in NaCl and 0·015 M in trisodium citrate at pH 7·0. Other concentrations will be denoted as follows: 10 SSC means a ten times more concentrated buffer, 0·01 SSC a 100-fold dilution, etc.) To make labelled DNA, the purification procedure is the same, but all safety precautions should be observed and all radioactive waste should be disposed off in proper containers.

V. THE DNA–AGAR METHOD

This method was developed by the Biophysics group at the Carnegie Institution in Washington D.C. The Carnegie Institution of Washington Year Books from 1962 on should be read in addition to the papers by Bolton and McCarthy (1962), Cowie and McCarthy (1963), McCarthy and Bolton (1963), Hoyer et al., (1964; excellent summary and discussion) and McCarthy and Hoyer (1964). It is by far the most successful and widely used method today for direct measurement of genetic relatedness amongst organisms.

The principles are as follows: high-molecular single stranded DNA is fixed in space in an agar gel, which prevents it from renaturation. It is mixed with a small amount of low-molecular-weight, single-stranded labelled DNA in solution. The mixture is incubated for a sufficient time (usually overnight) at an ionic strength and temperature allowing hybridization. Part, if not most, of the labelled DNA will bind to its homologous complement of agar-fixed DNA. All the heterologous and part of the homologous labelled DNA will renature in solution. The latter fraction is eluted at the same ionic strength and temperature. Upon increasing the temperature of the DNA-agar and decreasing the ionic strength of the eluent, the labelled DNA will "melt" from its agar-linked DNA counterpart; it is eluted and measured. In the direct method, the labelled DNA of strain A is mixed with DNA–agar from strains A, B, C, etc. The percentage hybridization of the labelled DNA with ordinary DNA is taken as 100% homology, and the percentage hybridizations with other DNA types are compared to that value. In the competition method, labelled and ordinary DNA–agar from the same strain is always used in the presence of a large excess of single-stranded, low-molecular-weight, ordinary DNA from the same (control) and other strains (see Section VC).

A. Preparation of stock DNA solutions

1. Stock solution for the preparation of DNA–agar

Dissolve the purified DNA from about 7·5 g of dry cells, after the isopropanol precipitation step according to Marmur (1961), in 8 ml of 01· SSC buffer. Add a drop of chloroform. The higher the molecular weight of DNA, the better for trapping in agar, and the more slowly it will dissolve. After several days the solution is homogeneously viscous (solution B). Stirring should be avoided in case the DNA molecules get broken. Dilute 1 ml of this solution with 5–10 ml of 0·1 SSC buffer; several days are required before complete solution (solution A). The concentration of solution A is measured at 260 nm (52 mg of pure sodium salt of DNA/ml has an absorbance of 1·0 at 260 nm). Solution A is adjusted to 1·5–2 mg of DNA/ml with 0·1 SSC buffer. All DNA should be completely dissolved before use. The adjusted solution A is used for preparing DNA–agar. At this stage it is important to check the molecular weight of the DNA by ultracentrifugation. The molecular weight should preferably be around 10×10^6. Samples with molecular weights of $1 \times 10^6 – 2 \times 10^6$ or less should be discarded, because DNA will not be trapped sufficiently firmly in the agar.

2. Stock solution of competing DNA

The concentration of DNA in solution B is determined at 260 nm after appropriate dilution. There should be 5–9 mg of DNA/ml of solution B.

Shear 5–10 ml in the French Pressure Cell (American Instrument Co., Silver Springs, Md., U.S.A.) at 12,000 p.s.i. Solution B is sometimes so viscous that it has to be put in the cell with a spoon. The sheared DNA is not very viscous. Its molecular weight should be 3×10^5–5×10^5. The exact concentration is determined after gentle mixing by measuring the absorbance at 260 nm. Competing DNA is denatured just before use by keeping it at 106°C in a closed screw-capped bottle for 6 min and quickly cooling it in ice. It does no harm to denature the sheared material several times.

3. [^{14}C]DNA stock solution

After the last isopropanol precipitation, dissolve the ^{14}C-DNA in 2–4 ml of 0·1 SSC buffer. Now adjust the solution to 1 SSC buffer, and take three 0·05 ml samples. One is used for determining the concentration at 260 nm, and the other two are used for radioactive counting. The specific activity can thus be calculated. It should be at least 500 counts/min/μg of DNA. After addition of a drop of chloroform the solution may be stored at 4°C. To prepare sheared and denatured ^{14}C-DNA for hydridization, add 1 ml of the above stock solution to 4 ml of distilled water in the French Pressure Cell. Stir with spatula and shear at 12,000 p.s.i. Denature the sheared solution in 0·2 SSC at 106°C in an oil or glycerol bath for 6 min in a sealed Pyrex tube and quickly cool it in an ice bath. The denaturation temperature does not always have to be so high: it depends on the mean molar guanine and cytosine percentage (%GC) of the DNA used and the ionic strength of the SSC solution. It is advisable to be at least 15°C above the melting temperature in the buffer in which the DNA is dissolved. The denaturation temperature can be estimated as follows. We found experimentally that the relation between Tm_x (melting point in x SSC buffer; e.g. for the normal buffer as given above, $x = 1$; for a 10-fold diluted buffer, $x = 0·1$) and the other parameters is—

$$Tm_x = 0·41 \times \text{percentage GC} + 69·3 + 15·0 \log x.$$

The denaturation temperature, Td_x has thus to be at least—

$$Td_x = 0·41 \times \text{percentage GC} + 84 + 15·0 \log x.$$

One can eliminate the calculation by always denaturing in 0·1 SSC at 106°C, no matter which strain is used. After denaturation, 0·05 ml is removed, dried on a planchet and counted to determine the concentration of DNA. The volume of the sheared, denatured DNA solution is determined and adjusted with distilled water and 10 SSC buffer to obtain a final concentration of 50μg of ^{14}C-DNA/ml of 2 SSC buffer. This solution usually keeps up to 2 months when stored at 4°C in the presence of a drop of chloroform. It is occasionally denatured again before use.

B. Preparation of DNA–agar

This is a modification of the method of Bolton and McCarthy (1962).

DNA–agar loses its hybridization capacity rapidly: after 1 week, hybridization values may drop several percent. It should be prepared just before use and only in quantities sufficient for 1 week's experiments.

To prepare the right amount of DNA–agar with the suitable concentration of DNA, a few simple calculations can be used. The total volume required is obtained by dividing the concentration of DNA in μg/ml of adjusted solution A by 650. The amount of solid agar (in mg) to be added is obtained by multiplying the result by 30. Example: the adjusted solution A contains 1,500 μg of DNA/ml. The total volume of liquid will be $1{,}500 \div 650 = 2{\cdot}305$ ml. The amount of agar to be added will be $2{\cdot}305 \times 30 = 69{\cdot}6$ mg. Weigh out the agar (Oxoid No. 3) in a 10–20 ml Pyrex screw-capped bottle, and add 1·305 ml of distilled water. Dissolve the agar by heating it in an oil or glycerol bath at 106°C for 10–15 min. Open the bottle slowly, because if it is opened too quickly, the solution will boil over, since it is under pressure. Add 1 ml of the adjusted DNA stock solution A. The total volume is now the required 2·305 and the buffer concentration is now 0·043 SSC. Shake vigorously 20–30 times. Denature for 6 min. at 106°C. Shake again 20–30 times. The bottle is opened carefully and the contents are poured into a large beaker which is standing in crushed ice. The empty bottle is closed and is also placed at once in the ice bath. After the agar has solidified, it is collected from both beaker and bottle. Press it twice through a 56 mesh stainless-steel screen. Transfer the agar particles to a water-jacketed column and wash them for about 2 h with 2 SSC at 60°C or at whatever temperature the hybridization will be carried out. The washing rate should be about 2 ml/min. All loose DNA is thus removed and the adhering DNA is now in 2 SSC. The last liquid is gently expressed with suction pear or compressed air.

The concentration of DNA in the agar is determined as follows. Weigh out 200 mg of agar in a 5 ml test tube. Add 1·8 ml of 5 M $NaClO_4$ and heat the mixture for 5 min in a boiling-water bath with a marble on top of the tube opening. Shake the mixture well, and spin it in a small centrifuge tube for 15 min at 12,000 g. The absorbance of the supernatant is read against a $NaClO_4$ blank—

$$(\text{Absorbance} - 0{\cdot}173)\, 500 = \text{mg of DNA/g of agar.}$$

where 0·173 is the absorbance due to agar alone. It should be determined separately for each new batch.

To make DNA–agar takes about 1 day.

C. The competition hybridization method

The principle is as follows. A certain amount of sheared, denatured ^{14}C-DNA is allowed to hybridize with a 10-fold excess of high-molecular-weight denatured DNA in agar in the presence or absence of sheared, denatured DNA from the same or different strains. We use 0, 20, 40, 60, 80 and sometimes 160 times the amount of ^{14}C-DNA as competing DNA Usually 40–60 times suffices. A competition curve is represented in Fig. 1.

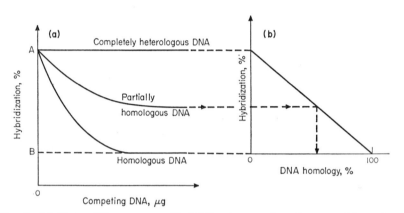

FIG. 1. (a) Competition curves with DNA of a varying degrees of homology. (b) Procedure for calculating the percentage DNA homology.

1. Preliminary estimation of the hybridization volume

With a specific activity of around 600 counts/min/μg of ^{14}C-DNA, as little as 20 μg of ^{14}C-DNA can be used. A good range for sufficient precision in counting and to avoid wasting labelled DNA is about 20–40 μg.

One has first to decide on the total volume to work with. Good hybridization is still obtained if the total volume of the system (in ml) is 1·6 times the weight of DNA agar, but this is not always feasible because of the large amounts of competing DNAs to be added, particularly when using two or more types of competing DNAs in the same mixture. The ratio of liquid volume/agar should be kept as low as possible.

2. Determination of the control curve

This is carried out in the presence of varying amounts of competing DNA of the reference strain itself. For example—

Stock solution of ^{14}C-DNA from reference strain A: 50 μg of DNA/ml in 2 SSC; specific activity, 600 counts/min/μg of DNA.
DNA agar from strain A: 500 μg of DNA/g of agar in 2 SSC.
Competing DNA from strain A, sheared, denatured, in 0·1 SSC, at concentration 5 mg/ml.

Hybridization systems

Volume of ^{14}C-DNA (25 μg), ml	0·5	0·5	0·5	0·5	0·5
Weight of agar (250 μg DNA), g	0·5	0·5	0·5	0·5	0·5
Competitive DNA					
in μg	0	500	1000	1500	2000
in ml	0	0·1	0·2	0·3	0·4
Buffer					
Volume of 10 SSC, ml	0	0·024	0·048	0·072	0·095
Volume of 2 SSC, ml	0·495	0·371	0·247	0·123	0

In this example the total volume of the liquid is thus 0·995 ml or twice the amount of agar. In these conditions the degree of hybridization will be less than optimal, but still workable. The smaller the ratio total volume/ amount of agar, the faster and higher the hybridization. It is thus obvious that the concentration of competing DNA in the stock solution should be as high as possible.

Each system is contained in a small tightly closed 5 ml screw-capped bottle, completely immersed in a water bath at the desired temperature (see below). When the liquid/agar ratio is rather high, hybridization is improved by rotating the vessel gently (60 rev/min).

3. Competition experiments with foreign DNA

Under the conditions described above, the competition curve usually levels off at about 1500 μg of the reference DNA and the same amount of foreign DNA. It is advisable though to test several levels of competing DNA.

4. Hybridization temperature

Following the method of McCarthy and Bolton (1963), hybridizations are carried out at 60°C. This temperature is only a guide. The higher the percentage GC, the higher the hybridization temperature should be. As a general guidance, the following table summarizes the maximal possible hybridization temperature in 2 SSC for organisms with different percentage GC—

GC, %	T_m in 2 SSC, °C	Approximate beginning of denaturation; highest possible temperature of hybridization, °C	Approximate optimal temperature of hybridization, °C
30	86	76	61
40	90	80	65
50	94	84	69
60	98	88	73
70	102	92	77

This aspect of the process is not yet completely understood.

5. *Duration of hybridization*

This depends on several factors, but largely on the concentration of DNA used. It can be estimated by using the C_0t values of Britten and Kohne (1966), in which C_0 is the total concentration of DNA expressed as moles of bound nucleotides (sodium salt; molecular weight, 331)/litre and t is time of hybridization (sec) after initiation of reaction. From Fig. 39 of Britten and Kohne (1966) one reads for *E. coli*—

C_0t in SSC buffer	Percentage hybridization
1	11
2	20
5	47
10	56
20	70
30	78
100	93

In 2 SSC buffer the reaction proceeds three times as fast.

In the experiment described above with 25 μg of ^{14}C-DNA and 250 μg of DNA in agar in 0·995 ml, C_0 is thus $M/1200$ A C_0t of 1 is thus reached after $1200/60 = 20$ min. By hybridizing for 16 h, $C_0t = 48$, and some 85% of the DNA will have hybridized. One can again see that the hybridization volume is important. By carrying out the above experiment in 2, 3 or 4 ml instead of 1 ml, the time to reach 85% hybridization would automatically double, triple or quadruple.

In the concomittant vessel with ca. 2000 μg of competing DNA, C_0 is $M/145$ and $C_0t = 1$ is reached after about 2·5 min. After 16 h hybridization is thus certainly completed.

D. The direct hybridization method

Labelled DNA from the reference strain is hybridized with an excess of DNA in agar as described before. There are two main differences between this and the competition method. First, obviously no competing DNA is used, and secondly the labelled reference DNA is hybridized not only with high-molecular-weight, single-stranded DNA of the reference strain in agar, but with similar DNAs from the other strains as well, likewise in agar. When for the competition method one has to prepare only one type of DNA–agar, for the direct method one has to prepare as many DNA–agars as there are strains to be studied. All the technical details are otherwise the same in both cases.

E. Collecting fractions

After the required incubation period, the contents from the hybridization bottles are quantitatively transfered with 2 SSC to a water-jacketed column, 16 mm i.d. × 18 cm long. We use a coarse fritted-glass disc, fused into the base of the column as a support. Before the column is filled, a disc of Schleicher & Schüll No. 589–2 filter paper and a layer of glass wool about 0·5 cm thick are put on top of the glass filter, in order to prevent the agar granules from being washed through. One can of course also use other devices, such as a hollow Teflon stopper with a fine silk, nylon or metal screen as column support. The column is kept at, say, 60°C with a circulating water bath. The agar is washed with 2 SSC solution at the same temperature at a rate of 10 ml/6 min, until 120 ml of solution is collected in 10 ml fractions. The fractions are collected in calibrated tubes and 1 ml of each fraction is dried on a planchet for counting. It is not advisable to use larger samples in order to prevent high self-absorption by the salts present during counting. To each planchet has been added already 0·05 ml of a 10% Teepol (Shell International Chemical Co., Ltd) detergent solution to allow more uniform drying. These fractions serve to remove all labelled DNA which is not bound on the agar. The agar is now washed with 0·01 SSC at 75°C to collect the hybridized fraction, at the same elution rate. Nine 10 ml fractions are collected, and 2 ml of each sample are dried on a planchet with the same amount of Teepol. The self-absorption of this small amount of salt is negligible. To save time, the samples of tubes 1, 2 and 9 go on separate planchets, and tubes 3 to 8 together onto 1 planchet. When the planchets are dry (infrared lamp) they are placed on the counter and each planchet is counted twice for 6 min.

The "tea bag" method (McCarthy and Hoyer, 1964)

Instead of the hybridization mixture being placed in a water-jacketted column, it is put in a silicone-treated glass tube 1 cm dia × 18 cm long, one end of which is capped by a 52 mesh Saran screen, attached by an O ring. tube is placed into another one 1·5 cm dia × 18 cm long and held some 3 cm from its bottom by some support. The contents of the hybridization vessel are quantitatively transferred to the inner tube with 15 ml of 2 SSC at 60°C. There are a total of 10 such 18 × 1·5 cm tubes, all in a water-bath at 60°C. Nine of them contain 15 ml 2 × SSC. After 15 min, the screen-capped tube with the agar is lifted from the first washing tube, drained and transferred to the next one. Nine such sequential transfers over a period of about 2 h will effectively remove all unbound nucleic acid. The temperature of the water bath is now increased to 70°C, and the screen-capped tube is now serially transferred to a set of five washing tubes with 15 ml of 0·01 × SSC to remove all hybridized DNA. Sampling for counting is carried out as described above.

F. Calculations

Let the total number of counts in the 12 wash planchets be w, the average background count be b and the total number of counts in the four hybrid planchets be h. Determine once and for all the correction factor (f) for self-absorption due to the dried 2 SSC solution in the washes (the factor is about 1·14). Now—

$$A = \tfrac{1}{6}\,(w - 12b) \times 10 \times f$$

where A is the total counts/min in the wash liquid, and —

$$B = \tfrac{1}{6}\,(h - 4b) \times 5$$

where B is total counts/min in the hybrid fraction. Further, $A + B = C$, where C is the total counts/min used, which should be the same as that calculated from the amount of labelled DNA used. The percentage hybridization in each particular experiment is $100\,\dfrac{B}{C}$.

To calculate the percentage DNA homology by the direct hybridization method, suppose that the percentage hybridization of the labelled DNA with the DNA–agar from the same reference strain is $r\%$, and that the percentage hybridization of the labelled DNA with the DNA–agar from an unknown strain is $x\%$. The percentage DNA homology is then $100\,\dfrac{r}{x}$. This means that $100\,\dfrac{r}{x}\%$ of the reference DNA is homologous to the unknown DNA, but the converse is not true. In order to know which fraction of the unknown DNA is homologous, one has to know the ratio between the total molecular weight of both types of chromosomal DNA.

To calculate the percentage DNA homology by the competition method, the percentage hybridization value in the absence of competing DNA is defined as 0% homology and the value of the flat, saturated part of the competition curve as 100% homology; the results are plotted on a graph as shown in Fig. 1. From the competition experiment with unknown DNA, the percentage homology can now be read off from Fig. 1 (*b*).

G. Comparison between the direct and the competition methods

The competition method offers the advantage that one always uses more labelled DNA and DNA-agar from the same batch; the only variable is the competing DNA. The reproducibility of this method is of the order of 1–2% homology. It has the technical disadvantage that large amounts of competing DNAs are required from every strain. In most cases the preparation of this DNA does not present difficulties. There are some organisms, however, that grow poorly, slowly or that are difficult to disrupt. In these cases it may be necessary to handle several hundred Roux flasks or up to 100 litres of culture medium in order to obtain sufficient DNA.

The direct method requires less DNA, but suffers from the disadvantage that the labelled DNA is hybridized each time with a DNA–agar from a different make. The quality of DNA agar may vary from one batch to another, depending on the size of the DNA embedded and on several other factors that are as yet not fully understood. The reproducibility of the direct method is usually 1–6%, sometimes it may be as high as 13%. In spite of the fact that sometimes more labour is involved for the competition method, it is at the present time our method of choice.

VI. THE MEMBRANE-FILTER TECHNIQUE

The technique of Gillespie and Spiegelman (1965) for detecting RNA complementary to DNA by hybridizing the RNA to DNA bound on nitrocellulose membrane filters has inspired some modifications to detect complementary DNA.

A. The albumin-coated filter method

This is the modification of Denhardt, (1966).

Denatured DNA sticks to nitrocellulose membranes. This can be prevented by coating them with albumin. The binding of denatured DNA to complementary denatured DNA, previously attached to the filter, is not hampered by the coating. We shall report here the details of the method as given by Gillespie and Spiegelman (1965) and Denhardt (1966).

Use Schleicher and Schüll Type B–6 coarse or Millipore HAWP 25 mm filters. Pre-wash the filter with 10 ml of 6 SSC buffer. Pass the solution of denatured DNA in 5 ml of 6 SSC through the filter at a rate of 5 ml/min. Wash the filter with 50 ml of 6 SSC, dry it overnight in a vacuum dessicator and further dry it at 80°C in a vacuum oven. About 95% of the DNA remains attached to the filter and less than 1% /day is washed out on subsequent use for hybridization.

Effect coating with 1 ml of a solution containing 0·02% Ficoll (Pharmacia, Uppsala, average molecular weight 400,000), 0·02% polyvinylpyrrolidone (Sigma, St. Louis, Mo., U.S.A., average molecular weight 360,000) and 0·02 Sigma bovine albumin, fraction V, in 3 SSC for 6 h at 65²C.

Prepare labelled DNA either with thymidine-requiring strains by growth on 1 μC of ^{3}H-thymidine/ml, with ^{32}P or with 8–[14]C-adenine; heat-denature it as described above, bring it up to 6 SSC and add it to the vials containing the filter. Continue the incubation for another 12 h. Everything that was said above on incubation time and temperature should also be applicable here, although it has not yet been tested in these conditions.

After incubation, wash each side of the filter with 40 ml of SSC. Dry the filters and count them in toluene–liquifluor in a scintillation spectrometer. In the best conditions reported, with a 500-fold excess of E. coli–DNA on the filter over the labelled DNA, only 18% of the latter was bound.

B. Removing non-hybridized DNA at low ionic strength and high pH

This is the modification of Warnaar and Cohen (1966).

Use Millipore filters HA or Membrane filters MF 30, 22 mm dia (Membrane Filter Gesellschaft, Göttingen). Prepare and load them with DNA as described by Gillespie and Spiegelman (1965). Effect hybridization in small screw-capped vials containing 3·2 ml of 1·25 SSC, pH 7·0, buffered with 10^{-2}M Tris–HCl and 1 μg of labelled, sheared and denatured DNA, together with a DNA-loaded filter. It is advisable to keep the volume as small as possible. Incubation time should be about 1 day. For good hybridizations, the fitter should contain at least 40 μg of DNA. An incubation temperature of 60°C is suitable, although—as pointed out above—it would be expected to vary with DNAs of different percentage GC. After incubation, remove the filters and rinse them in 3×10^{-3}M Tris–HCl at pH 9·4. Wash them on both sides by suction with 100 ml of the same buffer. This removes single-stranded DNA but not the hybridized one. Dry the filter and count it as usual.

Both modifications still need a thorough check with bacterial DNAs before they can be applied reliably to problems of bacterial taxonomy. Some aspects of the reaction rate between labelled DNA and filter-bound DNA and of the thermal stability of the duplexes have been studied by McCarthy (1967).

VII. THE ULTRACENTRIFUGAL CsCl GRADIENT METHOD

This technique was amongst the first to be used to detect DNA hybrids (Meselson et al., 1957; Schildkraut et al., 1961). It is now only rarely used for routine taxonomic studies, for reasons that will become obvious below. For completeness sake we shall present a brief account of the method.

The principle is that high-molecular-weight, single-stranded ordinary DNA of several strains are hybridized with high-molecular weight, single-stranded, heavy-labelled DNA of one or more reference DNAs. Since the hybrids have a specific gravity in between the heavy and ordinary DNA, they can be separated in a CsCl gradient by an analytical centrifuge.

Heavy-labelled organisms are usually grown in D_2O media with $(^{15}NH_4)_2 SO_4$ as nitrogen source (Marmur and Schildkraut, 1961). This works well with bacteria having simple nutritional requirements, but it is more difficult for more fastidious bacteria. One may also encounter all sorts of unexpected difficulties. As an example, we may cite our experience with *Xanthomonas* (Friedman and De Ley, 1965). Xanthomonads grow very well on a complex medium with 1 % yeast extract and 1% glucose. To prepare heavy DNA, a simple medium had to be developed with $(NH_4)_2SO_4$ as sole nitrogen source. The bacteria had to be adapted to increasing D_2O concentrations by transfer in media containing 0, 20, 40, 60, 80 and 100% D_2O. Not all the strains will grow and a selection has to be made. The more D_2O in the medium the more slowly they grow. Fully adapted organisms grow only on media solidified with agar; after one transfer in liquid D_2O medium they die. Full growth may take up to 2 weeks, whereas on ordinary media a few days suffice. A full adaptation period takes at least 1 month. Inoculation of Roux flasks has to be done from slants on agar with a loop. Upon growth, the entire surface has to be smeared with a sterile triangular spatula. Sterilization of all D_2O media has to be carried out in a pressure cooker containing D_2O. Preparation of the heavy DNA proceeds as described above. Another example of preparing deuterated, ^{15}N-labelled DNA from acetic acid bacteria is described by De Ley and Friedman (1964).

For hybridization, a modification of the method of Schildkraut et al. (1961) is used. Denature a mixture of heavy and ordinary DNA (6–20 μg each) in 1 ml of 1·9 SSC buffer in a small stoppered vial at 105°C for 10 min, Maintain it for 2 h in a water bath at 75°C and an additional hour at 70°C. after which the temperature should be lowered in steps of 5°C every 15 min to 25°C. Next, remove single strands by adding phosphodiesterase (Lehman, 1963). To do this, first dialyse the hybridization mixture against two changes of 500 ml of 0·07M glycine buffer, pH 9·2, once for 4 h and the second time overnight. Check the concentration by absorbance determination. To 0·7 ml of dialysed DNA solution add 0·05 ml of a 0·07M glycine buffer, pH 9·2 (containing 400 μg of $MgCl_2$ and 1/200 diluted mercaptoethanol) and 0·1 ml of phosphodiesterase solution (prepared according to Lehman, 1963) (10 units). Incubation is for 8 h at 37°C. The activity of the enzyme should be checked separately for each batch, since it declines with time. To 0·85 ml of the phosphodiesterase-treated DNA solution add ca. 1 μg in 0·01 ml of a reference DNA that is at least either 0·03ρ buoyant density heavier or lighter than the heaviest or lightest component of the hybridization mixture. Working with high percentage GC DNAs we always use DNA from *Cytophaga* sp. NCMB 292, which has a buoyant density ρ of 1·6931. Next add solid CsCl (No. 2041; E. Merck, Darmstadt, Germany) to obtain a suitable refractive index, measured with an Abbe 60 refractometer (Bellingham & Stanley, London) at 25°C. This brand of CsCl contains only traces of ultraviolet-absorbing substances and need not to be purified. The refractive index of the solution is adjusted so that all bands will be in the picture. The relation between buoyant density, ρ^{25}, and refractive index n_D^{25} is—

$$\rho^{25} = 10\cdot8601\,n - \tfrac{52}{D} - 13\cdot4974$$

Effect centrifugation in a model E Spinco Analytical ultracentrifuge at 44,770 rev/min for 22 h at 25°C, with rotor An-D, cells with Kel-F centrepiece and green anodized 1° negative wedge window. Before the runs, carefully clean and adjust the optics of the machine is, according to the Beckman Technical Bulletin TB 6101, August 1961. This is essential to obtain good pictures. The rotor chamber should be essentially free from oil. Take photographs with ultraviolet light on Kodak sheet film CF8 and develop them with Kodak Microdol. A cardboard mask below the Cl_2 and Br_2 filters is required to eliminate stray light and to obtain an even background. Make tracings of the pictures (enlarged 5 ×) with a Joyce-Loebl double beam recording densitometer MKIIIB with an effective slit width of 0·5 mm. The densities (ρ) of the DNA bands are calculated graphically by reference to the density ρ_0 of *Cytophaga* DNA using the formula—

$$\rho = \rho_0 + 4\cdot2\,\omega^2\,(r - r_0^2) \times 10^{-10} \text{g/cu.cm.}$$

where ω is the angular velocity in radians/sec and r and r_0 are the distances of the DNA bands from axis of rotation.

The method is expensive in investment of equipment, expensive in use, because of the large amounts of D_2O needed and the costs of the drives, is time consuming in calibration and frustrating since not all strains are amenable to growing in heavy media. The interpretation is sometimes difficult when the labelled DNA is not heavy enough and lies too close to the ordinary DNA, thus overlapping and covering the hybrid. Hybridization is also far less complete than in the agar method, because high-molecular-weight DNA is used and even small imperfections in nucleotide sequence can prevent extensive hybridizations. In our opinion and experience, this method cannot compete with the agar method.

VIII. INTRODUCTORY READING ON DNA HYBRIDIZATION WITH DIFFERENT ORGANISMS

Plants: Carnegie Institution Year Book 1964; Bendich and Bolton (1967).

Animals: Carnegie Institution Year Books from 1962 on; McLaren and Walker (1965); Walker and McLaren (1965); De Ley and Park (1966a).

Enterobacteriaceae: Schildkraut *et al.* (1961); Goodman and Rich (1962); McCarthy and Bolton (1963); Brenner *et al* (1967).

Pseudomonas, Xanthomonas: De Ley and Friedman (1965); Friedman and De Ley (1965); De Ley *et al.* (1966); Park and De Ley (1967).

Free-living nitrogen-fixing bacteria: De Ley and Park (1966b).

Acetic acid bacteria: De Ley and Friedman (1964).

Bacillus: Marmur *et al* (1963); Doi and Igarashi (1965; 1966); Takahashi *et al.* (1966).

Agrobacterium, Rhizobium, Chromobacterium: Heberlein *et al* (1967).

Streptococcus: Weissman *et al.* (1966).

Neisseria: Kingsbury and Duncan (1967).

Mycoplasma: McGee *et al.* (1965); Rogul *et al* (1965); Reich *et al.* (1966a, b).

Pasteurella: Ritter and Gerloff (1966).

Streptomyces: Tewfik and Bradley (1967).

Actinomycetes: Yamaguchi (1967).

Phages: Carnegie Institution Year Book from 1962 on; Hall and Spiegelman (1961); Schildkraut *et al.* (1962); Meinke and Jones (1967).

REFERENCES

Bendich, A. J., and Bolton, E. T. (1967). *Pl. Physiol.* **42**, 959–967.
Bolton, E. T., and McCarthy, B. J. (1962). *Proc. natn. Acad. Sci. U.S.A.*, **48**, 1390–1397.
Brenner, D. J., Martin, M. A., and Hoyer, B. H. (1967). *J. Bact.*, **94**, 486–487.
Britten, R. J., and Kohne, D. E. (1966). "Annual Report of the Director of the Department of Terrestrial Magnetism", Carnegie Institute Year Book 65, pp. 78–106.
Cowie, D. B., and McCarthy, B. J. (1963). *Proc. natn. Acad. Sci. U.S.A.*, **50**, 537–543.
De Ley, J., and Friedman, S. (1964). *J. Bact.*, **88**, 937–945.
De Ley, J., and Friedman, S. (1965). *J. Bact.*, **89**, 1306–1309.
De Ley, J., and Park, I. W. (1966a). *Nature, Lond.*, **211**, 1002.
De Ley, J., and Park, I. W. (1966b). *Antonie van Leeuwenhoek*, **32**, 6–16.
De Ley, J., Park, I. W., Tijtgat, R., and Van Ermengem, J. (1966). *J. gen. Microbiol.*, **42**, 43–56.
Denhardt, D. T. (1966). *Biochem. biophys. Res. Commun.*, **23**, 641–646.

Doi, R. H., and Igarashi, R. T. (1965). *J. Bact.*, **90**, 384–390.
Doi, R. H., and Igarashi, R. T. (1966). *J. Bact.*, **92**, 88–96.
Friedman, S., and De Ley, J. (1965). *J. Bact.*, **89**, 95–100.
Gillespie, D., and Spiegelman, S. (1965). *J. molec. Biol.*, **12**, 829–842.
Goodman, H. M., and Rich, A. (1962). *Proc. natn. Acad. Sci. U.S.A.*, **48**, 2101–2109.
Hall, D. B., and Spiegelman, S. (1961). *Proc. natn. Acad. Sci. U.S.A.*, **47**, 137–146.
Heberlein, G., De Ley, J. and Tijtgat, R. (1967). *J. Bact.*, **94**, 116–124.
Hoyer, B. H., McCarthy, B. J., and Bolton, E. T. (1964). *Science, N.Y.*, **144**, 959–967.
Kingsbury, D. T., and Duncan, J. F. (1967). *Bact. Proc.*, p. 40.
Kirby, K. S. (1964). *Prog. Nucleic Acid Res. molec. Biol.*, **3**, 1–31.
Kirby, K. S., Fox-Carter, E., and Guest, M. (1967). *Biochem. J.*, **104**, 258–262.
Lehman, I. R. (1963). In "Methods in Enzymology" (Ed. S. P. Colowick and M. O. Kaplan), 40–43. Academic Press, New York.
Marmur, J. (1961). *J. molec. Biol.*, **3**, 208–218.
Marmur, J., and Schildkrant, C. L. (1961). *Nature, Lond.*, **189**, 636–638.
Marmur, J., Seaman, E., and Levine, J. (1963). *J. Bact.*, **85**, 461–467.
McCarthy, B. J. (1967). *Bact. Rev.*, **31**, 215–229.
McCarthy, B. J. and Bolton, E. T. (1963). *Proc. natn. Acad. Sci. U.S.A.*, **50**, 156–162.
McCarthy, B. J., and Hoyer, B. H. (1964). *Proc. natn. Acad. Sci. U.S.A.*, **52**, 916.
McGee, Z. A., Rogul, M., Falkow, S. and Wittler, R. G. (1965). *Proc. natn. Acad. Sci. U.S.A.*, **54**, 457–461.
McLaren, A., and Walker, P. M. B. (1965). *Genet. Res.*, **6**, 230–247.
Meinke, W. J., and Jones, L. A. (1967). *Bact. Proc.*, p. 155.
Meselson, M., Stahl, F. W., and Vinograd, J. (1957). *Proc. natn. Acad. Sci. U.S.A.*, **43**, 581–584.
Park, I. W., and De Ley, J. (1967). *Antonie van Leeuwenhoek*, **33**, 1–16.
Riech, P. R., Somerson, N. L., Rose, J. A., and Weissman, S. M. (1966a). *J. Bact.*, **91**, 153–160.
Reich, P. R., Somerson, N. L., Hybner, C. J., Chanock, R. M., and Weissman, S. M. (1966b). *J. Bact.*, **92**, 302–310.
Ritter, D. B., and Gerloff, R. K. (1966). *J. Bact.*, **92**, 1838–1839.
Rogul, M., McGee, Z. A., Wittler, R. G., and Falkow, S. (1965). *J. Bact.*, **90**, 1200–1204.
Schildkraut, C. L., Marmur, J., and Doty, P. (1961). *J. molec. Biol.*, **3**, 595–617.
Schildkraut, C. L., Wierzchsowski, K. L., Marmur, J., Green, D. M., and Doty, P. (1962). *Virology*, **18**, 43–55.
Takahashi, H., Saito, H., and Ikeda, Y. (1966). *J. gen. appl. Microbiol.*, **12**, 113–118.
Tewfik, E., and Bradley, S. G. (1967). *Bact. Proc.*, p. 40.
Warnaar and Cohen (1966). *Biochem. biophys. Res. Commun.*, **24**, 554–558.
Walker, P. M. B., and McLaren, A. (1965). *J. molec. Biol.*, **12**, 394–409.
Weissman, S. M., Reich, P. R., Somerson, N. L., and Cole, R. M. (1966). *J. Bact.*, **92**, 1372–1377.
Yamaguchi, T. (1967). *J. gen. appl. Microbiol.*, **13**, 63–71.

CHAPTER XII

Hybridization of Microbial RNA and DNA

J. E. M. MIDGLEY

*Department of Biochemistry, University of Newcastle upon Tyne,
Newcastle upon Tyne, England*

I. General 331
II. Experimental Methods 333
    A. Preparation of DNA from microbial cultures . . . 333
    B. Preparation of RNA 336
III. Hybridization techniques 339
    A. Agar gels 339
    B. Cross-linked DNA gels 341
    C. Cellulose nitrate membrane filters 342
IV. Experimental Details of Hybridization (Gillespie and Spiegelman, 1965) 342
    A. Binding of DNA to filter 342
    B. Hybridization 343
V. Minimization by Ribonuclease of Nonspecific RNA Absorption . 347
VI. Arbitrary Nature of Filter Support 348
VII. Analytical Plans 348
    A. Messenger RNA 348
    B. Fractionation of messenger RNA 349
    C. Ribosomal RNA 350
VIII. Analysis of Hybridization Curves 350
    A. Saturation of DNA hybridizing sites with excess RNA . 351
    B. Estimations of the efficiency of formation of DNA–RNA hybrids 354
IX. Competition Experiments 355
Acknowledgments 359
References 359

I. GENERAL

The technique of hybridizing RNA to strands of denatured DNA has found widespread use in the estimation and characterisation of microbial RNAs. Early evidence demonstrating the feasibility of such methods came from Rich (1960), who showed that stable double stranded hybrids of

polyriboadenylic and polydeoxyribothymydylic acids could be formed in
solution from mixtures of single stranded molecules, in suitable conditions
of temperature and ionic strength. This work was extended to naturally
occurring nucleic acids, when hybridization was demonstrated between the
denatured single stranded DNA of the T2 bacteriophage and isotopically
labelled RNA formed after infection of *Escherichia coli* by T2 (Hall and
Spiegelman, 1961). The original method for hybrid detection depended
on the centrifugation of caesium chloride density gradients. This somewhat laborious technique involved the detection of the DNA–RNA hybrid
as a separate band after centrifugation of the density gradient to equilibrium. The strategy of hybridization of RNA to denatured DNA free in
solution was continued by Hayashi and Spiegelman (1961) and by Yankofsky
and Spiegelman (1963) to show the presence of messenger RNA in total
bacterial RNA and also the existence of genetic elements in bacterial DNA
responsible for the transcription of bacterial ribosomal RNA.

Because of the time-consuming manipulations involved in hybrid detection, more flexible methods were needed for the accurate analysis of
DNA–RNA hybrids. Several such methods were developed which require
the denaturation of DNA into single strands, its immobilisation in this
state upon an inert support, and the hybridization of the RNA to the
immobilised DNA in suitable conditions. Then thorough washing of DNA
+ support is sufficient to remove nonhybridized RNAs and the hybridized
RNA may be estimated. Supports which have been successfully employed
in this technique include phosphocellulose acetate (Bautz and Hall, 1962),
agar gels (Bolton and McCarthy, 1962), denatured DNA itself made into
an insoluble gel by crosslinkage of strands under the influence of ultraviolet irradiation (Britten, 1963), and cellulose nitrate membrane filters
(Gillespie and Spiegelman, 1965). Modifications of this technique have
been described in which DNA–RNA hybrids were first formed in free
solution and were then collected by filtration through cellulose nitrate
membrane filters (Nygaard and Hall, 1963; McConkey and Dubin, 1966).
In this case, the filters retained the DNA–RNA hybrids, whereas unhybridized RNA passed through.

It is proposed to examine briefly some of these methods to determine
which, in the author's opinion, are most flexible, accurate and uncomplicated. The technique best suited for measuring the degree of hybridization
of RNA to DNA must fulfil the following criteria:

(1) It must be applicable over a wide range of DNA/RNA weight ratios
used in tests, and must not depend on the identity of the DNA used.
(2) It must allow a ready analysis of a large number of samples simultaneously.

(3) It must give maximum reproducible efficiency of DNA–RNA hybridization, with minimal interference from DNA–DNA interactions.

(4) It must allow a reproducible accuracy of hybrid detection of at least 1–2% of the RNA added to the system.

(5) It must show minimal inaccuracies from the retention of RNA upon any material (including the support) other than DNA.

(6) It must be convenient to perform, with minimal losses and corrections for experimental mechanics etc.

Using these criteria, the author has chosen to disregard methods involving the formation of DNA–RNA hybrids in free solution as, although these methods are rapid, their success depends to a large extent on freedom from competition by DNA–DNA hybridization. In the literature (e.g. Mangiarotti and Schlessinger, 1968), lack of absolute reproducibility of results has been experienced using this technique, and in any case, the efficiencies of hybridization have often been considerably lower than quoted here. A recent detailed analysis of the methods of hybridization of DNA and RNA has emphasized the relative inflexibility of this approach (Kennell, 1968; Kennell and Kotoulas, 1968). Further, McConkey and Dubin (1966) reported high nonspecific absorption of RNA to Millipore filters used in their experiments. The method to be described below gives, at one and the same time, greater efficiency and reproducibility of hybridization, and lower nonspecific interference by RNA absorption to support over a wide range of DNA/RNA ratios. The immobilisation of the denatured DNA strands ensures that DNA–DNA interactions are of minimal importance.

II. EXPERIMENTAL METHODS

A. Preparation of DNA from microbial cultures

1. *Unlabelled DNA*

In some cases, it may be necessary to use unlabelled DNA in hybridization experiments owing to difficulties in labelling etc. It is important to use DNA of high molecular weight to obtain maximum efficiency and reproducibility of RNA hybridization. Thus, microbial DNA is prepared by a method derived mainly from those of Marmur (1961) and Kirby (1964). Harvested cultures are suspended in 0·4 M NaCl, 0·1 M EDTA, pH 8·0, at a concentration of 50–200 mg. wet weight cells per ml. In some cases, e.g. with the genus *Bacillus*, a preliminary incubation of cultures with lysozyme may be necessary to initiate cell lysis. 500 μg/ml lysozyme (EC 3.2.1.17) is added to cells suspended in EDTA–NaCl and the mixture is incubated for 1 h at 37°C. Sodium 4-aminosalicylate is added to a final concentration of 4% (w/v) and the mixture is made up to 2% sodium dodecyl sulphate (w/v). The mixture is immersed in a waterbath at 60°C for 10 min. Lysis of the organisms is completed, giving a marked increase in solution viscosity.

Phenol-cresol mixture (140 ml redistilled m-cresol, 110 ml water, 1 g 8-hydroxyquinoline, and 1 kg redistilled phenol) is added in equal volume to the lysate, prepared as described. The mixture is shaken for 30 min until all lumps have dispersed. The resulting emulsion is centrifuged at 10,000 g for 20 min at 2°C. The upper (aqueous) layer is removed, leaving behind denatured protein collected at the interface between the aqueous and phenolic layers. An equal volume of 2-ethoxyethanol (Cellosolve) is layered carefully on top of the aqueous phase. White strands of DNA collect at the interface. The layers are gently mixed with a glass rod and the precipitating fibres of DNA are spooled. Excess liquid is drained from the fibrous mass by pressure against the side of the vessel. For large amounts of DNA, precipitation is most conveniently performed in a measuring cylinder. If satisfactory yields of DNA are not obtained at this stage, much can be recovered by re-extraction of the denatured protein + phenolic layers by a volume of 0·4 M NaCl corresponding to 40% of the original aqueous phase. After shaking for 20 min, the DNA may be precipitated from the separated aqueous phase as described. Yields are then combined and allowed to dissolve in the minimum practical volume of 0·015 M NaCl, 1·5 mM sodium citrate (0·1 SSC) (about 1 ml solution per 5 mg DNA). Gentle stirring overnight in the cold room is necessary to bring all the DNA into solution. If a very opaque solution is formed at this stage, the solution may be centrifuged at 80,000 g for 30 min at 5°C, giving a clear supernatant fluid.

A 0·2% solution of pancreatic ribonuclease (EC 2.7.7.16) in 0·15 M NaCl is heated to 80°C for 10–15 min to destroy any deoxyribonuclease activity. Sufficient of a cooled solution is added to the redissolved DNA to bring the enzyme concentration to 50 μg/ml. The mixture is incubated at 37°C for 1 h. After incubation, Na 4-aminosalicylate is added to a final concentration of 4%, one volume of phenol-cresol mixture is added and the preparation is shaken for 15 min. The emulsion is centrifuged and the aqueous phase collected as previously described. To the aqueous layer is added a further 0·5 volumes of phenol-cresol and the cycle is repeated. To the separated aqueous phase 1/9 volume of a solution of 3 M sodium acetate in 1 mM EDTA, pH 7·0 is added, followed by a careful layering of 0·54 volumes of isopropyl alcohol on to the solution. The gelatinous precipitate of DNA formed on gently mixing the layers is spooled on to a glass rod. The precipitate is redissolved in 0·1 SSC solution and then precipitation and redissolution are repeated. The resulting solution is usually opalescent, indicating the presence of contaminating polysaccharide. The solution is adjusted to 0·5 M NaCl, and solid sodium benzoate is added to a final concentration of 20% (w/v). One volume of 2-butoxyethanol is layered on to the aqueous phase and the DNA, essentially free from contaminants, is spooled on to a glass rod. The final product is dissolved in 0·01 × SSC and dialysed against 30 times its own volume of the same buffer for 16 h at 3°C, with three buffer changes. The DNA is stored in solution at 5°C with the addition of a few drops of chloroform as an antifungal agent.

Storage in the deep freeze leads to considerable DNA degradation by shearing, and subsequent poor hybridization properties. Analysis of the DNA in the analytical ultracentrifuge indicates that most of the material has a sedimentation coefficient ($S°_{20, w}$) about 30S. Small amounts of degraded material are always noted. The yield of DNA is about 1–2 mg/g wet packed cells.

2. Radioactive DNA

For experiments involving double isotopic labelling for the detection of hybrids, [3H] or [14C]-labelled DNA is most conveniently prepared. Organisms may be specifically labelled in their DNA by the addition of [6-[3H]] or [2-[14C]] thymidine as growth medium supplement. In many cases, addition of thymidine to the growth medium causes induction of thymidine-destroying enzymes, giving rise to thymine which is often poorly incorporated into the cells. For example, thymine-requiring mutants of *E. coli* or species of the genus *Bacillus* are best used as a source of labelled *E. coli* or *Bacillus* DNA. *E. coli* strains employed in the author's laboratory include B3005* (an auxotroph requiring adenine and thymine) and 15TAU⁻* (requiring thymine, arginine and uracil). *B. subtilis*† 168 ind⁻thy⁻ (an auxotroph requiring tryptophan and thymine) may be used similarly. If similar mutants are not available in particular microbial species, it may be necessary to use [[32P]]orthophosphate, or nucleic acid precursors such as [2-[14C]]uracil or [8-[14C]]adenine. Specificity of isotope incorporation into DNA is, of course, lost in these cases, and this may lead to complications in the purification of DNA and exaggeration of minor contaminants in preparations.

---

* Obtained from Professor K. Burton, Department of Biochemistry, University of Newcastle upon Tyne.

† Obtained from Dr. J. D. Gross, Microbial Genetics Research Unit, Hammersmith Hospital, London, W.12.

B. Preparation of RNA

1. General

Whether or not labelled DNA is used for hybridization experiments, radioactively labelled RNA is necessary for this technique, owing to the small quantities usually bound in the hybridization process. The RNA types usually of interest in hybridization work are (i) messenger RNA (found as a fraction of "rapidly labelled" RNA), (ii) ribosomal RNA and (iii) transfer RNA.

2. Isolation of "rapidly labelled" RNA

In bacteria, "rapidly labelled" RNA is known to contain a high proportion of messenger RNA, with an average lifetime which is short compared with the cell division time, and which, therefore, is present as only a small proportion of the total bacterial RNA (Bolton and McCarthy, 1962; Levinthal, Keynan and Higa, 1962; Midgley and McCarthy, 1962). It can be preferentially labelled by permitting cultures to incorporate a radioactive RNA precursor for a short time. The most satisfactory protocol involves the radioactive labelling of RNA for a period not longer than 4% of the cell division time of the culture. RNA precursors used in this way can be either [^{32}P]orthophosphate or ^{3}H or ^{14}C-labelled uracil, adenine or guanine. It must be remembered that, however short the period of isotope incorporation into the cells, the RNA labelled is a mixture of messenger RNA and ribosomal RNA precursor (Bolton and McCarthy, 1962; Midgley and McCarthy, 1962), the latter predominating in steadily growing cultures. Thus a mere measurement of "rapidly labelled" RNA is not an accurate assessment of messenger RNA levels, making hybridization studies obligatory to permit valid estimations (Pigott and Midgley, 1968).

Microbial cultures are inoculated into fresh growth medium and are allowed to divide at a suitable temperature. When the exponential phase of growth has reached a point where the turbidity at 650 nm (1 cm light-path) is 0·6 (1 g wet weight cells/litre solution), radioactive nucleic acid precursor is added by vigorous injection. Incorporation of the isotope is allowed to proceed for the required short interval (e.g. 1 min at 37°C in an organism dividing once every 50 min). The culture is then poured on to approximately one volume of crushed frozen growth medium to arrest cellular metabolism as quickly as possible. The culture is harvested in the cold by centriguation at 10,000 g for 10 min. Cells may be crushed by grinding with alumina in a chilled mortar, or by a cooled Hughes (Hughes, 1951) or French Press (French and Milner, 1955) (see Hughes, Wimpenny and Lloyd, Volume 5B, page 1). Other more gentle procedures may be used to obtain RNA that is less degraded. These include lysis of spheroplasts

(Schaechter, 1963), treatment of suspended cells by EDTA-lysozyme (Kiho and Rich, 1964) or by freezing and thawing + treatment by lysozyme (Ron, Kohler and Davis, 1966).

From such extracts taken up in 10 mM $MgCl_2$, 10 mM Tris, pH 7·0, DNA is removed by incubation with 20 $\mu g/ml$ ribonuclease-free deoxyribonuclease (EC 3.1.4.5) purified by the method of Polatnick and Bachrach (1961). Incubation is continued for 20 min at 37°C. RNA may be prepared by treatment with one volume of phenol-cresol mixture (see preparation of DNA) followed by shaking at room temperature for 5 min. The emulsion is centrifuged at 10,000 g for 10 min, separating into two liquid phases and a creamy interface of denatured protein. The upper (aqueous) phase is removed and adjusted to 0·4 M NaCl. A further deproteinization step is carried out, using phenol-cresol mixture, and after centrifugation, the aqueous phase is precipitated by two volumes of cold ethanol. RNA is collected by centrifugation, after allowing the solution to stand for 1 h at −10°C. The RNA is then redissolved in 0·1 SSC solution; ethanol precipitation is repeated and the RNA recentrifuged and redissolved in the above buffer.

Some ribonuclease originally present in the crude cell extracts may have survived the phenol treatment described above. A 2·5 cm diameter column is prepared by layering a 2·5 cm depth of Dowex 50 × 8 on to a 15 cm depth of Sephadex G25, both equilibrated in 0·1 SSC, in a glass column. RNA solution (1 ml) is carefully layered on top of this dual column and is then washed through with 0·1 SSC. 3 ml fractions are collected and the first ultraviolet-absorbing peak is collected and pooled. This contains all the RNA, whereas any peak following this contains traces of phenol-cresol mixture and other contaminants. Any ribonuclease is removed by adsorption to the Dowex resin. The RNA is now in a state in which it will withstand the high temperature incubation for DNA–RNA hybridization without enzymic degradation to nonhybridizable small oligonucleotides.

3. Randomly labelled RNA

(a) *Ribosomal RNA*. Cultures are grown over several divisions in the continuing presence of exogenously supplied radioactive RNA precursor. The cells are broken, DNA is removed and the RNA is obtained as described for "rapidly labelled" RNA. However, gel filtration is carried out using Sephadex G200 under Dowex 50. In this case, ribosomal and messenger RNA elute as the first (high molecular weight) peak, whereas transfer RNA elutes later (Midgley, 1965). As the messenger RNA contamination accounts for only about 2% of the ribosomal RNA, this fraction is pure enough for many purposes to allow the assumption that it is essentially ribosomal RNA. However, if even this level of messenger RNA contamination is inadmissible, the following modifications can be used:

(i) After prolonged growth of the bacterial cultures in the presence of radioactive RNA precursors, the cells may be harvested and grown over 2–4 generations in growth medium containing the supplements in non-radioactive form. Messenger RNA will turn over many times during this period and will have eventually a specific activity much lower than the stable labelled ribosomal RNA due to "chasing" by unlabelled RNA precursors. Thus, isolation of ribosomal RNA as described above will give a mixture of heavily labelled ribosomal RNA and almost unlabelled messenger RNA.

(ii) If even the presence of unlabelled messenger RNA is inadmissible, the cultures, after isotope incorporation may be crushed and taken up in 10 mM $MgCl_2$, 10 mM Tris, pH 7·0 as described previously. The solution may be centrifuged at 100,000 g for 45 min, bringing down labelled messenger RNA with the ribosomal complexes. The pellet obtained after centrifugation may be taken up in 0·1 mM $MgCl_2$, 10 mM Tris, pH 7·0 and dialysed for 12 h against 30 times the solution volume of the same buffer. This splits the ribosome-messenger RNA complex into ribosomal subunits and free messenger RNA. Centrifugation at 100,000 g for 240 min at 2°C pellets the ribosomal subunits, leaving free messenter RNA in the supernatant fluid. The pellet is washed once in 0·1 mM $MgCl_2$, 10 mM Tris, pH 7·2, before being taken up in the same buffer and recentrifuged. After taking up the final pellet in a suitable buffer, deproteinisation is commenced as previously described.

(b) *Transfer and 5S RNA.* (i) Fairly pure transfer RNA is obtained from the second ultraviolet absorbing peak found during gel filtration of total cell randomly labelled RNA on Sephadex G200. It is, however, contaminated with 5S RNA and traces of phenol-cresol mixture and ribosomal RNA. Purification may be achieved by the use of DEAE-cellulose column chromatography. A 1 cm diameter, 10 cm long DEAE-cellulose column is prepared, equilibrated at 2°C with 0·01 M $MgCl_2$ 0·01 M Tris, 0·01 M NaCl pH 7·0. The Sephadex fractions containing tRNA are loaded on to this column and are eluted in a linear gradient (400 ml) of 0·01–1·0 M NaCl, in 0·01 $MgCl_2$, 0·01 M-Tris, pH 7·0. Pure transfer RNA (and 5S RNA) elutes as an ultraviolet absorbing peak at 0·5 M NaCl, whereas traces of ribosomal RNA are eluted (if at all) at much higher salt concentrations (Midgley, 1962). The transfer RNA fractions may be either ethanol precipitated or dialysed against 0·015 M NaCl, 0·0015 M sodium citrate to provide a suitable storage environment. All RNA fractions may be stored in the deep freeze at − 10°C.

(ii) If the presence of ribosomal 5S RNA is undesirable in the transfer RNA preparation, the crushed labelled cells may first be taken up in 10 mM Mg⁺⁺, 10 mM tris, pH 7·0 at 2°C. The suspension is then centrifuged at 20,000 g at 2°C for 5 min to pellet unbroken cells and large cell wall fragments. The supernatant fluid is then centrifuged at 100,000 g at 2°C for 60 min, pelleting the 70S ribosomes with the attached 5S RNA. The top two thirds of the supernatant fluid may now be deproteinised and purified as described previously. The RNA then obtained from DEAE-cellulose column chromatography is virtually pure transfer RNA.

III. HYBRIDIZATION TECHNIQUES

The choice of methods of DNA immobilisation for hybridization include the use of agar gels (Bolton and McCarthy, 1962), cross-linked insoluble DNA gels (Britten, 1963) or cellulose nitrate membrane filters (Gillespie and Spiegelman, 1965) as inert supports for denatured DNA. All methods discussed involve the prior immobilisation of single-stranded DNA upon the support before attempted hybridization.

A. Agar gels

The procedure is followed exactly as described by Bolton and McCarthy (1962). A useful review is given by Hoyer and Roberts (1967). Preparation of DNA-agar: 300 mg of agar (Ionagar No. 2, Oxoid) is dissolved in 5 ml of water at 100°C. 5 mg DNA in 5 ml 0·01 × SSC is heated to 100°C for 5 min and poured into the agar. The hot solutions are thoroughly mixed and poured into an empty 250 ml beaker in an ice water bath. When the gel has hardened, it is forced through a screen (35 mesh). The gel particles are washed with 2 × SSC at 60°C. Trapping of DNA is essentially quantitative at low concentrations. Hybridization of RNA with Immobilized DNA: RNA in 1 ml or less of 2 × SSC is incubated at 60°C with DNA-agar containing DNA, either in a thick slurry in a screw top vial, or in a chromatograph tube heated by circulating water. The preparations heated in vials are then transferred to chromatograph tubes for subsequent operations. The column is washed with ten 5-ml aliquots of 2 × SSC. Recovery of the remainder of the labelled RNA is effected by washing with up to 20 5-ml aliquots of 0·01 × SSC which are also individually collected. The total RNA is precipitated with 5% trichloroacetic acid. The precipitates are collected on membrane filters and assayed for radioactivity.

Reproducible high efficiencies of DNA–RNA hybridization are found with specimens of "rapidly labelled" bacterial RNA. Nonspecific adsorption of RNA to the gels is minimal, but the removal of hybridized RNA from the gels for counting is somewhat time consuming. Ionagar No. 2 is found to

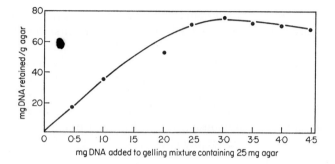

FIG. 1. Retention of denatured *E. coli* DNA by a granulated gel of 3% Ionagar No. 2.

be most suitable as a source of agar, but it must be stored in airtight bottles in the dark if its binding properties for denatured DNA are not to decline rapidly. The use of agar gels containing a higher percentage of agar (e.g. 6%) than quoted by Bolton and McCarthy (1962) gives no further experimental advantages.

An important drawback in this method is that the amount of denatured DNA which can be trapped in a fixed quantity of gel has a low finite limit. Figure 1 shows the effect of an addition of a fixed quantity of 3% agar to increasing quantities of DNA. At low DNA concentrations the entrapment is essentially quantitative, but a practical maximum of 75 mg of DNA per g agar powder was found to be trapped. This gave a limiting concentration of approximately 2·5 mg DNA per g drained gel. These limits are essentially practical ones, due to the difficulties of mixing viscous concentrated DNA solutions with the agar solution before gelling.

A further problem encountered is that of instability of the gels + DNA at the higher DNA concentrations employed. At higher DNA levels, leaching of DNA (possibly with hybridized RNA attached) becomes important during hybridization experiments. Leaching of DNA is especially significant during the elution of hybridized RNA at 70–75°C in solutions of lower ionic strength (Bolton and McCarthy, 1962). This results in the contamination of collected fractions by both DNA and agar in quantities often far in excess of the probable RNA content. A result of this is that production of membrane filters + precipitated labelled RNA for counting of hybridized material is impeded, and the sample thickness for counting is variable. If elution of hybridized material is attempted at 60°C, agar loss is smaller but the emerging labelled RNA may be spread over 200 ml of solution, giving rise to systematic errors in counting the many fractions resulting.

The practical difficulties in the agar-gel techniques of Bolton and

McCarthy (1962) may be summarized: (i) if the method is required as a routine assay for messenger RNA using low levels of DNA in a given weight of agar, it is generally satisfactory. A difficulty always encountered is the presence of leached agar particles with hybridized RNA after its removal from the DNA-agar gel. (ii) If a very wide range of DNA/RNA ratios is required in experiments (say from 1000/1 to 1/1) the DNA-agar technique has several drawbacks due to the necessity of using unstable DNA-agar gels with a high DNA content, to attain high DNA/RNA ratios combined with a reasonable level of radioactivity in the hybridized RNA.

B. Cross-linked DNA gels

Essentially the method is as described by Britten (1963). About 3 mg of microbial DNA are dissolved in 1–2 ml water and the solution is heated to 100°C for 10 min to separate the DNA strands. The solution is quickly cooled to 3°C, brought to the strength of SSC by adding concentrated 10 × SSC, and 100 mg of an inert supporting material is added. Two volumes of ethanol are added to precipitate DNA and the wet solid is exposed as a thin film to u.v. light (2 min with stirring, 5 cm from a 15 watt germicidal lamp). Only about 10% of the DNA is trapped as insoluble crosslinked gel or as a film on the inert support. Thorough washing of gel + support with SSC removes the unbound excess DNA.

Labelled RNA is incubated in a vial with the DNA-gel (either free or on the support) for 16 h at 60°C in 1 ml of 2 × SSC solution. After this period, the unhybridized RNA is removed at 40°C by placing the contents of the vial in a heated chromatography tube and washing the gel with 2 × SSC in 5 ml portions. The hybridized RNA may then be removed by raising the temperature to 80°C and continuing the washing procedure.

Alternatively, the gel + support may be centrifuged through 2 × SSC repeatedly at 5°C (10,000 g for 10 min). This frees the gel + support of unhybridized RNA and the hybridized RNA may then be freed by heating the gel + support in 2 ml 0·01 SSC at 50°C for 10 min. Recentrifuging leaves hybridized RNA in solution.

Supports used in the author's laboratory include paper powder (Whatman), kieselguhr (Hyflo Supercel), Geon 101 polyvinyl beads (Goodrich Co., Akron, Ohio, U.S.A.), Dowex 50 (20–50 mesh) (Dow Chemical Co., Midland, Mich. U.S.A.) and Zeokarb 225 (50–100 mesh) (Permutit Co. Ltd., U.K.). All these materials adsorb crosslinked DNA well, but upon forming columns, a poor flow rate of solution is noted through some of the materials. Channelling is common, giving rise to insufficient washing of the DNA + hybridized RNA.

A major drawback is that, during hybridization experiments at the necessary elevated temperatures, DNA may be eluted from the supports and lost during later washing to remove nonhybridized RNA. Solubilization of crosslinked DNA gels is also important if the insoluble gel itself is used without support. The degree of dissolution of gel varies from batch to batch, and leads to low efficiencies of binding of RNA together with uncertainties in collection and counting of the hybridized RNA after elution from the gels.

C. Cellulose nitrate membrane filters

The modification of Gillespie and Spiegelman (1965) is used. Several advantages are apparent in this method. Denatured DNA is bound to membrane filters, which are directly usable in scintillation counters after hybridization procedures and washing. Filters (2·5 cm diam.) generally adsorb up to 300 μg DNA before clogging, but multiple filters may be employed in a single experiment to raise the DNA levels to the required amount. After drying in vacuo, the filters, with DNA attached, are stable even at the highest DNA levels. No DNA is lost at any stage of the hybridization procedure. There is little nonspecific absorption of RNA to filters containing no DNA. A reproducible high efficiency of hybridization is noted. Hybridized RNA may be quantitatively removed from the filters in conditions of low ionic strength. The method is rapid, and has the added convenience that filters with bound DNA may be be prepared well in advance of experiments, and stored until needed at 3°C. This method is considered to possess sufficient potential to be selected as the most convenient and uncomplicated of all those studied.

A recent low-temperature modification of this method has been described by Bonner, Kung and Bekhor (1967). The addition of formamide-SSC to hybridization mixtures permits hybridization of RNA at 0–20°C. However, in our hands, the time taken to reach equilibrium between free and hybridized RNA is long and the reproducibility of hybridization of given RNA specimens is low.

IV. EXPERIMENTAL DETAILS OF HYBRIDIZATION (GILLESPIE AND SPIEGELMAN, 1965)

A. Binding of DNA to filters

A solution of microbial DNA of high molecular weight is diluted to a concentration of 100 μg/ml using 0·01 SSC (SSC = 0·15 M-NaCl, 0·015 M sodium citrate). It is then made 0·15 M with respect to NaOH by the calculated addition of 10 N alkali. The mixture is allowed to stand at room temperature, denaturation being monitored by measurement of ultraviolet absorbancy at 260 nm; this should increase by a factor of approx. 1·5 on completion. The denatured DNA solution is then cooled to 3°C in an icebath, before it is neutralized with 5 N acetic acid. Denatured DNA may be stored at 3°C in solution for several days without detectable renaturation.

Cellulose nitrate membrane filters (Type B6 coarse, 24 mm diam.), Schleicher and Schuell, Keene, New Hampshire, U.S.A., or Sartorius Membrane filter (25 mm diam.), Type MF 50, are presoaked for at least 1 min in 6 × SSC at 3°C and then transferred to a stainless steel filter apparatus of 2·5 cm diam. They are then washed with 10 ml 6 × SSC and the denatured DNA sample is allowed to filter through under gravity. If the binding of DNA is to be efficient, only slow rates of filtration are permissible. These manipulations must be carried out in the cold at 3°C, otherwise the efficiency of DNA retention upon the filters is lowered by at least a factor of ten. A rough guide to efficiencies obtained can be given: up to 100 μg DNA is bound with about 100% efficiency, 200 μg is bound with 70–80% efficiency, and 500 μg with 50–60% efficiency per filter (at 3°C).

The loaded filters are washed with 100 ml of 6 × SSC with the aid of a constant head device. The washed filters are allowed to dry at room temperature for at least 4 h followed by additional drying for 4 h at 80°C over P_2O_5 in vacuo. The preliminary drying of the filters at low temperature minimizes renaturation of DNA which may otherwise occur in the early stages of high temperature desiccation. The denatured DNA is now irreversibly attached to the membrane filter. Dry filters may be stored over P_2O_5 in an evacuated desiccator at 3°C for at least 28 days.

B. Hybridization

Dried filters containing DNA are placed in 25 ml widemouth conical flasks or scintillation vials and the RNA samples are added. The solution volume is made up to 1·5 ml and the salt concentration to 6 × SSC. This volume is just sufficient to cover the filter. If multiple filters are used, the volume may be increased until the filters are all covered with 6 × SSC. A volume increment of 0·3 ml per filter is sufficient.

The stoppered flasks are immersed in a waterbath at 66°C. After 16 h the flasks are removed and plunged in an icebath. A period of 16 h is necessary to ensure maximum hybridization at the usual DNA/RNA levels used (Fig. 2). Each side of the filter is washed with 50 ml 6 × SSC by suction filtration. The protocol following depends on the isotopic nature of the DNA and RNA used.

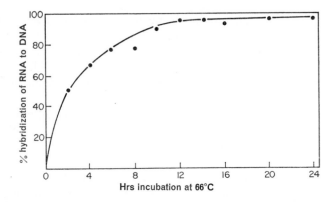

FIG. 2. Kinetics of hybridization of E. coli "rapidly labelled" RNA to denatured
E. coli DNA, bound to nitrocellulose membrane filters. For each point, the amount
of DNA bound to the filter was 200 μg, and 1 μg of RNA was hybridized. The
vertical scale has been normalised to give a maximum of 100% hybridization.

1. Using unlabelled DNA and labelled RNA

The total nucleic acid content of the filter is removed by the addition
of 2 ml 0·5 N HClO$_4$, followed by incubation in a boiling water bath for
30 min. The filters are removed from the liquid, and allowed to drain into
it for 10 min, after which time they are discarded. The cooled hydrolysate
is neutralised with 0·2 ml 5 N KOH and the resulting precipitate of KClO$_4$
is removed by centrifugation in the cold.

An aliquot of the nonhybridized RNA remaining in the hybridization
medium is precipitated by the addition of an equal volume of ice-cold
10% trichloroacetic acid on to a cellulose nitrate membrane filter, and the
filter is washed with 10% trichloroacetic acid (10 ml). Aliquots of the hydrolysate from the hybridization filter may be dried on membrane filters for
measurements of radioactivity to assess hybridized RNA. Other aliquots
assay the total nucleic acid by phosphate analysis. For valid results, the
sum of the radioactivity of filters containing non-hybridized RNA and
hybridized RNA should represent at least 90% of the total radioactivity in
the RNA used as initial challenge.

(a) *Assay of hybridized RNA.* Radioactive RNA hydrolysed as above is
estimated by scintillation counting using the system of Hall and Cocking
(1965). The scintillation mixture consists of: 4 g diphenyl oxazole (PPO),
0·1 g 1,4 bis(2-phenyl-oxazolyl) benzene (POPOP), 300 ml 2-ethoxy-
ethanol, and 700 ml. scintillation grade toluene. 1 ml samples of hydrolysate are pipetted into scintillation vials containing 0·1 ml 90% formic
acid and 2·6 ml 2-ethoxyethanol. 5 ml of the scintillation mixture, bubbled

with oxygen-free nitrogen for 10 min is then added and the vial is shaken until a clear liquid is obtained. Samples of nonhybridized RNA, similarly digested, are treated in the same way to give a measure of the stability of RNA during hybridization and a further hydrolysed sample of RNA gives an internal standard. The degree of hybridization of RNA may be obtained by reference to the standard RNA samples. The background rate is determined from a washed DNA-free filter incubated with the same quantity of RNA as the test sample, and subjected to identical hybridization, washing and hydrolysis procedure.

(b) *Assay of DNA bound to filters.* In the overwhelming majority of cases, the assay of unlabelled DNA by a colorimetric method contains negligible errors from RNA contributions, owing to the small amount of RNA hybridized. The quantity of DNA present cannot be accurately assayed by absorbancy at 260 nm or by the Burton (1956) diphenylamine test, since materials eluted from the nitrocellulose membrane filters may interfere considerably with either test. A suitable method for DNA analysis in these systems is that of Chen, Toribara and Warner (1956). This method is sensitive to < 0.5 μg phosphate.

The following reagents are prepared:

A. Ascorbic acid 10% (w/v) (stored at $5°C$ for no longer than 6 weeks).
B. 6 N H_2SO_4.
C. 2.5% (w/v) ammonium molybdate.
D. 1 vol $A + 1$ vol $B + 1$ vol $C + 2$ vol distilled water. (This is prepared immediately before use.)

Samples (0.5 ml) to be determined are ashed by the addition of 1 drop of conc. sulphuric acid and 2 drops of N $HClO_4$. The samples are heated until white fumes appear, and then more strongly until the bead remaining in the tube becomes clear (about 30 sec after first appearance of white fumes). The tube is cooled and approximately 1 ml of distilled water is added. The sample is then immersed in a boiling water bath for 10 min to ensure the complete solution of the sample. The contents of the tube are washed into a graduated test tube and the volume is made up to 4 ml with distilled water. Four ml of reagent D are added and the tubes are stoppered, shaken and incubated at $100°C$ for 2 h to allow maximum development of the colour. The tubes are then removed from the boiling water bath, and after cooling for 5 min at room temperature the absorbancy at 820 nm is read on a suitable spectrophotometer. Blanks are run in which 0.5 ml water replaces the DNA hydrolysate. The addition of a nitrocellulose membrane filter to the ashing procedure does not affect the calibration curve. Standard curves may be constructed in which sodium phosphate

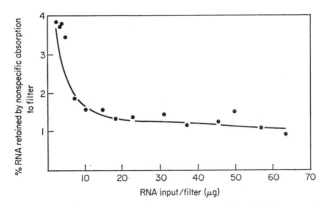

FIG. 3. Effect of RNA concentration on nonspecific binding to nitrocellulose membrane filters.

solutions of known strength substitute for the unknown samples. The calibration curve obeys Beers Law with samples containing up to at least 7–8 μg phosphate.

2. *Using ³H-labelled DNA and ¹⁴C-labelled RNA.*

The formation of filters and the hybridization technique is as described above.

(a) *Assay of labelled DNA and RNA.* The formation of doubly labelled DNA–RNA hybrids enables a considerable experimental simplification to be made. Membrane filters, after hybridization, are removed from the hybridization medium and washed as previously described. They are then placed in scintillation vials under an infrared lamp until dry. 6 ml of a mixture containing 6 g diphenyl oxazole (PPO) and 0·05 g 1,4 bis (2-phenyloxazolyl) benzene (POPOP) per litre scintillation grade toluene are added. The vial is left to stand in the dark for 24 h before counting in the scintillation counter, to permit equilibration of quenching materials slowly eluted from the filters. Usually, two channel counting is carried out to assay ³H-DNA and ¹⁴C-RNA. Standards are set up containing known amounts of ³H-DNA or ¹⁴C-RNA dried onto membrane filters. Further standards account for nonspecific absorption of RNA to the filter material by carrying filters with no DNA attached through the hybridization procedure.

(b) *Nonspecific absorption.* A problem in this procedure is that, contrary to the experiences of others (Gillespie and Spiegelman, 1965), nonspecific absorption of RNA to filters may depend on the weight of RNA offered per filter. Figure 3 indicates that nonspecific absorption increases to a limit of 4% of the input RNA. If the actual percentage hybridization is small, considerable errors may be introduced if the nonspecific absorption is underestimated.

V. MINIMIZATION BY RIBONUCLEASE OF NONSPECIFIC RNA ABSORPTION

Several authors (Hayashi and Spiegelman, 1961; Nygaard and Hall, 1963; Gillespie and Spiegelman, 1965; Friesen, 1966) have exploited the resistance of DNA–RNA hybrids to digestion by pancreatic ribonuclease as a means of reducing spurious hybridization and nonspecific absorption of RNA to supports. Using this method, Gillespie and Spiegelman (1965) claimed a reduction in nonspecific absorption of RNA to nitrocellulose membrane filters from 0·06% to 0·003% of RNA input. However, our observations give a maximum nonspecific binding of 4% of RNA input. Moreover, in many experiments, where the degree of RNA hybridization is considerable, other errors inherent in the method may make this refinement superfluous. An example of the effects of pancreatic ribonuclease on the hybridization of "rapidly labelled" RNA to DNA in E. coli is given in Table I. The treatment consists of immersion of washed filters containing hybrids for 1 h at room temperature in 5 ml 2×SSC with 20 μg/ml DNAase-free pancreatic ribonuclease (see section on preparation of DNA). Filters are then chilled in an icebath, and each side is washed with 50 ml 2×SSC. The filters are then ready for assay in the usual manner.

From Table I it can be seen that no advantages accrue if ribonuclease treatment is carried out in cases where the bulk of the RNA is hybridized

TABLE I

The effect of Ribonuclease Digestion on DNA-RNA Hybrids

% RNA hybridized		
Without RNase digestion	With RNase digestion	DNA/RNA ratio
23·5	23	10 : 1
31	30	50 : 1
43	43·5	75 : 1
50	49	100 : 1
66	64	150 : 1
78	79	230 : 1

Specimen of RNA hybridized: [14]C-"rapidly labelled" RNA from E. coli, contain 1/3 label in messenter RNA and 2/3 label in ribosomal RNA precursor.

at 66°C. This treatment will be of value if nonspecifically bound RNA is a significant fraction of the RNA correctly hybridized.

VI. ARBITRARY NATURE OF FILTER SUPPORT

It has frequently been observed that freshly obtained nitrocellulose membrane filters have a lowered efficiency of DNA adhesion and of maximal hybridization efficiency. Also, nonspecific absorption of RNA may be initially as high as 8% of the RNA input. For example, maximum efficiencies of RNA hybridization may be initially as low as 50–55% of input, whereas 70–75% is normally obtained. Efficiencies improve to the higher value after prolonged soaking of such filters at 3°C in SSC solutions for periods of not less than 24 h nonspecific absorption of RNA declining concomitantly. The reason for this phenomenon is quite obscure, but it emphasizes the necessity of adequately soaking and testing samples of membrane filters for hybridization efficiency before proceeding further.

VII. ANALYTICAL PLANS

In no case can RNA be hybridized to DNA in a single challenge with 100% efficiency, even though excess DNA sites are made available. Figures 4(a) and 4(b) show the maximum hybridization achieved for "rapidly labelled" and for ribosomal RNA from *E. coli*. Results obtained must be corrected by a factor of 100/70–80 (i.e. 1·26–1·42) to normalize to 100% theoretical efficiency of hybridization in any single test.

A. Messenger RNA

Microbial messenger RNA is present in total RNA as only a small percentage (Bolton and McCarthy, 1962; Midgley and McCarthy, 1962; Levinthal, Keynan and Higa, 1962; Armstrong and Boezi, 1965). It is representative of DNA cistrons from the bulk of the genome, and will hybridize with normal efficiency to these. As specific nucleotide sequences of longer than 12 residues are rarely repeated in microbial DNA (Thomas, 1966), we can expect that messenger RNA molecules will bind with minimum ambiguity to the correct DNA sites. Low DNA/RNA ratios used in hybridization tests will generally be adequate in providing enough sites for maximal binding of virtually all messenger RNA molecules. In practice, messenger RNA preparations are obtained mixed with total ribosomal RNA, and DNA/RNA ratios of 5/1 to 10/1 are suitable to ensure, at the same time, maximal binding of messenger RNA and minimal binding of ribosomal RNA. In several organisms tested, ribosomal RNA cistrons account for less than 0·5% of the DNA (Yankofsky and Spiegelman, 1963). "Rapidly

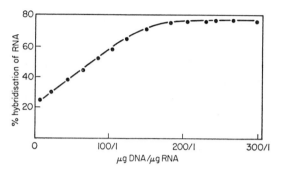

FIG. 4. (a) Hybridization of *E. coli* "rapidly labelled" RNA to DNA at various DNA : RNA ratios.

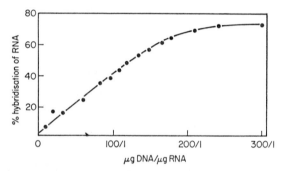

(b) Hybridization of *E. coli* ^{14}C-labelled ribosomal RNA with DNA at various DNA : RNA ratios.

labelled" RNA consists of 1/3 messenger RNA and 2/3 ribosomal RNA precursors (Midgley and McCarthy, 1962; Bolton and McCarthy, 1962). Messenger RNA amounts to about 1–2% of the total cell RNA and ribosomal RNA, 80%. As messenger and ribosomal RNA's are generally isolated together, then at a DNA/RNA ratio of 5/1, virtually all messenger RNA will be hybridized with 70–80% efficiency, whereas ribosomal RNA will be hybridized with $5 \times 0.5/100 \times 70{-}80/100$ (or 2% efficiency). A small correction must thus be made to take account of this factor. There is direct evidence that in *E. coli* DNA/RNA ratios of 10/1 bind all messenger RNA with maximum efficiency and no messenger RNA's are produced at rates which reduce this binding efficiency in any one instance (Pigott and Midgley, 1966; Pigott and Midgley, 1968).

B. Fractionation of messenger RNA

Reports have been made (Bolton and McCarthy 1964) that messenger RNA may be fractionated according to base composition by hybridization techniques. Messenger RNA of *E. coli*, hybridized at maximum efficiency, is removed in fractions by incubating the DNA–RNA hybrids in $0.01 \times$ SSC with a number of increments of increase in temperature. The first RNA hybrids removed at the lowest temperatures contain less $G + C$ than those lost at the higher temperatures and an elution profile can be developed, representing percentage messenger RNA eluted at a given temperature. The evidence indicates that for *E. coli*, messenger RNA base compositions may vary as much as 10% on either side of the mean $G + C$ content of the DNA.

C. Ribosomal RNA

The DNA cistrons responsible for the formation of ribosomal RNA in microbial cells are relatively few in number (Yankofsky and Spiegelman, 1963), whereas the RNA, being a stable end product in the cell, is a major component of the total ribonucleic acid. Large DNA/RNA ratios (see Fig. 4(b)) are necessary before ribosomal RNA will bind with a given amount of DNA with at least 70% efficiency (Avery, Midgley and Pigott, 1969). There is little evidence of nonspecific binding of ribosomal RNA to other DNA sites, but ribonuclease treatment may be used in studies of this kind to ensure minimization of spurious hybridizations.

4. Transfer RNA

The above comments on ribosomal RNA binding bear with even more weight on the binding of transfer RNA to DNA. DNA/RNA ratios of $1000 : 1$ to $5000 : 1$ are probably necessary to ensure the maximum binding of tRNA (Giacomoni and Spiegelman, 1962; Goodman and Rich, 1962). In addition, the considerable secondary structure of tRNA, and its relatively high $G + C$ nucleotide composition, make it probable that higher hybridization temperatures are required to effect efficient hybridization. Elevation of the hybridization temperatures may be necessary.

VIII. ANALYSIS OF HYBRIDIZATION CURVES

The formation of stable hybrids of RNA and DNA in solution may be considered primarily as a system that proceeds to equilibrium and that obeys the general laws of Mass Action. Thus, we can define DNA–RNA hybrid formation in solution as:

$$K_{equilm,} = \frac{[DNA_{specific, free}] \cdot [RNA_{specific, free}]}{[DNA\text{-}RNA\ Hybrid_{specific}]} \qquad 1$$

The exact value of K_{equilm}, in any given test depends, of course, on the conditions of solution employed (e.g. ionic strength), the temperature of incubation of the solutions, and the nature of the DNA and RNA used (e.g. their contents of G + C). Experiments are normally devised in conditions which strongly favour the formation of hybrids; a common protocol is to incubate the DNA and RNA at temperatures some 15°C below the melting out of DNA-RNA hybrids.

Owing to further complexities in this system, however, it is likely that Mass Action effects are not the only factors involved in determining the proportions of hybridized and nonhybridized RNA at equilibrium. For example, prolonged incubation of nucleic acids at the elevated temperatures needed for successful hybridizations leads to shearing of certainly the RNA molecules (Attardi, Huang and Kabat, 1965) and probably also some of the DNA molecules on the filters. This may lead to damage of DNA regions capable of binding the RNA and also to the formation of a randomly sized fraction of small pieces of RNA that may hybridize with less stability with the DNA (Niyogi, 1969). Further, the probability of some interference of hybrid formation on immobilised DNA by close association of denatured strands on the filters or by hindrance from the inert support itself, are also potentially important factors. We thus consider that Mass Action effects are probably not the sole arbiters of the extent or efficiency of the hybridization processes in any given test, although they are probably the most important.

A. Saturation of DNA hybridizing sites with excess RNA

This approach to semi-quantitative analysis of DNA–RNA hybridization is the one most frequently encountered. The basis of this technique is that increasing amounts of RNA are offered to a fixed amount of denatured DNA, until no more RNA will hybridize. Thus, the maximum weight of RNA that will hybridize with unit weight of DNA, using a large excess of RNA in a challenge, is a measure of the percentage of the DNA specifically capable of forming hybrids.

The curves obtained by graphical analysis may be described by mathematical functions, derived from equation (1) giving the proportions of hybridized and nonhybridized RNA and DNA at equilibrium, (Avery and Midgley, 1969). Let us suppose that we have a fraction of RNA composed of several different species that hybridize independently with different regions of the DNA. Let us also assume that the RNA species are present in the mixture in the same weight ratios as the DNA regions to which they hybridize. Then, we can describe the hybridization curve obtained by plotting Weight of RNA Hybridized/Weight of DNA against Weight of

RNA Offered/Weight of DNA as:

$$\frac{Y}{\mu D} = \frac{(1+1/C+\beta)+/-\sqrt{(1+1/C+\beta)^2-4/C}}{2} \qquad (2)$$

where Y = weight of RNA hybridized, D = weight of DNA in the test, μ = the fraction of the DNA taking part in the specific hybridization process, $\beta = K_{equilm,}/\mu D$ and $C = \mu D/X$, where X = the weight of RNA offered in the hybridization. Plotting $Y/\mu D$ values (as ordinate) against $1/C$ (or $X/\mu D$) values (as abscissa), we obtain hybridization curves as shown in Fig. 5, for values of β of 0, 0·01, 0·1, 0·2 and 0·4 (Avery and Midgley, 1969). For the case where $\beta = 0$, the equation (2) simplifies to:

$$\left(\frac{Y}{\mu D}-1\right)\cdot\left(\frac{Y}{\mu D}-\frac{1}{C}\right) = 0.$$

This gives two straight lines, as shown in the Fig., for the solutions $Y/\mu D = 1$ or $1/C$. At the intersection of these two lines, the value of $1/C$ from the abscissa corresponds to the fraction of the DNA taking part in the hybridization process. This value is also obtained by extrapolation of the line given by $Y/\mu D = 1$ to the ordinate. In practice, of course, $\beta > 0$, and therefore, as shown in Fig. 5, a considerable excess of RNA is necessary to ensure that the hybridization curve has reached the value given by $Y/\mu D = 1$. However, if $\beta \leqslant 0·01$, the curve is a good approximation to that obtained when assuming that $\beta = 0$, with a maximum 10% error at the inflexion point of the hybridization curve.

When this simple situation no longer holds, i.e. when the RNA species are present in a mixture in weight ratios different from those of the corres-

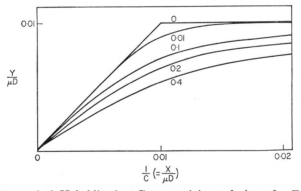

FIG. 5. Theoretical Hybridization Curves, giving solutions for Equation (2) (see text), for $\beta = 0, 0·01, 0·1, 0·2$ and 0·4. The RNA specimen comprises products from 1% of the total DNA, the weights of the RNA species being equal to the weights of the DNA cistrons from which they were transcribed.

ponding regions of DNA to which they hybridize, the saturation curve is further complicated by breaks in slope of the early part of the curve, as the various RNA species saturate their specific binding sites at different values of $1/C$. Indeed, if some specific RNA species are present as minor components in a mixture, then extremely high RNA–DNA ratios may be required to ensure that the corresponding DNA sites are saturated. In this case, the hybridization curves can be represented in a different way, to allow a simplification in estimating the amount of DNA taking part in the hybridization. When the amounts of RNA offered for binding with the DNA are greatly in excess of the available DNA sites, it may be assumed that the amount of RNA bound with the DNA is a negligible fraction of the total. Thus, $[RNA_{free}]$ in (1) is equivalent to X, the amount of RNA added. Using the symbols in (2), therefore, at large excesses of RNA;

$$\frac{1}{Y} = \frac{K_{equilm,}}{\mu D}\frac{1}{X} + \frac{1}{\mu D} \tag{3}$$

If a graph is plotted of $1/Y$ against $1/X$, in these conditions, a straight line results, with slope $K_{equilm,}/\mu D$ (β). If this line is extrapolated to $1/X = 0$ then $1/Y = 1/\mu D$, giving the amount of DNA involved in the specific hybridization process. Considerable caution has to be exercised in the interpretation of these hybridization curves. Since (3) only holds when the excess of RNA over available DNA sites is of the order of twentyfold or greater, the contribution of minor contaminating RNA species in any fraction is much exaggerated, as these are bound with high efficiency to their DNA sites. For example, a contaminant contributing only 1% to the original RNA will, at a twenty-fold excess of RNA over DNA sites, now contribute some 20% to the hybridized material. This will affect the extrapolated value of $1/\mu D$ considerably. Close attention to purity of RNA specimens is therefore crucial for this kind of analysis.

Practical difficulties also add to the uncertainties in the accuracy of estimations by this method. Owing to the fact that only a small proportion of the offered RNA will bind with the DNA, the amount of RNA that hybridizes with other regions of the DNA nonspecifically, or that binds with the support filter itself, becomes comparable with the amounts of RNA correctly hybridized. Thus, treatment of support+DNA–RNA hybrids with ribonuclease is essential, to remove these spuriously bound fractions. It is not certain that the hybrids are absolutely stable to the action of ribonuclease, so that estimations by this method must be considered as lower limits.

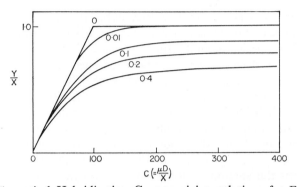

FIG. 6. Theoretical Hybridization Curves, giving solutions for Equation (4) (see text), for $\beta = 0$, $0\cdot01$, $0\cdot1$, $0\cdot2$ and $0\cdot4$. The RNA specimen was assumed to be identical with that used in Fig. 5.

B. Estimations of the efficiency of formation of DNA–RNA hybrids

An alternative approach to the analysis of DNA–RNA hybridization has been described by Avery and Midgley (1969) and by Avery, Midgley and Pigott (1969). By this method, the efficiency of hybridization of RNA with DNA is measured, over a range of DNA–RNA ratios where the bulk of the RNA offered is hybridized. This method has the advantage that ribonuclease digestion of spuriously bound RNA is no longer necessary, and that at least some of the problems arising from damage to both DNA and RNA during the hybridization processes are minimized.

Let us consider the hybridization of an RNA specimen with DNA, the RNA being constituted as described for the saturation analysis described previously. Then, using the symbols as described for equations (2) and (3):

$$\frac{Y}{X} = \frac{(1+C+\beta C)+/-\sqrt{(1+C+\beta C)^2-4C}}{2} \tag{4}$$

Thus, plotting values of Y/X (ordinate) against C (abscissa), we obtain hybridization efficiency curves as shown in Fig. 6, for a range of values of β from 0 to $0\cdot4$. For the case where $\beta = 0$, (4) simplifies to:

$$\left(\frac{Y}{X}-1\right)\cdot\left(\frac{Y}{X}-C\right) = 0.$$

The hybridization curve thus becomes two straight lines, for solutions Y/X = 1 or C. At the intersection of these lines, the value of C from the abscissa corresponds to the reciprocal of the fraction of the DNA taking part in hybridization. Since, however, $\beta > 0$, the hybridization curves give apparent inflexion points at values of C that are greater than is the case when $\beta = 0$. Thus, although values of $\beta \leqslant 0.01$ give curves closely approximating to that for $\beta = 0$, higher values underestimate the fraction of DNA taking part in RNA binding.

Examples of this type of analysis quoted in the literature (Avery and Midgley, 1969; Avery, Midgley and Pigott, 1969) indicate that for bacterial ribosomal RNA–DNA hybridization, the value of $\beta \leqslant 0.01$, as close agreement is obtained between results from saturation of DNA sites with excess RNA (i.e. values of $T/\mu D$) and those from analysis of efficiency of RNA hybridization. Indeed, a fairly accurate analysis can be made of the proportions of two RNA species in a mixture, if the proportions by weight are different from those of the specific DNA sites to which they hybridize (Avery and Midgley, 1969).

IX. COMPETITION EXPERIMENTS

As in the case of hybridization curve analysis, there are several alternative methods of studying the competition of RNAs for hybridization at DNA sites. The basis of competition experiments is that, given a limited availability of DNA sites for the acceptance of RNA hybrids, the addition of excess unlabelled RNA may reduce the hybridization of a labelled sample. If progressive additions of unlabelled RNA to fixed amounts of labelled RNA diminish the hybridized label, then fractions are present in both unlabelled and labelled RNA with a considerable degree of nucleotide sequence homology. If the labelled RNA is completely diluted out by a large excess of unlabelled material, then RNA species in the labelled fraction must all be present in the unlabelled competitor. However, if a percentage of the labelled RNA is hybridized with undiminished efficiency in the presence of excess unlabelled competitor, then this fraction cannot be present in the unlabelled RNA. Examples of competition curves for *E. coli* and *B. subtilis* 16S and 23S ribosomal RNAs are given in Figs 7 and 8 to show this phenomenon (Avery and Midgley, 1969; Avery, Midgley and Pigott, 1969). These curves indicate that, whereas in *B. subtilis* the ribosomal RNAs of 16S and 23S do not compete for each others' binding sites upon the DNA (Yankofsky and Spiegelman, 1963; Oishi and Sueoka, 1965; Mangiarotti, Apirion, Schlessinger and Silengo, 1968; Smith, Morell, Dubnau and Marmur, 1968; Avery and Midgley, 1969), this is not true of the corresponding RNAs from *E. coli* (Attardi, Huang and Kabat, 1965; Mangiarotti *et al.*, 1968; Avery, Midgley and Pigott, 1969).

The choice of amounts of DNA and labelled RNA in experiments of this kind is critical. Two approaches can be made. In the first, the initial amount of labelled RNA offered to DNA is high enough to saturate all appropriate DNA sites with labelled RNA. The unlabelled RNA competitor

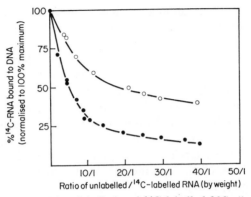

FIG. 7. Competition of (a) unlabelled and ^{14}C-labelled 23S ribosomal RNA and (b) unlabelled 16S and ^{14}C-labelled 23S ribosomal RNA, from *E. coli*, for DNA sites in denatured DNA bound to nitrocellulose membrane filters. The initial DNA/[14]C-23S RNA ratio was 300 : 1 (just sufficient to allow maximum hybridization of the RNA).

●———● ^{14}C-labelled 23S RNA v. unlabelled 23S RNA.

○———○ ^{14}C-labelled 23S RNA v. unlabelled 16S RNA.

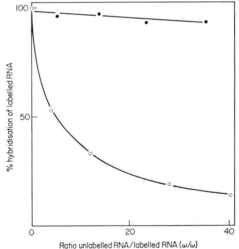

FIG. 8. Competition of (a) unlabelled and ^{3}H-labelled 16S ribosomal RNA and (b) unlabelled 23S and ^{3}H-labelled 16S ribosomal RNA, from *B. subtilis*, for DNA sites in denatured DNA bound to nitrocellulose membrane filters. The initial DNA/[3]H-16S RNA ratio was 650 : 1 (just sufficient to allow maximum hybridization of the RNA to the DNA).

○———○ ^{3}H-labelled 16S RNA v. unlabelled 16S RNA.

●———● ^{3}H-labelled 16S RNA v. unlabelled 23S RNA.

is now added in subsequent experiments with the same DNA : labelled RNA ratio, so that the hybridization competition curve commences from a point at which the DNA cistrons are already saturated with RNA, having a high specific activity at first, but falling as more unlabelled RNA competitor is added. In the second case, a DNA : RNA ratio is chosen for the initial hybridization of the labelled RNA that is just sufficient to bind all the RNA offered with maximum efficiency. Then competition curves may be constructed as before, using unlabelled competing RNA.

If the hybridizationally distinct RNA species are present in any fraction in a weight ratio much different from the weight ratios of the specific DNA cistrons, these two methods will give different competition curves. Only in the case where the competing RNA fractions are present in the same weight ratio as that of their accepting DNA cistrons will the two competition curves coincide.

Let us take a simple example. Suppose a labelled RNA fraction to be tested by competition analysis consists of 80% by weight an RNA from cistron 1 and 20% by weight from cistron 2 in the DNA. Further, let us suppose that the molecular weight of the RNA produced from each cistron is the same. In the first set of experiments, the labelled RNA is hybridized with DNA so that cistrons 1 and 2 are both saturated. This has only been achieved by alteration of the amounts of the two RNA species from 80 : 20 in the offered sample to 50 : 50 in the hybrid. If competition is now carried out by offering excess unlabelled RNA with species in the weight ratio 80 : 20, results will be obtained as shown in Table II for several different ratios of labelled RNA/unlabelled RNA competitor. In the second set of experiments, where the DNA : RNA ratio is chosen so that, initially, all the labelled RNA will just bind, then the ratio of RNAs in the hybridized material will be the same as in the offered RNA. However, this will give rise to unsaturation of cistron 2 responsible for the RNA present as 20% of the total by weight. This has first to be saturated by the addition of unlabelled RNA before true competition can begin at this site. Table II shows the results obtained for competition experiments using the same ratios of labelled/unlabelled RNA as in the previous method. These simple examples show the differences obtained in two experimental protocols using competition of identical mixtures of labelled and unlabelled RNAs. Obviously, if two different RNA fractions are to be compared, the experimental difficulties in analysing the results by the second method increase considerably, and a careful choice of the ratio of DNA : labelled RNA has to be made. In this case, it is usual to analyze the RNA by using sufficient labelled RNA to saturate the available DNA sites before beginning competition.

However, where identical labelled and unlabelled RNA fractions are

TABLE II

Competition of labelled and unlabelled RNA for DNA sites

	% Labelled RNA Hybridized	
$\dfrac{\text{Unlabelled RNA}}{\text{Labelled RNA}}$ (Initially Hybridized)	(i) DNA cistrons initially saturated with labelled RNA	(ii) All labelled RNA bound to DNA at minimum DNA/RNA ratio
0	100	100
$\frac{1}{1}$	50	60
$\frac{2}{1}$	33	47
$\frac{5}{1}$	16·7	30
$\frac{10}{1}$	9·1	16
$\frac{20}{1}$	4·9	8·7
$\frac{50}{1}$	2·0	3·5

RNA from two cistrons of equal size, in weight ratio 80/20.

competed, the second method can give information about the relative amounts of individual RNA species present in the mixture. Only in the special case where all the DNA cistrons are just saturated at the minimum DNA : RNA ratio for maximum hybridization of the RNA will the competition curves using both methods coincide at all points. In all other cases, the curves will differ, and with simple mixtures of (say) two RNA species, such as 16S and 23S ribosomal RNAs, the comparison of the two curves allows us to estimate their proportions.

Hybridization analyses by these methods are not, of course, restricted to the study of mixtures of RNAs. The purity of the RNA product from a single genetic site can also be checked by competition analysis. If the RNA product is 100% pure and derived from one DNA cistron, then its hybridization and competition curves will behave exactly as described for RNA mixtures containing individual species in the same weight ratios as those of the DNA cistrons from which they were transcribed. If it is impure, then, as long as it comprises the major fraction of the specimen the degree of purity can be estimated by these techniques.

X. HYBRIDIZATION OF RNA AND DNA FROM HETEROLOGOUS SOURCES

In view of the complexities of analysis of genetically homologous RNAs and DNAs, the information to be obtained by heterologous challenges must be more limited. Valid DNA : RNA hybridization, involving long stretches of near-perfect sequence homology may shade almost imperceptibly into less and less stable hybridizations as the homology of individual gene products with the DNA decreases. In practice, hybridization and competition curve analysis is carried out at elevated temperature (e.g. 66°) to ensure that any hybrids formed are those with most stability and maximum homology. As the experimenter's choice of conditions for DNA–RNA hybridization is usually quite arbitrary, statements concerning the extent of homology of RNAs and DNAs from different organisms must be treated with caution. Nevertheless, evidence of genetic relatedness between micro-organisms has been obtained in this way, both for messenger RNAs of various species (McCarthy and Bolton, 1963) and for ribosomal RNAs (e.g. Doi and Igarashi, 1965, 1966; Dubnau, Smith, Morell and Marmur, 1965; Moore and McCarthy, 1967; Takahashi, Saito and Ikeda, 1967).

ACKNOWLEDGEMENTS

The author is M.R.C. Senior Research Associate in the M.R.C. Research Group in the Structure and Biosynthesis of Macromolecules, University of Newcastle-upon-Tyne. Much of the work given in the experimental figures was carried out as part of the studies of the Research Group by Dr. R. J. Avery. I am indebted to both him and Dr. G. H. Pigott for discussions in the preparation of this manuscript, and to Professor K. Burton for his helpful comments.

REFERENCES

Armstrong, R. L., and Boezi, J. A. (1965). *Biochim. Biophys. Acta*, 103, 60.
Attardi, G., Huang, P. C., and Kabat, S. (1965). *Proc. Nat. Acad. Sci U.S.*, 53, 1490.
Avery, R. J., and Midgley, J. E. M. (1969). *Biochem. J.*, 115, 383.
Avery, R. J., Midgley, J. E. M., and Pigott, G. H. (1969). *Biochem. J.*, 115, 395.
Bautz, E. K. F., and Hall, B. D. (1962). *Proc. Nat. Acad. Sci. U.S.*, 48, 400.
Bolton, E. T., and McCarthy, B. J. (1962). *Proc. natn. Acad. Sci. U.S.A.*, 48, 1390.
Bolton, E. T., and McCarthy, B. J. (1964). *J. molec. Biol.*, 8, 201.
Bonner, J., Kung, G., and Bekhor, I. (1967). *Biochemistry*, 6, 3650.
Britten, R. J. (1963). *Science*, 142, 963.
Burton, K. (1956). *Biochem. J.*, 62, 315.
Chen, P. S. Jr., Toribara, T. Y., and Warner, H. (1956). *Analyt. Chem.*, 28, 1756.
Doi, R. H., and Igarashi, R. T. (1965). *J. Bact.*, 90, 384.
Doi, R. H., and Igarashi, R. T. (1966). *J. Bact.*, 92, 88.
Dubnau, D., Smith, I., Morell, P., and Marmur, J. (1965). *Proc. natn. Acad. Sci. U.S.A.*, 54, 491.

French, C. S., and Milner, H. W. (1955). *In* "Methods in Enzymology". (Ed. N. Kaplan and S. Colowick), Vol. 2, Academic Press, New York. 65.

Friesen, J. D. (1966). *J. molec. Biol.*, **20**, 559.

Giacomoni, D., and Spiegelman, S. (1962). *Science*, **138**, 1328.

Gillespie, D., and Spiegelman, S. (1965). *J. molec. Biol.*, **12**, 829.

Goodman, H. M., and Rich, A. (1963). *Nature, Lond.*, **199**, 318.

Hall, B. D., and Spiegelman, S. (1961). *Proc. natn. Acad. Sci. U.S.A.*, **47**, 137.

Hayashi, M., and Spiegelman, S. (1961). *Proc. natn. Acad. Sci. U.S.A.*, **47**, 1564.

Hoyer, B. H., and Roberts, R. B. (1967). *In* "Molecular Genetics", Part II (Ed. J. H. Taylor). Academic Press, New York. 425.

Hughes, D. E. (1951). *Brit. J. exp. Pathol.*, **32**, 97.

Kennell, D. (1968). *J. molec. Biol.*, **34**, 85.

Kennell, D., and Kotoulas, A. (1968). *J. molec. Biol.*, **34**, 71.

Kiho, Y., and Rich, A. (1964). *Proc. natn. Acad. Sci. U.S.A.*, **51**, 111.

Kirby, K. S. (1964). *In* "Progress in Nucleic Acid Research and Molecular Biology" (Ed. J. N. Davidson and W. E. Cohn). Academic Press, New York. 3, 1.

Levinthal, C., Keynan, A., and Higa, (1962). *Proc. natn. Acad. Sci. U.S.A.*, **48**, 1631.

McCarthy, B. J., and Bolton, E. T. (1963). *Proc. natn. Acad. Sci. U.S.A.*, 156.

McCarthy, B. J., and Bolton, E. T. (1964). *J. molec. Biol.*, **8**, 184.

McConkey, E. H., and Dubin, D. T. (1966). *J. molec. Biol.*, **15**, 102.

Mangiarotti, G., Apirion, D., Schlessinger, D., and Silengo, L. (1968). *Biochemistry*, **6**, 456.

Marmur, J. (1961). *J. molec. Biol.*, **3**, 208.

Midgley, J. E. M. (1962). *Biochim. biophys. Acta*, **61**, 513.

Midgley, J. E. M. (1965). *Biochim. biophys. Acta*, **108**, 340.

Midgley, J. E. M., and McCarthy, B. J. (1962). *Biochim. biophys. Acta*, **61**, 696.

Moore, R. L., and McCarthy, B. J. (1967). *J. Bact.*, **94**, 1066.

Niyogi, S. K. (1969). *J. biol. Chem.*, **244**, 1576.

Nygaard, A. P., and Hall, B. D. (1963). *Biochem. biophys. Res. Commun.*, **12**, 98.

Oishi, M., and Sueoka, N. (1965). *Proc. natn. Acad. Sci. U.S.A.*, **54**, 483.

Pigott, G. H., and Midgley, J. E. M. (1966). *Biochem. J.*, **101**, 9P.

Pigott, G. H., and Midgley, J. E. M. (1968). *Biochem. J.*, **110**, 251.

Polatnick, J., and Bachrach, H. L., (1961). *Analyt. Biochem.*, **2**, 161.

Rich, A. (1960). *Proc. natn. Acad. Sci. U.S.A.*, **46**, 1044.

Ron, E. Z., Kohler, R. E., and Davis, B. D. (1966). *Proc. natn. Acad. Sci. U.S.A.*, **56**, 471.

Schaechter, M. (1963). *J. molec. Biol.*, **7**, 561.

Smith, I., Dubnau, D., Morell, P., and Marmur, J. (1968). *J. molec. Biol.*, **33**, 123.

Takahashi, H., Saito, H., and Ikeda, Y. (1967). *Biochim. biophys. Acta*, **134**, 124.

Thomas, C. A. Jr. (1966). *In* "Progress in Nucleic Acid Research and Molecular Biology (Ed. J. N. Davidson and W. E. Cohn), Vol. 5. Academic Press, New York. 315.

Yankofsky, S. A., and Spiegelman, S. (1963). *Proc. natn. Acad. Sci. U.S.A.*, **49**, 538.

CHAPTER XIII

Cell Walls

Elizabeth Work

*Department of Biochemistry, Imperial College of Science and Technology,
London, S.W.7*

I. Definition 361
II. Preparation of Cell Walls 362
   A. Preparation of cells 362
   B. Disintegration of cells 363
   C. Recovery and purification of walls from broken cell suspensions 368
III. Examination of Walls 372
   A. Criteria of purity 372
   B. Separation of polymers 373
   C. Techniques for examination of constituents of separated
      polymers or of whole walls 378
   D. Enzymic Methods · · · · · · · · · 400
IV. Muralytic Enzymes 408
   A. Lysis 408
   B. Linkage split 408
   C. Fractionation of soluble products 413
References 414

I. DEFINITION

The cell wall of a microbe may be defined as the principal structure of
the cell responsible for mechanical rigidity and shape, and essential to the
normal functioning of the cell.

In most organisms, with the exception of Gram-negative bacteria, there
are clear morphological, functional and chemical differences between the
cytoplasmic (plasma) membrane and the cell wall which lies immediately
outside this membrane. In these cases the cell wall can usually be mechanically separated from the cytoplasmic membrane and therefore may be
considered as a separate morphological entity.

The cells of Gram-negative bacteria have more complicated multi-
layered walls, which have external lipoprotein-containing layers having a
morphological similarity to a "unit membrane", in addition to the internal
"unit" membrane bordering the cytoplasm. It is not yet certain whether
these two membrane systems are separate entities with different structures
and functions. Since conventional preparations of walls from Gram-
negative bacteria usually contain portions of both these membranes, it has
been suggested that such preparations should be referred to as "envelope"
(i.e. wall and membrane) (Salton, 1967).

II. PREPARATION OF CELL WALLS

A. Preparation of cells

The first requirement in the preparation of cell walls being an adequate amount of cells, consideration must first be given to the volume of culture medium required. An average yield of bacterial cells is approximately 1 g per litre of culture medium; since the yield of wall is about 15–30% of the dry weight of the cell, one can expect a yield of walls of 150–300 mg per litre of culture medium. If the aim of the work is a chemical investigation of wall composition, it is advisable to have a stock of at least 1 g of wall, implying the use of up to 5 litres of medium. Other types of study may require much smaller amounts of walls and more manageable volumes of culture.

The advisability of handling a large bulk of live cells must be considered, bearing in mind the likelihood of aerosol formation during the preparation of walls. Where there is danger of pathogenicity, suitable precautions must be taken, either by killing the cells before harvesting (2% formalin) or by adopting strict safety precautions.

Control of homogeniety of the culture is very necessary, and the culture should be examined microscopically at all stages for contaminants. If the final growth stage is carried out in numerous vessels, the contents of each vessel should be checked before bulking.

Attention should be given to the composition of the culture medium, to the speed of growth and to the phase of vegetative growth at which the cells are harvested. In the past it has often been customary to harvest both slow-growing and fast-growing organisms early in the stationary phase of growth. This practice may ensure repeatability from one preparation to another, provided cell death and consequent autolysis is avoided. The process of autolysis may often involve enzymic attack on cell wall polymers with consequent change in composition, and should therefore be avoided. Autolytic activity inherent in many types of cells is much decreased in stationary cultures (Shockman, 1965); on the other hand, cells from stationary cultures often have thickened walls and are therefore more difficult to break than logarithmically growing cells. Alteration in growth rates due to changes in composition of media is known to alter overall cell wall composition through change in the proportions of various polymers (Ellwood and Tempest, 1969). Ideally, cultures should be grown under constant conditions in a chemostat; this ideal not being generally attainable, comparisons should be made on walls from cells in as similar a physiological state as possible.

It is customary to wash harvested cells once or twice with cold water or 0·9% NaCl; if however, strongly autolytic cells are to be examined, this step may be omitted. Some degree of undetected wall breakdown may occur in species of bacterial cells where autolysis is not expected, even in the short time it takes to spin down organisms from culture medium, resuspend them and treat so as to inhibit autolytic enzymes. To obtain completely undegraded mucopeptide of walls from *Escherichia coli*, it was necessary immediately after stopping aeration to squirt the culture into boiling water or to pass it through a coil immersed in boiling water (Weidel *et al.*, 1963). This procedure, although stabilizing the mucopeptide component of the wall, may have a detrimental effect on wall ultra-structures, and may also prevent subsequent removal of all cytoplasmic contents from the walls. Alternative methods of preventing lytic enzyme attack, where it may occur readily, is to keep cells at all times at 2°C or to suspend the cells in formalin (Pickering, 1966) or in 1–4% sodium dodecyl sulphate solution (Thompson, private commun.): the latter method will affect ultra-structure of Gram-negative bacterial cell walls. Undegraded "soluble mucopeptide" is prepared as described later (III.B.2).

Walls may be made from fresh or lyophilized cells; acetone-dried cells have also been used, but generally with less success. If wet fresh bacterial cells have to be stored before breakage, they should be resuspended, not kept as a pad in the bottom of a centrifuge tube, as this encourages autolysis, even at 0°C.

Certain types of cells may require preliminary treatment to get them into suspensions suitable for mechanical disintegration. Fungal hyphae may have to be shortened; this can be done by treatment for 5 min in a high-speed blender (Crook and Johnson, 1962); this step usually serves also to separate spores, most of which are in the filtrate when the broken hyphae are recovered by filtration. Cells which contain large amounts of lipid are difficult to disintegrate because of clumping, and are best subjected to preliminary solvent extraction at room temperature. Thus, bacillus species which contain poly-β-hydroxybutyric acid can be extracted with chloroform (Humphrey and Vincent, 1962), while the lipids of mycobacteria and Actinomycetales can be removed by extraction with alcohol, ether and chloroform (de Wij and Jolles, 1964) or with alkaline ethanol (Cummins and Harris, 1958; Yamaguchi, 1965). The gelatinous sheath of many blue-green algae also prevents disintegration; it can be removed by preliminary treatment with DNAase (0·1 μg/ml) (Höcht *et al.*, 1965).

B. Disintegration of cells

1. *General*

Wall fractions from microbial cells are usually prepared from mechanically broken cells by removal of extraneous material by washing, differential centrifugation and enzymic digestion. There are many different ways of carrying out each step; the methods selected will depend on the organism, the facilities available, the amount of wall required and the final aim of the investigation. Thus, it may be desired to produce a complete envelope fraction, as free as possible from cytoplasmic contamination, but which has retained its original biological, chemical and morphological characteristics; on the other hand, the object may be to examine the overall chemical composition of the mechanically and chemically stable components of the walls.

Cells for disintegration are suspended in a suitable medium such as water, buffer or salt solution; in some cases 1–4% dodecylsulphate can be present; this inhibits autolytic activity (Thompson, private commun.). Of the many methods available for disintegrating microbial cells (see Hughes, Wimpenny and Lloyd, Volume 5B, page 1), only certain ones invariably give reliable, pure, preparations of walls from bacteria and related organisms. The few techniques described here are chosen because there is minimum fragmentation of walls and optimal removal of extraneous materials during subsequent steps. Various other methods have been used successfully in selected cases, but cannot be relied on to produce clean homogeneous undegraded wall preparations from most types of microbial cells. Thus, the Hughes press gives "envelope" fractions rather than walls, even of Gram-positive bacteria; sonic or ultrasonic machines lead to solubilization of certain wall constituents, while stirring with small glass beads in a Waring blender or Virtis homogenizer abrades the metal blades and produces walls contaminated with fine metal particles. Autolysis under toluene is not recommended as it produces considerable degradation of walls.

Methods for disintegration of algae, yeast and fungi are usually similar to those used for bacteria, although it may be necessary to experiment with several methods (including some not advised for bacteria) to find the one best suited to the species under investigation (see Novaes et al. (1967), where a different method of disintegration was used for each of 3 species of Phycomycete). Bacterial spores are more resistant to disintegration than are vegetative cells and need more vigorous treatment.

Breakage of the cell wall of micro-organisms involves the expenditure of considerable amounts of energy, with consequent heat production. It is essential that preparations are kept cold during disruption to avoid cytoplasmic materials getting denatured and adhering to the inner wall surfaces; the more successful types of apparatus provide efficient cooling during the process. Shaking the cells with small glass beads (Ballotini) 0·1–0·2 mm diam (grade 12) at high speeds is a technique used in several methods of disintegration. The Mickle disintegrator (1948) is widely used on a small scale (up to 400 mg dry wt of cells per cycle of approximately 20 min for making walls of bacteria); it has to be operated in a cold room, with frequent stops to cool the pots. Disintegration on a larger scale (6 g dry wt bacteria per cycle) is carried out at 0°C in a shaker head designed to run in a refrigerated International Centrifuge model PR 2 (Shockman et al., 1957; Shockman, 1962). This apparatus proved very useful in the author's laboratory for preparation of bacterial cell walls (e.g. Allsop and Work, 1963); unfortunately it is now out of production, if however a model is available it should be used (taking precautions to protect centrifuge against undue vibration). It consists of 4 steel pots (60 ml capacity) fixed to horizontal arms; when fitted over the centrifuge spindle, it is subjected to a complicated motion resulting in rapid shaking of the mixture in the pots when the centrifuge is run at a speed of 1100–2000 rpm. Disintegration usually takes about 20 min, but longer periods are required for certain organisms.

The Braun tissue disintegrator (Merkenschlager et al., 1957) also provides a satisfactory method of cell disruption using Ballotini beads (Bleiweis et al., 1964). The cell suspension is shaken in a glass bottle (capacity, 65 ml) placed horizontally in a chamber connected to an eccentric cam; the chamber is shaken at 2000–4000 oscillations per min. A stream of liquid CO_2 delivered to the chamber cools the assembly, but care must be taken not to use too much cooling and so freeze the mixture. Disintegration is extremely rapid (3–5 min) and the temperature remains below 4°C (Huff et al., 1964); an air space equivalent to 20% of the total volume of the bottle is essential to disintegration. The instrument (manufactured by B. Braun Melsungen Apparatebau, Melsungen, Germany) is relatively cheap compared with most other cell disintegrators considered here; it can even be made in a well-equipped laboratory workshop.

If it is suspected that the cell walls contain alkali-labile groups, glass beads must be carefully selected, since many of them liberate alkali when violently shaken. After use, the beads are filtered, washed and cleaned with nitric acid or cleaning fluid before being used again.

The pressure cell of Milner, Lawrence and French (1950) (see Hughes, Wimpenny and Lloyd, Volume 5B, page 1), designed for breaking chloroplasts and first used for breaking bacteria for enzymic studies (Hoare and Work, 1955), has proved to produce good preparations of walls of bacteria and other organisms. The Aminco model produced by American Instrument Company, Silver Springs, Md, U.S.A., has a capacity between 5 and 40 ml. The press can be pre-cooled by standing overnight at 2°C, and the cell suspension is not heated appreciably during disruption as the heat capacity of the press is large; however it cannot be used on more than two batches in succession because the press will have warmed up appreciably by the

TABLE I

Conditions for breakage of various microbial cells in the Ribi refrigerated cell fractionator. Data from manufacturer's leaflet (Ivan Sorvall Inc., Norwalk, Connecticut, U.S.A.)

Organism	Species	Cell pressure (p.s.i. on 1·5 in. pressure cell)	Dry wt. of suspension of organism (mg/ml)
Fungi	*Penicillium chrysogenum*	20,000–30,000	100
Yeast	*Saccharomyces cerevisiae*	40,000	—
Bacteria	*Streptococcus pyogenes*	30,000–40,000	
	Streptococcus lactis	50,000	
	Escherichia coli	20,000	up to 50
	Salmonellas	15,000–20,000	
	Mycobacteria	20,000–25,000	
Bacterial spores		50,000	50
Rickettsiae	*Coxiella burnetii*	50,000	5

third batch. For larger quantities of cells, an adaptation of this press providing cooling at the nozzle (Ribi *et al.*, 1959) was developed. A commercial model is the Ribi Refrigerated Cell Fractionator (Ivan Sorvall Inc., Norwalk, Connecticut, U.S.A.); continuous cooling with liquid nitrogen is applied and dissipates the heat generated at the orifice. As many batches as desired can be run through without dismantling the apparatus, by means of a single-acting hydraulic valve. The apparatus is however very costly, and should not be used by the inexperienced as the needle valve can be easily damaged. Table I gives some of the applications and working conditions recommended by the manufacturers. It will be seen that very high working pressures are recommended for certain organisms, although the usual pressure is about 20,000 psi; above 25,000 psi a gradual fragmentation of walls is reported. With pressures of less than 7000 psi, the extracts are very viscous, owing to the fact that DNA is not depolymerized under those conditions; above 15,000 psi, DNA is depolymerized and viscosity is no problem.

In all methods of cell disintegration, no standard conditions can be given; the time or pressure varies from one type of organism to another, and even with change in growth conditions. Only by microscopical examination of the cell suspension at intervals during the course of disruption can the right conditions be assessed.

2. Practical details of cell disintegration

(a) *Resuspension of cells.* Efficient disruption, no matter which method is used, is hindered unless the cells have been homogeneously suspended.

This can be one of the most tedious processes in the preparation of walls from bacterial cells when using the usual method for resuspension of pads of micro-organisms in centrifuge tubes (sucking and blowing with a pipette). If this method is employed, care must be taken to avoid contamination of the preparation with saliva, which contains lysozyme: a cotton wool plug is essential. Mechanical aids to resuspension are recommended, especially when large volumes of cells are to be broken. The Silverson Laboratory Mixer (Silverson Machines Ltd, 55 Tower Bridge Road, London, S.E.1, England) is very efficient in resuspending bacterial cells or walls. A small mixing head is immersed in liquid just above the cell pad; the liquid is drawn into the base of the mixing head and expelled at high speed from the sides, thus setting up a pattern of circulation which in a few minutes produces a homogeneous cell suspension with no appreciable heating effect.

Lyophilized cells should be suspended in stages, the first of which is to make a smooth paste by adding liquid slowly to the cells with hand stirring, the paste can then be further diluted by mechanical means.

(b) *Mickle disintegrator.* Cold cell suspension (10 ml), containing 10–20 mg dry wt of cells/ml, is mixed with 10 ml of Ballotini beads No. 12 in each of two plain flat-bottomed glass cups closed with rubber stoppers protected by cellophane sheet. The cups are screwed into the machine in the cold room, and shaken at the maximum amplitude of about 5 cm and at mains frequency for 5 min: the cups are then removed and recooled in ice. The process is repeated for 10–30 min, according to species, until disruption is complete. Foaming can be largely eliminated by addition of one drop of octanol to each cup; if lipid structures are to be examined in detail it might be wiser to omit this, but disintegration will probably take longer.

(c) *Braun disintegrator.* Cell suspension (30 ml), 20–50 mg dry wt/ml, is mixed with 20 ml of Ballotini beads No. 12 in the stoppered glass bottle. The bottle is secured in the inner chamber of the machine, and compressed CO_2 is passed through for about 0·5 min (if the time is too long, liquid in the bottle may freeze). The bottle is shaken at 3000 strokes/min with continued use of cooling liquid, for $3\frac{1}{2}$–5 min according to species.

(d) *French press.* The cold suspension (5–40 ml), containing 33 mg dry wt cells/ml, is placed in the well of the pressure cell held upside down on a special stand supplied by the manufacturers, the pressure cell having been previously cooled by standing overnight at 2°C. The needle valve is closed and the cold piston is inserted, the unit is placed on the platform of a hydraulic press standing on a home-made disc to fit under the curved base, and a home-made collar is slipped over the top to position the piston (both necessary to prevent the cell from upsetting when pressure is applied). The press is pumped and the top of the piston is carefully directed into the recess in the top plate of the press. When the pressure (indicated on dial) has reached 5 tons psi, the needle valve is carefully opened so as to allow the passage of a slow stream of liquid without fall in internal pressure. Pumping and regulation of the valve are continued until the apparatus is emptied of liquid. Only one more batch of suspension can be treated without recooling the cell, otherwise the apparatus will not be sufficiently cold to prevent heating of liquid emerging from the orifice.

(e) *Ribi press.* Manufacturer's instructions should be followed, see also Table I.

C. Recovery and purification of walls from broken cell suspensions

1. *Inactivation of autolytic enzymes*

If the whole cells have not been treated to inactivate autolytic enzymes, the broken cell suspension should be kept at 2°C and treated as soon as possible, even before removal of glass beads. A short heat treatment is most often used; 10–15 min at temperatures varying from 60–100°C is advocated by various workers, the lower temperature should be selected if morphological studies are to follow. Sodium dodecylsulphate (2% w/v) is known to inactivate certain autolytic enzymes (Shockman, Thompson and Conover, 1967; Thompson, private commun.) and might therefore be a useful reagent for preventing autolysis in broken cell suspensions of organisms other than Gram-negative bacteria (whose lipid layers will be affected by the detergent).

2. *Removal of cytoplasmic membrane and cytoplasm*

After filtration through coarse grade sintered glass to remove beads, unbroken cells are separated from the filtrate and washings by a low-speed centrifugation step (usually 500–1000 g for 7–10 min) and the crude walls are sedimented at higher speeds. Such walls are contaminated by membrane fragments and cytoplasmic contents. These can sometimes be removed simply by multiple washings at the centrifuge with saline and buffers (Barkulis and Jones, 1957). In other cases, particularly after heat treatment which probably precipitates proteins on the walls, digestion with proteolytic enzymes and DNAase or RNAase is necessary. Trypsin is the most widely used proteolytic enzyme (Cummins and Harris, 1965); it has the advantage that it acts at pH not far removed from neutrality, important where acid or alkali-labile groups are present in wall polymers, as with several bacterial species. Washing with saline and/or buffer is still important, even after enzyme treatment. Phosphate buffer (0·1 M) is commonly used, but where detailed analysis of phosphorus-containing components is to be carried out, tris buffer should be used instead, as inorganic phosphate has proved to be impossible to remove completely from some bacterial wall preparations (Knox and Hall, 1965).

Although many proteins of cell walls are resistant to digestion with trypsin, it should always be realized that there may be trypsin-sensitive proteins, as indeed is the case with streptococcal M protein (Barkulis and Jones, 1957), a protein in staphylococcal walls (Grov and Rude, 1967) and a protein in *E. coli* buried deep in the wall (Leutgeb *et al.*, 1963).

The use of sodium dodecylsulphate to inactivate autolytic enzymes (see p. 368) may help in subsequent isolation of walls through disaggregation of the cytoplasmic membrane (Shafa and Salton, 1960); treatment of the walls of Gram-positive bacteria with this detergent was found to remove protein contaminants which may have originated from the membrane (Sutow and Welker, 1967).

There are many variations in the details of the methods used in separating walls from the rest of the cellular constituents; some are determined by the type of cell under investigation and the characteristics to be examined, others are derived from personal preferences of the investigators. Bacterial contamination of cell walls during preparation can be a source of trouble, and must be avoided by the use of a bacteriostatic agent in all liquids. A convenient agent is chloroform, all reagents being made up in chloroform-saturated water; this should not be used where studies on ultrastructure of lipid-containing walls are to be made, in this case sodium azide or some other bacteriostat should be used. Simple reduction in working temperature should not be relied on to prevent growth of contaminants since microorganisms frequently grow on wall preparations stored at $+2°C$.

Density gradient centrifugation (sucrose or glycerol gradients) has sometimes been used as a step in preparation of cell walls of bacteria or related organisms. This technique has particular advantages in the case of small organisms, such as Rickettsiae, where intact organisms are difficult to separate from crude walls (Ribi and Hoyer, 1960; Wood and Wisseman, 1967). Some studies have been made on density gradient centrifugation of separated bacterial wall preparations (Roberson and Schwab, 1960; Yoshida *et al.*, 1961; Allsop and Work, 1963; Huff *et al.*, 1964; Work and Griffiths, 1968). A pad of clumped walls or whole cells nearly always appeared at the bottom of the gradient; in some cases there were discrete bands (Yoshida *et al.*, 1961) with different chemical compositions (Allsop and Work, 1963; Work and Griffiths, 1968); in other cases (*Staphylococcus aureus*, Huff *et al.*, 1964; *E. coli*, Work and Knivett, unpub.) broad bands were apparent which were obviously polydisperse and may have been due to fragmentation of the walls. A preparative zonal centrifuge head (Anderson, 1967) should be useful for preparing large amounts of homogeneous walls.

3. Practical details

(a) *Typical preparation of 1 g quantity of bacterial walls* (*Allsop and Work,* 1963). See Table 2. A suspension of bacterial cells (5 g dry wt), broken in the Shockman head of the International centrifuge or by any of the methods described in section II B 2, is kept cold until disruption of the batch is complete, and is then heated at 60°C for 10–15 min. The Ballotini beads, if present, are removed by filtration through a sintered glass filter and washed three times with an equal volume of water. The filtrate and washings are combined and centrifuged at 1000 g for 7–10 min to remove unbroken cells. The crude walls are sedimented from the supernatant liquid at 20,000–25,000 g for 20 min, suspended (II B 2(a)) in 200 ml of 0·1 M phosphate buffer* pH 7·8 containing 0·1% (w/v) crystalline trypsin and a few mg of DNAase and shaken at 37°C until the turbidity ceases to fall (6–18 h). Any material subsequently sedimenting at 1000 g is discarded, and the walls in the supernatant fraction are sedimented at 15,000–20,000 g for 15–20 min, and then washed at the centrifuge successively with three washes each of 0·9% NaCl*, 0·1 M phosphate buffer pH 7·0 and water*. The final preparation can be stored as a pad in the cold (caution contaminants!) or, preferably, is lyophilized.

Essentially the same methods apply to walls of other types of organisms; sedimentation of walls from larger cells does not usually require such high speeds.

(b) *Density gradient centrifugation.* Linear gradients (0–40% w/v sucrose in 1 M NaCl) are prepared in centrifuge tubes of any convenient size. The following example uses 250 ml glass bottles which have good capacities.

The gradients are made as follows (Ribi and Hoyer, 1960); 80 ml of 1 M NaCl and 80 ml of 40% (w/v) sucrose in 1 M NaCl (both previously filtered through fine paper or Millipore HA filters) are each placed in two flat bottomed containers A and B respectively. A has an outlet at the base leading into B through rubber tubing supplied with a screw clip. B also has an outlet tube at the base opposite to the inflow; it leads to a short length of capillary tubing directing the flow of liquid to the side of the neck of the tilted centrifuge bottle. A screw clip controls the outflow from B; first it is opened sufficiently to allow liquid to emerge drop by drop. After 5 ml of sucrose has run out of B, the clip between A and B is opened and the NaCl runs in and merges with sucrose at the bottom of B where the solutions are stirred by a vibromixer. The drops emerging from B slide

* All reagents are made up in chloroform-saturated water.

TABLE II
Scheme for typical preparation of bacterial cell walls

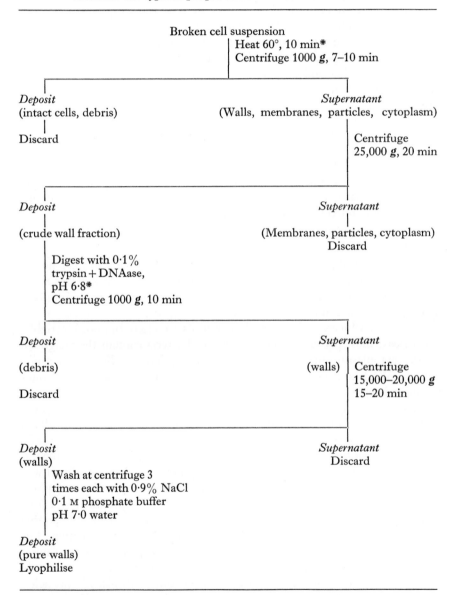

Broken cell suspension
Heat 60°, 10 min*
Centrifuge 1000 g, 7–10 min

Deposit
(intact cells, debris)

Discard

Supernatant
(Walls, membranes, particles, cytoplasm)

Centrifuge
25,000 g, 20 min

Deposit

(crude wall fraction)

Digest with 0·1%
trypsin + DNAase,
pH 6·8*
Centrifuge 1000 g, 10 min

Supernatant

(Membranes, particles, cytoplasm)
Discard

Deposit

(debris)

Discard

Supernatant

(walls) Centrifuge
15,000–20,000 g
15–20 min

Deposit
(walls)
Wash at centrifuge 3
times each with 0·9% NaCl
0·1 M phosphate buffer
pH 7·0 water

Supernatant
Discard

Deposit
(pure walls)
Lyophilise

* These steps are omitted in certain circumstances, e.g. where morphological studies are to be carried out.

down the side of the tilted centrifuge bottle and should cause no turbulence of liquid already present. Finally the bottle is stood upright for a few minutes. Gradients thus produced can be kept (in the cold) for several days.

Walls (100–200 mg), evenly suspended in 10 ml of 1 M NaCl, are layered carefully on to the top of the gradient. The bottles are centrifuged for 30–60 min in a swing-out head at 1400 *g*. The individual bands are collected by pipette, pooled and centrifuged, and the pads are either resuspended in M NaCl and subjected to another gradient centrifugation, or dialysed free of sucrose.

III. EXAMINATION OF WALLS

A. Criteria of purity

The main contaminants likely to be present in cell wall preparations are cytoplasmic particles, fragments of cytoplasmic membranes, or debris from external sources.

1. *Appearance*

A good, typical preparation of bacterial cell walls usually forms a homogeneous, colourless translucent pad in a centrifuge tube. Dark debris at the bottom of the pad usually originates from ingredients of the wash liquids or from washers of centrifuge tubes; it can be separated by scraping off the overlying translucent wall material. The walls are often, but not invariably, colourless; most pigmented Gram-positive bacteria contain their pigment in the cytoplasmic membrane (with the exception of *Micrococcus radiodurans*, Work and Griffiths, 1968). The location of pigments in Gram-negative bacteria is not yet clearly defined.

Macroscopic appearance is not a good criterion of homogeneity, nor is the appearance in the light microscope, except to detect unbroken cells or extraneous bacterial contamination. The electron microscope is a useful tool for assessing purity of wall preparations. The simplest type of preparation is a negatively-stained one, prepared by mixing a drop of diluted wall suspension on a carbon-coated microscope grid with one drop of phosphotungstic acid (2% w/v, adjusted to pH 7·0). After approximately 1 min, most of the liquid is withdrawn from the grid by a Pasteur pipette, so that a thin film remains on the grid. This is left to dry before examination. The resulting preparation should have a homogeneous appearance in the microscope and consist entirely of empty collapsed rods or spheres. There should be no extraneous material such as unbroken cells, cytoplasmic particles, fragments of membrane or flagellae; such inhomogeneity if observed, indicates that further purification is required. Walls from Gram-negative bacteria may not appear as homogeneous as those from other types of micro-organisms. This is because the membranous layers from these walls may partly peel off and round up into small spheres which will then be seen associated with the larger walls. In fact, the purity of walls from this type of organism is very difficult to establish.

2. Composition

The presence of nucleic acids in wall preparations should indicate cytoplasmic contamination, but is not easy to establish by direct spectroscopic examination owing to the light scattering produced by the wall suspensions. Examination of phosphate buffer washings from walls for absorption at 260 nm could establish the presence of nucleic acids in the preparation (Barkulis and Jones, 1957). Examination of the amino-acids can be very informative in hydrolysates of walls of Gram-positive bacteria prepared as Table II. Since these wall preparations are usually substantially devoid of proteins, the presence of major proportions of many amino-acids other than the 3 to 5 mucopeptide amino-acids (see C 2(d)) will indicate contamination with proteins from membrane or cytoplasm. (Minor proportions of protein amino-acids are frequently found in such wall preparations but their significance is not known.) This is not the case with walls from Gram-negative bacteria and other micro-organisms, where trypsin-resistant proteins are major wall components.

B. Separation of polymers

Cell walls of micro-organisms provide the necessary rigidity and mechanical strength required by cells capable of living an independent existence; this rigidity is produced by one or more insoluble polymers. The type of polymer present depends on the type of organism. In many algae, cellulose provides the main rigid unit of the wall; in other algae, there are xylans or mannans; occasionally chitin is found. Yeasts appear to contain both mannans and glucans bound to glycoproteins, and sometimes have chitin; most filamentous fungi contain chitin, cellulose being less commonly found. Bacteria (other than certain halophiles) and the related blue green algae, rickettsiae and organisms of the psittacosis group, all have a unique type of polymer not so far found anywhere else in nature. This is mucopeptide (known also as glycopeptide, peptidoglycan, glucosaminopeptide or murein). This insoluble polymer has a repeating backbone of N-acetyl-amino sugars cross-linked by short peptide bridges. Walls from all types or organisms may also contain other associated macromolecules such as polysaccharides, proteins and lipids; in addition, bacterial walls may have other specific polymers such as teichoic acids and lipopolysaccharides.

1. Yeast, algae and filamentous fungi

The lyophilized walls are extracted with hot ether or with 1/1 (v/v) chloroform/methanol at 4°C under nitrogen to remove lipid. They are then subjected to a series of acid and alkaline extractions which will remove various polymers (Table III); the soluble polymers are precipitated from solution with ethanol; they can be further treated with Pronase to remove protein contaminants (Sentandreu and Northcote, 1968). A preliminary hydrolysis of the wall residue and qualitative examination of its sugar content will suggest which insoluble polymers to expect (e.g. preponderance of glucosamine suggests chitin, preponderance of glucose suggests α-cellulose). Final positive identification of cellulose or chitin in insoluble fractions can be made by means of enzymes. Cellulose is hydrolysed by the

TABLE III

Methods for solubilization of polymers from walls of yeasts, algae and filamentous fungi

Treatment	Reference	Polymers dissolved	Polymers not dissolved
3% w/v NaOH, 6 h, 100°C	1	mannans proteins part of glucans	glucans hemicellulose α-cellulose
Ethylenediamine, 3 days, 37°C	2	mannan-protein part of glucan-protein	glucan-proteins
24% KOH 2 h, 25°C under N₂	3	xylans mannans hemicelluloses	α-cellulose
10% NaOH 4 h, 100°C under N₂	3		α-cellulose
N-HCl 100°C	4	chitosan	chitin
conc. HCl 10 min 4°C	4	chitin	
72% H₂SO₄ 48 h, 25°C	5	chitin	
Chitinase (Cal Biochemicals)	6	chitin	
Pronase	6	some proteins	
β-glucanase	7	α-cellulose	

References: 1. Northcote and Horne, 1952.
          2. Korn and Northcote, 1960.
          3. Northcote, Goulding and Horne, 1960.
          4. Applegarth, 1967.
          5. Grisaro et al., 1968.
          6. Chattaway et al., 1968.
          7. Anderson and Millbank, 1966.

β-glucanase from a crude preparation of snail digestive juice (Suc digestif d'*Helix pomatia*, Industrie Biologique Française, Gennevilliers, Seine, France). The purified glucanase is prepared from this source by the method of Anderson and Millbank (1966), using fraction A which should be devoid of chitinase activity. Alkali-insoluble wall fraction (10 mg) suspended in 50 ml of McIlvaine citrate-phosphate buffer pH 5·8 is shaken with 2·5 mg of β-glucanase at 30°C for 1 h. Glucose will be liberated and can be estimated.

Chitinase (California Biochemicals Ltd) can be used as positive identification of chitin in the insoluble residue, possibly after previous treatment with pronase and β-glucanase (Chattaway *et al.*, 1968). Insoluble wall fraction (5 mg) suspended in 5 ml of 0·5 M phosphate buffer pH 6·3 is shaken with chitinase (10 mg) at 36°C for 22 h: glucosamine is liberated. Chitin may also be identified by the chitosan sulphate reaction (Roelofson and Hoette, 1951). The dried alkali-insoluble residue of the walls is placed on microscope slides, mixed with one drop of 15% H_2SO_4 and covered with a cover slip. The slides are warmed to 80°C on a sand bath and then slowly raised just to boiling point over a microburner and immediately returned to the sand bath to cool down to 80°C. The presence of chitin in the original material is indicated if there are minute crystals of chitosan sulphate, similar to those formed from a standard chitin sample. A polarizing microscope is an advantage in identifying the crystals which are usually about 3–10 μm. The crystals may be characterized by staining with fuchsin (1% aqueous) or picric acid (1%); when examined under low magnification with crossed nicols, they appear as "stars" coloured red or yellow according to the stain used. Since dust particles are also stained, examination under high power should also be carried out.

2. Bacteria

Teichoic acids and various polysaccharides can often be extracted from bacterial walls with trichloroacetic acid solutions or hot formamide. Lyophilized wall preparation (20 mg) is stirred in a glass or metal centrifuge tube with 10 ml of formamide in an oil bath at 150°C for 15 min (Perkins, 1963). The mixture is cooled and mixed with 25 ml of acid-ethanol (2 N HCl : ethanol, 1 : 19, by vol). The insoluble residue is removed by centrifuging and washed with acid-ethanol, ethanol and ether and is finally dried *in vacuo*. Extractable materials are obtained from the formamide supernatant fluid by precipitation with 10 vol of acetone, dissolving in water and dialysis; the formamide insoluble wall residue, on shaking vigorously with water at room temperature, also yields soluble polymers.

Shaking walls with trichloroacetic acid (10% w/v, 60 ml per g of walls) at 2°C for periods up to 4 days usually removes most but not all of the teichoic acids; extracted teichoic acid is precipitated with 4 vol of ethanol (Armstrong et al., 1958). With certain organisms (e.g. Lactobacillus arabinosus) hot 10% (w/v) trichloroacetic acid may be required to remove all the teichoic acids (Archibald et al., 1961). It should be pointed out that total removal of teichoic acids will probably degrade them to shorter polymers in the process through contact with acid (Sanderson et al., 1962; Strominger and Ghuysen, 1963). If undegraded teichoic acids are required, the extraction time should be shorter (e.g. 16 h). Addition of ethanol to such extracts usually produces a white precipitate of teichoic acid, which can be purified by redissolving in 10% trichloroacetic acid and reprecipitating with ethanol.

For quick, qualitative examination of teichoic acids, extraction with 2 N HCl for 24 h at 20–22°C has been used in screening for major differences in components: the extracts, being already acid, are hydrolysed directly (Wolin et al., 1966).

Other polysaccharides, e.g. teichuronic acids, containing acidic sugar residues (glucuronic acid, or other uronic acids, Janczura et al., 1961; Perkins, 1963) can be removed by repeated extractions with 5% w/v trichloroacetic acid (50 ml/g of wall) at 35°C for 3–4 days. The trichloroacetic acid is removed from the extracts by repeated extraction with ether, and the ether is removed by gassing with N_2. Teichoic acids, but not teichuronic acids, are removed by 0·1 N NaOH at 35°C under N_2 (Hughes and Tanner, 1968).

Identification of the polysaccharides is carried out after hydrolysis and examination of the constituent sugars. Teichoic acids are identified by their high phosphorus content and by the presence of polyols. Any of these polymers may have serological activity. Mucopeptides being insoluble and usually less reactive biologically, can only be identified after hydrolysis and analysis of their specific constituents (Section III.C.2).

These methods of extracting polymers from cell walls are all relatively drastic and may cause breakage of covalent bonds either in the extracted polymers, or in the insoluble mucopeptide, or between polymer and mucopeptide. The mucopeptide may also become formylated by formamide, either on the sugar hydroxyl groups, or on free amino groups (Heymann et al., 1964; Perkins, 1965b). The susceptibility of the mucopeptide residue to digestion by lysozyme may be quite different from that of the original wall preparation, due either to removal of blocking polymers, or through substitution on hydroxyl or amino groups (Perkins, 1967; Work, 1967).

Less drastic methods of isolation of polymers from cell walls of Grampositive bacteria involve the use of various muralytic enzymes which degrade and solubilize the mucopeptide and at the same time release the associated polymers (see Section IV) (Krause and McCarty, 1961; Haukenes 1962; Ghuysen et al., 1965; Hughes, 1965; Knox and Hall, 1965; Bricas *et al.*, 1967). It is a common finding that a certain proportion of the poly-saccharides or teichoic acids so prepared are covalently linked to muco-peptide. Indeed, this has also been the case with certain solvent-extracted polymers (Smith *et al.*, 1956; White, 1968). This suggests that in the cell wall the polymers may be interlinked to some extent, probably through mucopeptides.

Preparations of "soluble mucopeptide" are often used to obtain homo-geneous starting material for structural studies. They are made by digesting the cell walls, suitably treated to remove other polymers, with lysozyme (Schleifer and Kandler, 1967) or some other muralytic enzyme to which the mucopeptide is susceptible (Ghuysen *et al.*, 1968). In a typical preparation (Hughes, 1968), trichloroacetic acid-extracted walls of *Bacillus licheni-formis* (1·0 g) are suspended in water and pH is adjusted to 6·5 with dilute ammonia; chloroform or toluene (1 ml) and lysozyme (10 mg) are added and the suspension is shaken overnight at 37°C. The mixture is heated at 100°C for 10 min, and the insoluble residue is removed by centrifuging, washed and discarded.

Soluble mucopeptide can be prepared (Hughes, 1968) directly from grow-ing cultures of Gram-positive bacteria so as to avoid any danger of break-down by autolysis during harvesting. Trichloroacetic acid (50%, w/v) is added to the culture (100 ml per 1 of culture) and the mixture is shaken at 35°C overnight. The insoluble residue obtained after centrifugation is again shaken 16 h at 35°C with trichloroacetic acid (50%, w/v) and then washed with water. The solid, suspended in 10 mM ammonium acetate (25 ml) containing 5 mg of lysozyme and chloroform, is digested overnight at 35°C and treated as above.

The main polymers of walls of Gram-negative bacteria are different from those of the Gram-positive organisms. There is only 3–10% of mucopeptide (compared with 40–90% in Gram-positive); the remainder of the wall consists of lipoproteins, lipopolysaccharides and polysaccharides. In the intact cell and the freshly isolated wall, the mucopeptide is protected against attack by added muralytic enzymes by layers of other polymers, probably lipoproteins. The mucopeptide is however very susceptible to the cells' own autolytic enzymes which must be inactivated immediately on harvesting if undegraded mucopeptide is required (Weidel *et al.*, 1963). Unlike the majority of wall polymers of Gram-positive bacteria, the poly-mers of Gram-negative cell walls are located in layers, many of which can be clearly seen in the electron microscope and which, in one case, have been specifically and successively stripped by various reagents (Weidel *et al.*, 1960; Leutgeb *et al.*, 1963). Various methods have been used for the preparation of mucopeptides from walls of Gram-negative bacteria (Leutgeb *et al.*, 1963; Mandelstam, 1962; Schocher *et al.*, 1962; Graham and May,

1965), but none can be recommended as yielding an undegraded preparation free from other polymers.

Treatment with hot 45% (w/v) aqueous phenol (see Sutherland and Wilkinson, Volume 5B, p. 345) will extract most of the lipopolysaccharide, phospholipids, lipoprotein, protein and polysaccharide components of the walls. On cooling, the aqueous phenol separates into two layers, the top one contains lipopolysaccharide and polysaccharides, while the lower layer and the interface yield proteins and phospholipids. A mild technique for specifically extracting lipopolysaccharides from rough forms of bacteria uses a monophasic solvent consisting of liquid phenol-chloroform-petroleum ether at temperatures below 20°C (Galanos et al., 1969). The method should be applicable to the extraction of R-type lipopolysaccharides from cell walls.

Sugars may also be identified by gas chromatography combined with mass spectrometry, a technique devised by Hellerquist et al. (1969); if knowledgeable collaborators are available, this technique should be considered as it is very sensitive. Proteins may be recovered from phenol extracts of walls by removal of phenol by dialysis. If it is desired to examine phospholipids separately, they are best extracted with methanol and then chloroform from the lyophilized walls before phenol extraction.

While it is probable that the lipopolysaccharide is derived from an exterior layer of the wall (Knox et al., 1966; Mergenhagen et al., 1966), it is unclear whether the lipoprotein component originates only from an exterior layer, or whether it also comes from the cytoplasmic membrane which may be present in the wall preparations. These polymers are all serologically active.

C. Techniques for examination of constituents of separated polymers or of whole walls

1. Hydrolysis of polymers

Constituents of polymers can only be identified and estimated after the polymers have been broken down into their monomeric building blocks. This usually involves a hydrolytic step, which may cause damage to some of the more labile components; this damage is particularly likely to occur in the presence of mixtures of different types of constituents. Ideally, the optimal hydrolysis conditions should be determined for release of each component in walls of each species of organism (Salton and Pavlik, 1960; Sutow and Welker, 1967). In practice, it is usually satisfactory to adopt a routine method of hydrolysis for each type of polymer; if accurate determination of a specific constituent is desired, corrections can be made by extrapolation to true values from a time-course hydrolysis plot, the constituent being estimated after various hydrolysis times.

(a) *Polysaccharides*. A mild acid hydrolysis is usually used to liberate neutral sugars (1–2 N acid (HCl or H_2SO_4), 2–6 h at 100–110°C, 1 ml acid/5 mg of wall, in a sealed tube). It is not advisable to remove HCl by evaporation, even at room temperature, since concentration of the acid will occur and some sugars may be destroyed. Either of the acids may be removed by treatment with an anion exchange resin (e.g. Dowex 2) in the bicarbonate form; or H_2SO_4 can be removed by addition of solid barium carbonate until the hydrolysate is neutral, then removing solids by centrifuging or filtering. The neutralized hydrolysate is concentrated by evaporation *in vacuo*. For quantitative work, it is advisable to set up a control hydrolysis with a known amount of the component to be estimated added to the walls. Carboxylated sugars (uronic acids) are more labile than neutral sugars, and hydrolysis with 0·5–1 N HCl for 1–3 h at 100°C is usual (Janczura *et al.*, 1961). Amino sugars are more resistant to acid hydrolysis than neutral sugars and need somewhat stronger conditions of hydrolysis (minimum conditions are 3 N HCl for 4 h at 95°); they are however destroyed to some extent.

(b) *Mucopeptides*. Amino-acids are usually released from mucopeptides by hydrolysis with 4 N–6 N HCl for 16 h at 105°C. HCl is removed by drying *in vacuo* over NaOH, with three successive additions and removals of water. Before the final drying, black humin may be removed by filtration. It is usual to examine the amino sugars of mucopeptides (glucosamine and muramic acid) in the hydrolysate prepared for amino-acids analysis; but, if HCl is removed by evaporation, there is a risk that some muramic acid may be destroyed. For quantitative work, exact neutralization with NaOH is recommended (Ghuysen *et al.*, 1966) and milder conditions of hydrolysis (see para 1 (a)). It should be remembered that the amount of amino sugar present in a hydrolysate represents the sum of release and degradation reactions. When using the stronger hydrolysis conditions used for amino-acid release, a correction can be applied for destruction, either specifically by extrapolation to zero time, or more generally using the figures 1–5% per hour (Young *et. al.*, 1963). For examination of peptide fragments from mucopeptides, partial hydrolysis is carried out in 4 N HCl at 100°C for 0·5–2 h (Schleifer and Kandler, 1967).

(c) *Proteins*. Similar conditions to mucopeptides.

(d) *Lipopolysaccharides*. The lipid A portion of the molecule is split by hydrolysis in 1 N H_2SO_4 for 30 min at 100°C; it separates out from solution. The oligosaccharides remaining in solution are hydrolysed for a further 15 h at 100°C to liberate heptose and other sugars. To test for the presence of 3-deoxy-2-oxo-octonoic acid, the lipopolysaccharide is hydrolysed for 25 min at 100°C with 0·25 NH_2SO_4 (Taylor et al., 1966), or by an even milder method, in 25 mM ammonium acetate buffer pH 4·6, for 1 h at 100°C (Knox, 1966).

(e) *Teichoic Acids*. Ester-linked D-alanine is removed from teichoic acids by very mild alkaline hydrolysis, for example using 2 N ammonia for 3 h at 37°C or N NaOH or LiOH (2 ml per 6·5 mg teichoic acid) for 3 h at room temperature (Ellwood et al., 1963). Ammonia is removed by evaporation to dryness *in vacuo*, keeping the temperature low; Dowex 50 (NH_4^+ form) is used to remove other bases. This type of hydrolysis does not cleave any phosphodiester bonds either in the main polymeric chain or in the linkage to muramic acid (Sanderson et al., 1962).

The hydrolysis procedures of most use in determining the general structure of teichoic acids are acid hydrolysis and a combination of alkali hydrolysis and dephosphorylation of the products with a phosphatase. Acid hydrolysis is carried out in 2 N HCl (0·2 ml per 2 mg teichoic acid) for 3 h at 100°C. HCl is removed *in vacuo* over NaOH or by chromatography through a column (0·8 × 3 cm) of Dowex 3 (OH-form). Alkaline hydrolysis uses N NaOH (0·2 ml per 2 mg teichoic acid) for 3 h at 100°C; sodium ions are removed by passage of the hydrolysate through a short column (2 cm × 1 cm) of NH_4^+ form of Dowex 50 or Zeocarb 225. Hydrolysis with phosphomonoesterase is carried out on the eluate after concentration *in vacuo*.

Teichoic acids can also be cleaved with HF without significant hydrolysis of glycosides (Glaser and Burger 1964); this can be a useful additional hydrolytic method since there are no degradative products or phosphates to be identified. Teichoic acid is dried overnight in polyethene tubes in a vacuum desiccator and incubated with 0·2 ml of 60% HF at 0°C for 3–5 h. The sample is cooled to − 60°C lithium hydroxide sufficient to neutralize HF is added and the solution is allowed to warm to 0°C. The pH is adjusted to 7 with lithium carbonate and the insoluble LiF is removed by centrifugation and washed with water. The supernatant and washings are concentrated and put on a column prepared from 1 g of Darco G 60 charcoal and 1 g of Celite. The column is washed with water and eluted with 10% ethanol.

2. Identification of constituents of polymers

(a) *Sugars*. For identification of neutral sugars found in polysaccharides see Herbert, Phipps and Strange, Volume 5B, p. 209.

The sugars of lipopolysaccharides are described by Sutherland and

Wilkinson, Volume 5B, p. 345). The characteristic heptose of lipopoly-saccharides gives a colour with the modified Dische reaction (Osborn, 1963), and affords an easy way of identifying this type of polymer. Carboxylated sugars give a positive carbazole (Dische, 1947) reaction. Their identification is for specialists.

(b) *Hexosamines*. The 2-amino-2-deoxyhexoses, D-glucosamine and D-galactosamine are the most widely distributed amino sugars; glucosamine and its derivative muramic acid, are constituents of all mucopeptides. Other rarer amino sugars found in polysaccharides associated with bacterial walls are 2-amino-6-deoxyhexoses, D-quinovosamine and D- and L-fucosamine; monoamino-di-deoxy and diamino-tri-deoxy-hexosamines, mannosamine; uronic acid derivatives of glucosamine, galactosamine and mannosamine. Identification of these rarer amino sugars cannot be dealt with in this review, but is described by Crumpton (1959) and Wheat (1966) who cover paper electrophoresis, paper chromatography and ion exchange chromatography.

The common hexosamines, glucosamine, muramic acid and galactosamine, can be identified by paper chromatography or electrophoresis; they give purple spots with ninhydrin, and rather weak red spots with Ehrlich's reagent. To carry out the Ehrlich reaction (Partridge, 1948), spray the chromatograms with 1% acetylacetone in butanol freshly mixed with 1/20 volume of a mixture of 4 vol ethanol + 1 vol 50% KOH; heat at 100°C for 5 min. Then spray with p-dimethylaminobenzaldehyde (1 g in 30 ml ethanol + 30 ml conc HCl + 30 ml n-butanol). Heat at 105°C.

Chromatograms are best run on paper, previously washed with 0·1 M BaCl₂, irrigated 48 h with butanol : pyridine : water (6 : 4 : 3, by vol.) (Heyworth *et al.*, 1961). Paper electrophoresis (7–10 v/cm) can be carried out in M-acetic acid. By these means glucosamine, galactosamine and muramic can be separated and identified against markers.

Final identification of these hexosamines can be carried out, if necessary by conversion to the corresponding pentose by ninhydrin treatment (Perkins, 1963). The hexosamines as spots on paper separated by chromatography or electrophoresis, are eluted and heated for 30 min at 100°C in a capillary tube with an equal volume of 2% (w/v) ninhydrin in a 4% (v/v) solution of pyridine in water. The reaction mixture is then examined by ascending paper chromatography in butanol-acetic acid-saturated boric acid (9 : 1 : 1 by vol) and the pentose spots are revealed by spraying with aniline phthalate (Partridge, 1949). The mobilities of the conversion products are compared with those of standard marker pentoses (R_{ribose} values = arabinose, 0·27; xylose, 0·39; lyxose, 0·44). Arabinose and lyxose are derived respectively from glucosamine and galactosamine (Stoffyn and

Jeanloz, 1954), while an acidic pentose with an Rf less than any of the marker pentoses is produced by muramic acid (Strange and Kent, 1959).

(c) *Amino-acids.* The presence of significant amounts of proteins in cell walls is immediately apparent after examination of the amino-acid content of the hydrolysate by any of the methods commonly in use. A preparation having mucopeptide as its main amino-acid-containing constituent, will have no more than five major amino-acids; these are invariably aline, glutamic acid and one dibasic amino-acid such as diaminopimelic acid, lysine, ornithine or diaminobutyric acid; sometimes glycine, aspartic acid, serine or threonine may also be present. A chromatogram from 2 mg of a hydrolysed wall preparation from a species of Gram-positive bacterium (containing no protein), presents a relatively simple picture with at the most seven major spots represented by the aforementioned amino-acids and glucosamine and muramic acid; relatively faint spots corresponding to some or all the protein amino-acids may be present. A protein-containing cell wall (e.g. from a Gram-negative bacterial species) presents a more complicated picture, with 12 or more spots of the predominant protein amino-acids and diaminopimelic acid. Two-dimensional chromatography with the systems phenol-water (ammonia atmosphere) and butanol-acetic acid-water (63–10–27 by vol) is a useful means of examining the wall for gross content of amino-acids and amino sugars (Allsop and Work, 1963). Other systems use the solvents isopropanol-acetic acid-water (75 : 10 : 15 by vol) followed by α-picoline-conc. ammonia-water (70 : 2 : 28 by vol) (Schleifer and Kandler, 1967), or butanol-pyridine-acetic acid-water (60 : 45 : 4 : 30 by vol) (Esser, 1965) or ethanol: ammonia: water (85 : 2 : 13) and ethanol: water (85 : 15) on Kodak K5 IIV chromatographic papers (Tinelli, 1966). The dibasic amino-acids can be identified in the hydrolysate by chromatography on Whatman No. 1 paper in methanol-water-pyridine-12 N HCl (32 : 7 : 4 : 1 by vol) (Rhuland *et al.*, 1955). On dipping the dried chromatogram into ninhydrin dissolved in acetone (0·1% w/v)* and heating at 105°C for 5 min, the different dibasic amino-acids give permanent spots of characteristic colours (Table IV), while other amino-acids give transitory purple spots. (*Caution*—spots do not appear if chromatograms have been in contact with collidine or lutidine vapours.) The mobilities of all the diabasic amino-acids (Table IV) are lower than those of other naturally-occurring amino-acids, with the exception of cystine (identical with *meso*-diaminopimelic acid) and glycine (identical with lysine). Cystine is not found in mucopeptides, while the glycine colour fades rapidly. Hexosamines do not give coloured spots under these conditions, so do not interfere. Paper electrophoresis at pH 11·5 (0·025 M sodium

* Store at +2°C.

TABLE IV

Mobilities and colours of spots given by the dibasic amino-acids on chromatograms on Whatman No. 1 paper. Solvent, methanol : water : pyridine : 12 N HCl (32 : 7 : 4 : 1, by vol); chromatograms dipped in ninhydrin (0·1% w/v) in acetone, heated 5 min at 105°C. On storage at 100m temperature, colour changes occur, as shown.

Amino-acid	^{R}lysine	Colour with ninhydrin
3-OH-Diamopimelic	0·32	Green → yellow
Meso-Diaminopimelic	0·48	Green → yellow
DD-Diaminopimelic	0·42–0·48	Green → yellow
Diaminopropionic	0·5	Grey-brown
LL-Diaminopimelic	0·65	Green → yellow
2-4-Diaminobutyric	0·71	Grey
Ornithine	0·83	Red-brown
Lysine	1·00	Purple → brown

(From Perkins and Cummins, 1964.)

TABLE V

Order of elution of mucopeptide constituents on Amino-acid Autoanalyser (Technicon). Operating conditions as in Schmidt, 1966; gradient 1, temperature 60°C throughout.

Constituent	Postition relative to known peaks	Separation from neighbouring peaks	Colour value
Aspartic			4·98
Threonine			5·12
Serine			5·56
Muramic	Bet'ser and glu	Just separated from glu	3·08
Homoserine	Bet'muramic and glu	Some overlap with both	3·50
Glutamic			5·12
Glycine			5·56
Alanine			5·38
Glucosamine			5·16
3-OH-Diaminopimelic	Just faster than met	Not fully separated	6·04
*Diaminopimelic	Between leu and norleu	Just separated	6·40
Diaminobutyric			4·40
Ornithine			6·42
Lysine			6·22

* Norleucine cannot be used as an internal standard when diaminopimelic acid is present.

carbonate) will also distinguish between the basic amino-acids (lysine, ornithine and diaminobutyric acid), whose mobility varies directly with chain length (Perkins and Cummins, 1964).

All the amino-acids and amino sugars of mucopeptides can be identified, and also estimated, on the amino-acid auto analyser (Table V). If the conditions are right, there is little or no interference from the other natural amino-acids, so the specific amino-acids can be detected in the presence of proteins.

Some other unusual amino-acids which have been identified in muco-peptides are homoserine, 3-hydroxy-diaminopimelic acid (Perkins, 1965a and c), ε-(aminosuccinoyl) lysine (Swallow and Abraham, 1958; Cummins and Harris, 1958) and threo-3-hydroxy-glutamic acid (Schleifer *et al.*, 1967). Homoserine is present in acid hydrolysates partly as the lactone, which behaves as a base and on paper stains yellow with ninhydrin. The lactone can be converted to the open-chain form by heating for 5 min at 100°C with 10 N ammonia. ε-(Aminosuccinoyl) lysine is a stable neutral cyclic peptide originating, during hydrolysis, from the ε-(aspartyl)-lysine linkage present in the mucopeptides of several species of Gram-positive bacteria (Swallow and Abraham, 1958). Prolonged acid hydrolysis converts it to aspartic acid and lysine. On automated amino-acid analysers it appears as a peak near the tryptophan position.

(d) *Lipids*. The separation and identification of extractable lipids is described by Sutherland and Wilkinson, Volume 5B, p. 345. The bound lipids of lipopolysaccharides are only released after saponification (see Sutherland and Wilkinson, Volume 5B, p. 345).

(e) *Teichoic acid*. The teichoic acids most commonly encountered as major components of walls of Gram-positive Eubacteriales are phosphodiester-linked polymers of either glycerol or ribitol carrying glycosidically-linked sugars and ester-linked D-alanine (Baddiley, 1968; Archibald and Baddiley, 1966). They usually contain molar proportions of phosphorus and polyols, with equal or lesser proportions of sugars and alanine. The sugars are usually either glucose of N-acetyl glucosamine and the polymer chains may vary from 7 to 20 units in length, but may be broken during extraction procedures.

Exceptional teichoic acids, as for example the ribitol teichoic acids from various streptomycetes, may lack D-alanine but succinate (as mono ester) or acetate may be present. In others, the sugar is present in the main polymer chain, attached to the polyol by phosphodiesters or glycosidic linkages, as in teichoic acids from *Bacillus subtilis*, *Staphylococcus lactis*, *Pneumococcus*; other sugars may be present (e.g. N-acetyl galactosamine and a diamino-trideoxy hexose). D-Alanine is not always ester-linked to the polyol, it can be bound to a side chain sugar residue by a less labile link. While it may not be possible for the inexperienced worker to arrive at the exact constitution of a certain teichoic acid, a good deal of information as to constituents and chain length can be obtained by using a few standard techniques, such as acid and alkaline hydrolysis (IIIC.1.b) and splitting by phosphatases and other enzymes (described on p. 387), usually followed by paper chromatography and estimation of total and inorganic phosphorus (see Herbert, Phipps and Strange, Volume 5B, page 209), or periodate oxidation (see p. 392).

Hydrolysis of teichoic acids with dilute ammonia or normal alkalis removes ester-linked D-alanine which appears in the hydrolysate as the free amino-acid and its amide. A more specific test for ester-linked alanine is its release as alanine hydroxamate by reaction with hydroxylamine (Ellwood et al., 1963). Teichoic acid (1·0 mg) and 0·5 M hydroxylamine (freshly prepared from the hydrochloride by the treatment with Dowex 1, OH⁻ form) are heated at 37°C for 2 h; alanine hydroxamate is identified in the products by paper chromatography in n-propanol-water-0·5 M hydroxylamine (pH 7·4)–0·1 M EDTA (8 : 17 : 2 : 1, by vol). A spot (Rf = 0·43) is revealed by spraying with ferric chloride reagent for hydroxamates (a saturated solution of ferric chloride in n-butyl alcohol saturated with water). Kinetic studies on the rate of release of alanine from teichoic acid can be carried out using alanine methyl ester as a standard (Archibald et al., 1968). A series of tubes containing 1 ml of 0·05 M hydroxylamine and 1 mg of teichoic acid are heated at 45°C. Samples (0·5 ml) are removed at intervals, mixed with N HCl (1 ml) and M FeCl$_3$ (1 ml). The concentration of hydroxamate is determined colorimetrically at 540 nm.

Acid hydrolysis of ribitol teichoic acid gives, among other compounds, inorganic phosphate, 1,4-anhydroribitol, ribitol mono- and diphosphates, 1,4-anhydroribitol phosphates and sugars (Table VI). Anhydroribitol and inorganic phosphate are produced by acid decomposition of ribitol phosphates; while formation of anhydroribitol phosphates and ribitol diphosphates indicates the presence in the polymer of ribitol residues joined to each other through phosphodiester linkages. Glycerol teichoic acids yield on acid hydrolysis D-alanine, glycerol and its mono- and di-phosphates, inorganic phosphate and a sugar; the formation of both glycerol and its diphosphates indicates that here again the polymer contains glycerol units joined by phosphodiester groups.

Alkaline hydrolysis of ribitol teichoic acids (e.g. one containing glucose, as in B. subtilis) produces mainly isomeric glycosylribitol monophosphates; there are small amounts of glycosylribitol, and of diphosphates which arise from the ends of the chain (Table VI). Subsequent dephosphorylation with alkaline phosphatase produces a large amount of glycosylribitol,

TABLE VI

Some products of hydrolysis of teichoic acids, and their behaviour on paper chromatograms

	Present in hydrolysate		Mobility in solvent[1]			Periodate Schiff Reaction[2]	Ninhydrin reaction
	Acid	Alkali	A (Rf)	B (Rf)	C (R glucose)		
Inorganic phosphates	+	trace	0·18	..	..	–	–
Alanine	+	+	0·59	..	..	–	+
Alanine amide	–	–	0·70	..	..	–	+
Glucose	+	–	0·56	0·27	1·0	S, W	–
Galactose	+	–	..	..	0·9	S, W	–
Glucosamine	+	–	0·56	0·21	0·73	S, W	+
Galactosamine	+	–	0·51	0·21	0·62	S, W	+
Ribitol	+	–	0·65	0·30	..	R	–
1,4-Anhydroribitol	+	+	0·75	..	0·47	S	–
1,4-Anhydroribitol-5-phosphate	+	..	0·29	..	..	S	–
Ribitol phosphates	+	..	0·33	..	..	R(Y)[3]	–
Ribitol glucosaminide	+	+	0·56	..	..	R	+
Ribitol monoglucoside	–	+	0·55	..	..	R	–
Ribitol diglucoside	–	+	0·40	..	..	R	–
Glycerol	+	+	0·75	0·49	..	R	–
Glycerol monophosphates	+	+	0·32	0	..	R	–
Glycerol diphosphates	+	+	0·13	0	..	–	–
Glucosyl glycerol	–	+	..	..	..	S, R	–
Glucosaminyl glycerol	–	+	..	0·21	..	S, R[4]	+
Galactosaminyl glycerol	–	+	low	0·21	..	S, R[4]	+

1. See text p. 388 for description of solvents.
2. S = slow reaction,   W = weak reaction,   R = rapid reaction.
3. Y = yellow colour given by 3 isomer, which is always present in acid hydrolysates.
4. Speed and colour depends on whether glycosyl residue is on 1 or 2 position on glycerol.

inorganic phosphate and some ribitol. Similar types of results are obtained with glycerol teichoic acids.

Hydrolysis of teichoic acids or their degradation products with enzymes can also give much useful information. Phosphodiesterases do not attack teichoic acids. Phosphomonoesterases can be used to remove and estimate monoesterified phosphate end-groups from intact teichoic acids, or to degrade further the products of alkaline hydrolysis. Either calf intestinal phosphatase or *E. coli* alkaline phosphomonoesterase can be used (both obtainable from Sigma Chemical Co., St. Louis, Mo). The reaction with calf enzyme is carried out in either ethanolamine-HCl or 0·1 M ammonium carbonate buffers at pH 9·5; *E. coli* enzyme is used at pH 8·5. Usually 0·1 mg enzyme is used per mg teichoic acid. Incubation is continued at 37°C until the level of inorganic phosphate is constant; total and inorganic phosphate are then estimated. The dephosphorylated products are then hydrolysed either in 2 N HCl for 2 h at 100°C or by specific enzymes and examined for sugars and polyols.

Digestion with α and β-glucosidases (Worthington Biochemical Corp) can show the type of linkage of glucose residues in whole polymers or in fragments; either or both α and β-links being found. α-Glucosidase digestion is carried out in 0·06 M phosphate buffer pH 6·8 containing 10^{-3} M mercaptoethanol; β-glucosidase is used in 0·05 M acetate buffer, pH 5·0. The type of link between hexosamines and the polymer chain may be found by using acetyl-hexosaminidases; here teichoic acid can be used directly, but its hydrolysis products will have to be re-acetylated before digestion. This is done with 6 mg of product in the presence of Dowex 2(CO_3^- form) (1 ml) + methanol (0·1 ml) in 2 ml of water. The mixture is cooled to 0°C, acetic anhydride (0·03 ml) is added and the mixture is shaken at 4°C for 90 min. The resin is removed and washed in water; supernatant and washings are concentrated and passed through a small column of Dowex 50 (H^+ form). Washings and eluate are concentrated and a fraction, containing about 0·5 mg, is treated at 38°C for 18 h with β-acetyl-glucosaminidase from pig epididymis (Levvy and Conchie, 1966) in the presence of 0·1 M citrate buffer pH 4·2, 0·1 M NaCl and 0·01% bovine serum albumin. The digest is freed from salts by passage through small columns of Dowex 50 (NH_4^+ form) and then Dowex 2 (HCO_3^- form). Eluates and washings are concentrated and examined by paper chromatography. Both glucosamine and galactosamine are split off if they are present in β-linkages. Enzyme from rat epididymis (Roseman and Dorfman, 1951) contains both α- and β-acetylglucosaminidases and therefore will free all hexosamines. In this way the proportion of α- and β-linked hexosamines may be found (Sanderson *et al.*, 1962). The proportions are known to vary in different strains of *Staphylococcus aureus* (Torii *et al.*, 1964).

Hydrolysis products of teichoic acids are a complex mixture of substances which can be often identified by paper chromatography (Table VI) (Armstrong et al., 1958; Ellwood et al., 1963; Archibald et al., 1961). General separation is effected in solvent A (n-propanol-aq. ammonia (sp. gr. 0·88)-water (6 : 3 : 1, by vol)) by ascending chromatography on Whatman No. 4 paper. Polyols may be separated on Whatman No. 1 in solvent B (n-butanol-ethanol-water-aq. ammonia (sp. gr. 0·88) (40 : 10 : 49 : 1, by vol); organic phase). Polyols are detected by periodate-Schiff reagent as follows: dried paper is sprayed with 1% aqueous sodium metaperiodate, left for 5–10 min, treated with SO_2 and then sprayed with Schiff's reagent (1% aqueous suspension of pararosaniline hydrochloride treated with SO_2 until a pale straw-coloured solution results). Colours are allowed to develop at room temperature (see Table VI for results). Phosphate esters and inorganic phosphate are revealed by perchloric acid molybdate reagent, as follows: the dried paper is sprayed with a solution containing 60% (w/w) perchloric acid (5 ml), N HCl (10 ml), 4% (w/v) ammonium molybdate (25 ml), water to 100 ml. After removing excess water in a current of warm air, the paper is heated at 85°C for 7 min, is then allowed to regain moisture from the air, after which it is hung for 5–10 min in a jar containing dilute H_2S gas. Phosphate esters now appear as intensely blue spots. There is a buff background, which gradually turns blue owing to atmospheric ammonia; the buff background can be regenerated by exposure to HCl fumes. Sugars and amino-acids are separated by descending chromatography (Whatman No. 1) in solvent C (pyridine-ethyl acetate-water-acetic acid 5 : 5 : 3 : 1, by vol) and revealed by alkaline silver nitrate or ninhydrin as appropriate.

It should also be possible to identify and estimate ribitol and glycerol in teichoic acid hydrolysates by thin layer chromatography of their trimethylsilyl esters. Gregory (1968) has separated ribitol, anhydroribitol (produced by acid hydrolysis of ribitol) and glycerol as follows. The hydrolysate dried by addition and evaporation in vacuo of chloroform (redistilled and stabilized by addition of 1% absolute ethanol) was dissolved in pyridine (10·2 ml); to this was added successively hexamethyl disilazane (0·1 ml) and trimethylsilyl chloride (0·05 ml). The sample was stoppered, shaken for a few minutes, and centrifuged. The supernatant was separated, pyridine was removed from it by distillation in vacuo at 45°C, addition of chloroform and re-evaporation. Finally the product was dissolved in 0·2 ml chloroform. Thin layer chromatography in benzene on plates precoated with Merk Silica Gel F 254, gave spots which were visualized by U.V. fluorescence as the plate dried, and by spraying 50% H_2SO_4 followed by charring at 150°C. Rf values were as follows: ribitol 0·92; anhydroribitol, 0·34; glycerol 0·67.

Gas chromatography on a 5 ft, 3·5% SE 52 on 85–100 mesh silanized

Diatomate-C glass column at 130°C, with carrier gas flow rate of 45 ml/min gave the following retention times—ribitol (2 peaks) 32·8 and 38·5 min, anhydroribitol 11·7 min, glycerol 2·7 min. Hexoses run very much more slowly and do not interfere.

Gas chromatography of silanized glucosyl-glycerols has also been described (Brundish and Baddiley, 1968).

Constituents of hydrolysis products from teichoic acids can be separated into groups by ion exchange chromatography. In one method (Ellwood *et al.*, 1963) the acidic (phosphorylated) products are separated from the non-acidic products. Teichoic acid (217 mg) is hydrolysed with N KOH (5 ml) and adjusted to pH 7·0 with 72% (w/v) perchloric acid then cooled to 0°C. Insoluble potassium perchlorate is centrifuged off and washed with a little ice-water. Supernatant and washings, concentrated to 1 ml, are applied to a column of Dowex 2 (HCO_3^- form, 50 ml). The column is washed with water (500 ml); all the "non-acidic" fragments will be in the percolate and washings. Acidic material is eluted from the column with 10% (w/v) ammonium carbonate (250 ml) followed by water (100 ml). Combined eluate and washings are evaporated to dryness *in vacuo*; the residue is dissolved in water (10 ml) and again evaporated. This is repeated five times to remove ammonium carbonate.

Chromatography of alkaline hydrolysates on Dowex 1 (Cl^- form) will separate some of the phosphate esters (Sanderson *et al.*, 1962). Alkaline hydrolysate of teichoic acid (5 mg) is adjusted to pH 9·0 with N HCl and then passed over a column of Dowex 1 Cl^- (1×30 cm 200–400 mesh, 8% cross-linked) followed by 50 ml of water. The column is then eluted at a rate of 0·35 ml/min with a linear gradient of HCl (reservoir contains 0·02 N HCl (300 ml) and mixing flask contains water (300 ml)). Phosphate esters may also be separated on DEAE-cellulose (CO_3^{2-} form) eluted with a linear concentration gradient (0–0·15 M) of $(NH_4)_2$ CO_3 pH 9·0. A column 60 cm × 2 cm was used for 109 mg teichoic acid which had been hydrolysed with N NaOH, the mixing vessel contained water (21) and the reservoir contained 0·15 M $(NH_4)_2$ CO_3 (21), flow rate was 2 ml/min (Archibald *et al.*, 1968).

3. *Analysis of cell walls and polymers*

Cell walls, and some of the separated polymers, are not soluble in water, and sampling for analysis is usually carried out on aqueous suspensions. It is essential that these suspensions are absolutely homogeneous; this is sometimes difficult to attain owing to the tendency of some walls to clump. In such cases, it may be necessary to shake the suspension with large glass beads (not Ballotini), or even to sonicate.

The following analyses are of value when examining whole walls, particularly at different stages of extraction; total N, total and organic P, total carbohydrate (see below), protein, extractable and bound lipid. Changes in these substances as a result of various extraction procedures will often give information as to which polymers are being removed. For details of procedures see Herbert, Phipps and Strange, Volume 5B, page 209; and Sutherland and Wilkinson, Volume 5B, page 345. Other analyses are carried out after preliminary hydrolysis (III.C.I.4).

The amount of cross-linking of the peptide chains of mucopeptides is also important to assess. This can be done by determining the content of diamino-acid in the preparation before and after dinitrophenylation of the wall (III.C.3.b), the difference between the two figures representing the amount of diamino-acid with a free end amino group; this can be confirmed by direct estimation of the mono-DNP amino-acid separated from the dinitrophenylated walls (VI.B.2.b). Similarly, the alanine released on hydrazinolysis of the walls (see IV.B.2.c) will give a measure of the carboxy-terminal groups. Cross-linking of the peptide chains is usually between the terminal carboxyl group of D-alanine and the free amino group of the diamino-acid, so measurement of these groups will give information on the degree of cross linkage. Since glutamic acid, diaminopimelic acid and aspartic acid are frequently present as amides in mucopeptides, determination of amide ammonia is important. Acetyl residues, bound to $-OH$ or $-NH_2$ groups of amino sugars or amino-acids, may also have to be estimated (III.C.3.c).

(a) *Estimation of total carbohydrate* (Dubois *et al.*, 1956; Allsop and Work, 1963). Walls (400 μg) in 2 ml of water are mixed (in a test-tube of diameter 16–20 mm) with 50 μl of aqueous phenol (80% w/v), and 5·0 ml of conc. H_2SO_4 is added rapidly from a fast-flowing pipette directly on to the surface. The mixture heats up sufficiently for the colour to develop, and after standing for 5 min, the tubes are cooled in tap water. The absorbance at 480 nm is read against a reagent blank; the colour is stable for several hours. A glucose solution is used as a standard, and the results can be expressed as glucose equivalent.

This reaction is relatively little affected by proteins; hexosamines do not react.

(b) *Dinitrophenylation of walls and estimation of free amino groups* (Ingram and Salton, 1957). Walls (50 mg) suspended in 5 ml of water containing 500 mg sodium bicarbonate are shaken with 10 ml of 5% (v/v) ethanolic solution of 1-fluor-2,4-dinitrobenzene for 5 h at 37°C in the dark. The dinitrophenylated walls are deposited by centrifugation, washed successively four times each with ether, ethanol and water, and are dried. The walls are hydrolysed (4 N HCl, 16 h, 100°C) and HCl is removed in the usual way. The hydrolysate is shaken with three batches of equal volumes of ether, the ether extract is evaporated to a small volume. The DNP-amino-acids are chromatographed on paper using 1·5 M phosphate buffer pH 6·0, and n-propanol : 0·2% ammonia (8 : 2 by vol) as solvents, or on thin layer plates of silica gel G as described later (p. 399). Substitution at the free amino groups has caused the DNP-amino-acids to loose their dipolar ionic characteristics and to become soluble in ether, but mono-DNP-derivatives of diamino-acids still maintain their polar character and are water-soluble; these can be identified on chromatography both by their yellow colour and by the fact that they give a purple colour with ninhydrin. DNP-amino-acids are estimated as described on p. 399.

(c) *Estimation of acetyl groups. Total (O and N) acetyl* (Lodowieg and Dorfman, 1960). The sample is deacetylated with methanolic HCl; the acetyl groups are converted to methyl acetate which is distilled off and estimated colorimetrically as a hydroxamic-ferric complex.

Reagents. Alkaline hydroxylamine (prepared just before use) by mixing equal volumes of 0·35 M hydroxylamine HCl and 1·5 N NaOH.

Perchloric acid, 0·75 M.

Ferric-perchloric solution, 1·9 g of $FeCl_3.6H_2O$ dissolved in 5 ml of conc HCl, and mixed with 5 ml of 70% perchloric acid. Solution is evaporated almost to dryness and diluted to 100 ml with water; acidity is between 0·55 and 0·65 M (Stable 1 month in cold).

Methanolic HCl (2 N). Absorb dry HCl into cold anhydrous methanol. Titrate against standard alkali and dilute to 2 N with methanol. Protected from moisture, it is stable for a few months in the cold.

Procedure. Sample containing 1·0–10 μmoles of acetyl residues is evaporated in a Pyrex tube to dryness over anhydrous $CaCl_2$ using an oil pump. Methanolic HCl (0·5 ml) is added, the solution is cooled to −70°C and the tube is sealed. The mixture is heated in boiling water for 4 h and cooled. The tube is opened and placed at the bottom of a distillation tube with a side arm leading down through a bung or ground joint to near the bottom of a receiving tube with a side arm connected to a $CaCl_2$ tube and a water pump. The receiving tube is immersed in solid CO_2-methanol and the apparatus is evacuated. The distillation tube is immersed in a water bath at 35–40°C and all the liquid is distilled from the sample tube which is then washed with 0·5 ml methanol and the distillation is repeated. The receiving tube is removed and placed in a bath at 20–25°C. After 2 min, 1·0 ml of water is added, followed by 2·0 ml of alkaline hydroxylamine. The tube is shaken, and left for 10 min; 2 ml of 0·75 M perchloric acid is added, and mixed by shaking, then 1·0 ml of ferric-perchloric solution.

The optical density is measured at 520 nm against an HCl-methanol blank, and the amount of methyl acetate formed is estimated by comparison with standard ethyl acetate (0·005 M ethyl acetate in methanol: water (1 : 1 v/v), diluted suitably with methanol : water) treated in the same way as the sample.

N-acetyl groups. Walls are stirred for 18 h at room temperature with 0·01 N NaOH to remove O-acetyl groups and then washed three times with water. The washed walls are dried and weighed before estimating acetyl groups as before. The difference between the two readings represents O-acetyl. An alternative method of estimating O-acetyl groups can be used without prior formation of methyl acetate by direct reaction of walls with alkaline hydroxylamine, and estimation *in situ* of aceto-hydroxamic acid (Hestrin, 1949).

Formaldehydogenic end-groups (Periodate oxidation). The non-phosphory-lated end of ribitol teichoic acids is a 4-O-glycosyl ribitol-5-phosphate which may also be substituted on the 2- or 3-position of the ribitol by an alanine ester. If alanine is removed by dilute ammonia, periodate will rapidly oxidize the C_1–C_2 diglycol of the terminal ribitol unit, producing formalde-hyde. No other primary-OH groups which could yield formaldehyde are present in teichoic acids. Thus measurement of the formaldehydogenic groups in relation to total phosphorus content has been used to estimate chain length of teichoic acids (the other method is to measure the propor-tion of total phosphate present as phosphomono ester by measuring the amount released by phosphomonoesterase treatment (p. 387).

The oxidation is carried out as follows (Ellwood *et al.*, 1963); teichoic acid without alanine residues (p. 384) (47 mg/ml water) is mixed with an excess of 0·1 M sodium metaperiodate (1 ml) and kept at 4°C in the dark; control solutions without teichoic acid are also included. Samples (0·05 ml) are withdrawn at intervals to check periodate consumption (change in absorbance at 223 nm of solution diluted 250 times). When there is no further decrease in periodate concentration (90 h), samples are removed for determination of total phosphorus and formaldehyde. To determine formaldehyde, 0·5 ml of reaction mixture is mixed with 10% sodium bisulphite (0·5 ml) to remove residual periodate. To 1 ml of this is added 5 ml of chromotropic acid reagent (4,5,-dihydroxy-2,7-naphthalene sulfonic acid, 1 gm/100 ml water plus 450 ml of 24 N H_2SO_4 stable in dark for 2 weeks). The mixture is heated for 30 min at 100°C, cooled and treated with 0·5 saturated thiourea (0·5 ml) which destroys the dark colour of the reagent. Absorbance is read at 570 nm against that of a reagent blank. Standard formaldehyde is provided by a sample of pure hexamethylene tetramine (5 mg) hydrolysed in 2 N H_2SO_4 (5 ml).

If difficulty is experienced in estimating formaldehyde directly in the oxidation mixture (Hay *et al.*, 1965), it can be estimated by the Conway diffusion technique (Conway, 1962). 0·4 ml sample is pipetted into the outer chamber of a Conway unit and 0·8 ml of chromotropic acid reagent is placed in the inner well. After greasing and sealing the units, distillation is allowed to proceed in the dark for 15–18 h. The contents of the inner well are then transferred to a stoppered test tube and heated in boiling water in the dark for 30 min. The tube is then cooled in ice, allowed to come to room temperature and the absorbance is read at 570 nm.

4. *Quantitative analyses on hydrolysates of walls and polymers*

(a) *Sugars*, see Herbert, Phipps and Strange, Volume 5B, p. 209.

(b) *Hexosamines*. Prior to most colorimetric estimations, the hexosamines should be separated from neutral sugars, if present, by absorption on Dowex 50, to avoid the possible interaction of neutral sugars with the colour reagent. Dowex 50 or Zeocarb 225- × 4, 200–400 mesh (H^+ form) is washed successively on a Buchner funnel with 2 N NaOH, water, 3 N HCl, water, and excess moisture is removed by suction. 5 ml of 1 : 1 (w/v) aqueous suspension of the washed resin is pipetted into a column 10 mm in diam. The hydrolysate, after removal of HCl, is placed on the column which is washed with 15 ml of water; the effluent and wash are discarded. The hexosamines are eluted with 2 N HCl (10 ml) and the eluate is dried *in vacuo*, or HCl is removed by ion exchange (p. 389).

Total hexosamines. The most satisfactory method of estimating total amino sugars is by the Morgan–Elson reaction in which N-acetylamino sugar is assayed. Preliminary conversion of free amino sugar to the N-acetyl derivative is quantitative.

Micro estimation of hexosamines by Morgan–Elson reaction (Ghuysen *et al.*, 1966). Hydrolysates (3 N HCl, 4 h at 95°C) containing 10–100 mμ moles of total hexosamines are exactly neutralized with an equivalent amount of 3 N NaOH. 30 μl are transferred to 3 ml tubes, saturated $NaHCO_3$ (10 μl) and freshly prepared 5% (v/v) acetic anhydride (10 μl) are added. The tubes are mixed and stood at room temperature for 10 min, when N-acetylation occurs. Excess acetic anhydride is destroyed by heating the tubes in boiling water for exactly 3 min and then immersing them in cold water. 5% (w/v) potassium tetraborate† (50 μl) is added. Samples are mixed and heated in boiling water for 7 min, cooled, and mixed with 600 μl of diluted Morgan–Elson colour reagent (*p*-Dimethylaminobenzaldehyde (16 g) dissolved in acetic acid to give a total volume of 95 ml; conc. HCl (5 ml) is added. This mixture is termed the colour reagent. Before use dilute 2 vols with 5 vols acetic acid). The solutions are mixed and the colour is developed in

† See p. 409 for preparation of potassium tetraborate.

20 min at 37°C and measured at 585 nm. If the heating times are strictly adhered to, glucosamine and muramic acid have the same molar extinction coefficient of 11,500.

Individual hexosamines can also be determined colorimetrically or by paper chromatography or on the amino-acid analyser (Table V) or more specifically on a micro scale by enzymic reactions.

Colorimetric estimations of individual hexosamines. The following colorimetric method estimates glucosamine, galactosamine and muramic acid separately in crude cell wall hydrolysates (Stewart-Tull, 1968):

Reagents. Acetylacetone reagents. Made immediately before use. I. Redistilled acetylacetone (b.p. 138–140°C), 1 ml, dissolved in 100 ml of bicarbonate buffer pH 9·8 (23·02 g Na_2CO_3, 2·76 g $NaHCO_3$, 5·84 g NaCl in 1 litre water). pH readjusted if necessary. II. Methanol-triethylamine-acetylacetone-pyridine (7·3 : 2·0 : 0·6 : 0·1 by vol.).

Colour reagents. A. Recrystallized *p*-dimethylaminobenzaldehyde (0·8 g) dissolved in 100 ml of ethanol containing 3·5 ml conc. HCl. B. Dimethylaminobenzaldehyde (0·8 g) dissolved in 30 ml of ethanol and 30 ml conc. HCl. Both reagents are colourless and stable at 2°C for 14 days.

Procedure The hydrolysate of cell walls (containing 20–100 μg hexosamines) in 2 ml of water is mixed with 5·5 ml of acetylacetone reagent I and heated in a stoppered tube in boiling water for 20 min. After cooling, the mixture is transferred quantitatively (3 washes of 2 ml) into a 100 ml distillation flask connected to a 20 cm condenser (Cessi and Piliego, 1960); the contents are boiled over a Bunsen flame and 4 ml of distillate is collected over 4 min in a 10 ml cylinder containing 6 ml of colour reagent A. The extinction is measured 30 min later at 545 nm to estimate glucosamine and galactosamine. The solution (9 ml) remaining in the distillation flask is cooled to 50°C, and added to 1 ml of colour reagent B. The mixture is stood for 24 h and then extinction at 510 nm is measured to determine muramic acid.

If galactosamine is known to be present it is estimated as follows: cell wall hydrolysate (0·1–0·5 ml) in a test tube is dried *in vacuo* and incubated in the stoppered tube for 16 h at 55°C with 0·5 ml of acetylacetone reagent II. It is then rapidly dried *in vacuo* and dissolved in 6 ml of borate buffer pH 8·0 at 35°C; the stoppered tube is heated in boiling water for 20 min and cooled to 35°C. The contents are quantitatively transferred to the distillation apparatus. 4 ml is distilled as before into 10 ml of colour reagent A. After standing 30 min, extinction is measured at 550 nm to determine galactosamine.

Glucosamine and muramic acid can be separated before estimation as follows (Hughes, 1968); the hydrolysate, containing 0·2–1·0 μmole of each hexosamine, is evaporated to dryness in a small tube *in vacuo* over NaOH.

A suspension of Norit SX Plus charcoal (2·5 ml, 50 mg/ml) is added, and the tube is shaken for 10 min at room temperature. The suspension is transferred to a sintered glass filter, and the tube and charcoal are washed with small portions of water until 10 ml (containing glucosamine) has been collected. Then the charcoal is washed with 10 ml of 10% (v/v) ethanol to elute muramic acid. Both fractions are evaporated to dryness *in vacuo* and dissolved in a suitable amount of water for analysis.

Specific enzymic estimations of hexosamines. Estimations using specific enzyme reactions are carried out as follows. Glucosamine is determined in hydrolysates by conversion to N-acetylglucosamine phosphate by a mixture of enzymes from yeast; the N-acetylglucosamine phosphate is determined by the Morgan–Elson reaction (Luderitz *et al.*, 1964). The enzyme is made from 20 g of fresh yeast ground with 60 g aluminium oxide in a mortar at 4°C for 15–30 min. To the paste, 25 ml of phosphate buffer (0·1 M, pH 7·5) is added, and the mixture is centrifuged for 30 min at 0°C at 10,000 g, and the supernatant is centrifuged at 100,000 g for 70 min. To the supernatant (about 20 ml) protamine sulphate (2% w/v, 1·6 ml) is added and the mixture stood 30 min at 4°C and then centrifuged at 10,000 g. The clear supernatant is dialyzed against 1 litre of 0·002 M EDTA (pH 7·0) for 12 h and clarified by centrifuging. This preparation, which contains hexokinase, acetate-activating enzyme, and D-glucosamine-6-phosphate acetylase is stable at −15°C. The reagent mixture (0°C) consists of coenzyme A (1·2 mg), glutathione (1·5 mg), 0·02 M tris buffer pH 8 (250 μl), 0·1 M potassium acetate (50 μl), 0·25 M MgCl$_2$ (30 μl), 0·1 M ATP (90 μl) and enzyme preparation 400 μl. Samples and glucosamine standards (2·5–35 millimicromoles) are dried in a desiccator over CaCl$_2$ and NaOH. To each tube, and to a blank, 25 μl of the above reagent mixture is added. After mixing, the tubes are corked and heated at 38°C for 1 h. Then 1·3% sodium tetraborate (150 μl) is added, and the tubes are heated for 7 min at 100°C. Morgan–Elson colour reagent (p. 393) diluted with 8 vol of acetic acid is added and after heating 20 min at 37°C, extinction is read at 585 nm.

Muramic acid can be determined specifically in hydrolysates by treatment with alkali which eliminates the D-lactyl moiety; D-lactate dehydrogenase is then used to estimate the D-lactate liberated (Tipper, 1968). Cell wall hydrolysate (18 h in 4 N HCl at 100°C) containing about 50 millimicromoles of muramic acid is neutralized and incubated in 0·04 M sodium phosphate, pH 12·5 (60 μl) for 1 h at 37°C; D-lactate is determined in 20 μl. D-Lactate dehydrogenase is obtained from *Leuconostoc mesenteroides* cells grown for 20 h at 30°C in AC-1 medium. The washed cells are sonicated in 0·002 M phosphate buffer pH 7·5 (3 : 1 w/v) for 20 min. The soluble supernatant fraction obtained after centrifugation at 20,000 g for 20 min is fractionated with ammonium sulphate; the fraction precipitating between 50–75% saturation is collected by centrifugation and dissolved in water to give a solution containing 1·7 mg/ml (Dennis, 1962). The solution to be estimated (20 μl) is diluted to 75 μl with water and mixed with 1·0 M tris-HCl buffer, pH 9·0 (10 μl) and 10 μl of the acetylpyridine analogue of NAD (10 mg/ml). 4 μl of D-lactate dehydrogenase is added and the change in extinction at 363 nm is read to estimate the amount of reduced 3-acetyl pyridine-NAD formed. Standards of muramic acid alone (Cyclo Chemical Corp. Los Angeles) and of muramic acid added to the cell walls before and after hydrolysis should be set up.

(c) *Amino-acids*. All the amino-acids in a hydrolysate of cell walls can be determined on the amino-acid analyser (Table V). For separation of certain mucopeptide constituents, special conditions may have to be used. These may have to be found by trial and error. Diamino-acids can be estimated in the hydrolysate colorimetrically; since these amino-acids—with the exception of lysine—are usually confined to mucopeptides, it follows that their levels in cells may be taken as a measure of the mucopeptide content of the cells (15–20% of mucopeptide is diamono-acid).

If an amino-acid analyser is not available, the amino-acids of mucopeptides can be determined by estimation of the colour given by ninhydrin on paper chromatograms or paper electrophoresis. Quadruplicate samples of the hydrolysate are run in parallel with known quantities of amino-acids; the separation system depends on the amino-acids present; chromatography in butanol-acetic acid-water (63 : 10 : 27 by vol) or butanol-pyridine-water (6 : 4 : 3 by vol), or electrophoresis in M formic acid have been used (Mandelstam and Rogers, 1959; Perkins and Cummins, 1964; Perkins, 1965, b and c). The sheets are very thoroughly dried at 60°C, sprayed with 0·6% ninhydrin in butanol and heated at 50°C for 1 h. The amino-acid spots and several blank areas are cut out and transferred to test tubes. The purple colour is eluted by shaking with 70% (v/v) ethanol and the absorbancy is ready at 570 nm.

An alternative method uses thin layer chromatography (Esser, 1967).

Estimation of diaminopimelic acid is carried out as follows (Work, 1957, 1963):—test solution (0·5 ml), containing 5–100 μg of diaminopimelic acid is mixed with acetic acid (0·5 ml) and 0·5 ml ninhydrin reagent (250 mg ninhydrin in 6 ml of acetic acid and 4 ml of 0·6 M phosphoric acid). The solutions are covered and heated for 5 min in boiling water, then cooled in tap water, and diluted to 5 ml with 3·5 ml of acetic acid. The absorption of the yellow solutions is read at 345 or 440 nm against a reagent blank. HCl does not interfere. Other diamino-acids give similar colours except ornithine, which gives pink solutions absorbing at 515 nm. If only small amounts of lysine are present, interference with diaminopimelate measurements can be eliminated by reducing the heating time to 2 min, during which lysine does not react. In the presence of excess (4 moles excess) of lysine or of ornithine, proline or tryptophan, the reaction conditions can be changed to render the reaction specific to diaminopimelic acid, but only in the absence of HCl. The heating is carried out in the presence of 4 m of acetic acid, for 1·5 h at 37°C; absorption is read at 440 nm. The quantities can all be scaled down if necessary.

An alternative relatively specific method for diaminopimelic acid is that of Gilvarg (1958). To a sample volume of 0·13 ml, containing 5–80 μg diaminopimelic acid, are added conc. HCl (0·07 ml) and ninhydrin (5% w/v in methyl cellosolve, 0·40 ml). The mixture is heated 20 min in a boiling water bath, propanol : water (1 : 1, by vol) (4·40 ml) is added and the absorption is read at 420 nm.

(d) *Amide ammonia.* Amide ammonia is estimated after hydrolysis with 4 N HCl at 100°C for 4 h. Under these conditions production of ammonia from isoglutamine and isoasparagine is complete, while β 1,4-N-acetylglucosaminyl-N-acetylmuramic acid yields no trace of ammonia. Ammonia can be estimated on certain amino-acid auto-analysers, but not on the Technicon model. It may also be determined colorimetrically as follows, after diffusion from the alkaline hydrolysate (Teinberg and Hershey, 1960).

Reagents. Ammonia-free water, prepared by distillation from acid, is used for all solutions except saturated sodium carbonate. When stored for more than 1 week blanks of the solutions may rise.

1. Sodium carbonate, saturated solution (50 g Na_2CO_3 boiled with 100 ml of water for 10 min).

2. *Colour reagents*

   (a) Sodium carbonate (0·05 M).
   (b) Sodium nitroprusside (0·005% w/v), dilute a stock 0·5% solution 1 : 100 daily.
   (c) Sodium phenolate (2·5% w/v) (5 g phenol + 2·5 g NaOH in 200 ml water). Stand 48 h before use.
   (d) Sodium hypochlorite (0·05–0·07 N). Dilute 1 : 20 a stock solution containing 5–6% available chlorine. Titrate with sodium thiosulphate to check normality.

Semi-micro determination of ammonia. Diffusion takes place in a stoppered Warburg vessel without the centre well with 1·0 ml saturated Na_2CO_3 in the side arm and 1·0 ml of hydrolysed sample (0·03–0·4 μmole of ammonia) in the flask. A piece of filter paper moistened with 1 N H_2SO_4 is held in the flask above the sample by means of a glass tube passing through a stopper. The flask is shaken for 30 min at a constant temperature after tipping in alkali; the filter paper is transferred to a test-tube and 1 ml of water is added, followed by 1 ml of each of the colour reagents, added in the order, a, b, c, d. Stand 30 min, then add 5 ml of water or, if necessary, still more water. Read absorption at 625 nm.

Micro determination of ammonia (Tipper *et al.*, 1967). Hydrolysed sample (200 μl containing 5–20 mμ moles of ammonia) is placed in a glass vial (5 ml capacity) and partially neutralized with 80 μl of 4 N-NaOH. The vial is provided with a glass rod passing through a stopper and extending about half way down; 5 μl of 2 N H_2SO_4 is placed on the end of the rod, 200 μl of saturated K_2CO_3 is added to the hydrolysate and the vial is immediately stoppered. The vials are held by clips to a slowly rotating board (10 rpm) so that the rods are horizontal. After 1 h the rod is washed into a tube with water (5 × 20 μl), and the washings are partially neutralized with 4 μl of 2 N NaOH. 20 μl of this solution is then mixed with 20 μl of each of the colour reagents (in the order a, b, c, d, final volume 100 μl). After 1 h at room temperature, water (300 μl) is added; the absorption is read at 625 nm.

(e) *Amino-acid stereo-isomers.* Mucopeptides are characterized by their content of D-amino-acids. Glutamic acid is usually present only in the D-form, while D-aspartic acid has also been found. At least 50% of the alanine is D-alanine; ornithine and diaminobutyric acid may be either the D- or the L-isomers. Diaminopimelic acid is most frequently found as the *meso-* (internally compensated D,L) isomer, especially in the Gram-negative bacteria, where it is the only diamino-acid to have been identified in the mucopeptide. LL- or DD- diaminopimelic acid occur in some Gram-positive organisms and Actinomycetales, sometimes together or in conjunction with the *meso*-isomer.

The methods most commonly employed for identification and estimation of amino-acid isomers have been, until recently, dependent on stereo-specific enzyme reactions (Table VII), but physical methods are now also available for estimation and separation of stereo-isomers.

Physical methods. Direct measurement of optical configuration of amino-acids in amounts low enough to be applicable to cell wall analyses is possible after converting them to their dinitrophenyl derivatives, thereby enhancing their optical rotations. Measurements on 1 ml quantities containing 0·5–1·5 μmoles are feasible in the polarimeter, while optical rotatory dispersions measured in a spectropolarimeter can give information on 0·25 μmoles in 1 ml (Bricas *et al.*, 1967); Perkins, private communication). These methods can only be applied to the pure dinitrophenyl derivatives; separation from other amino-acids can be effected either before or after dinitrophenylation, or by a combination of the two procedures.

Recently, Manning and Moore (1968) have described a method whereby

D and L amino-acids can be separated and determined by ion exchange chromatography as dipeptides. The dipeptides are made rapidly and almost quantitatively by coupling the amino-acids in the acid hydrolysate with an L-amino-acid N-carboxy-anhydride. The resulting pairs of dipeptides (LL and DD) are separable on an amino-acid analyser. One part of D-amino-acid can be detected in the presence of 1000 parts of L-isomer with 2 μmole samples. The method has been worked out for protein amino-acids and is therefore directly applicable to some of the mucopeptide amino-acids; it has not yet been used with ornithine, diaminobutyric acid or diamino-pimelic acid, but Moore (private commun.) believes it should be applicable to these amino-acids.

The three isomers of diaminopimelic acid can be identified by sequential chromatography, first of the free amino-acid and then of the di-dinitro-phenyl (DNP) derivatives prepared on eluates from the first chromatograms. On Whatman No. 1 paper, the LL-isomer of the free amino-acid has a higher mobility than the other two isomers in methanol-water-pyridine-12 N HCl (32 : 7 : 4 : 1, by vol) (Table IV and Hoare and Work, 1957), or methanol-water-pyridine-formic acid, 98% w/v (80 : 19 : 10 : 1, by vol, Perkins, 1965). The latter solvent is to be recommended if it is desired to elute the slower moving meso and DD isomers as it produces less trailing. The di-DNP derivative of *meso*-diaminopimelic acid can be distinguished and separated from the di-DNP derivatives of the other isomers (Bricas *et al.*, 1967; Perkins, 1969). The systems used are either (a) Whatman No. 4 paper, previously dipped in 0·1 M phthalate buffer pH6 and dried, irrigated with t-amyl alcohol saturated with the same buffer; (b) Whatman No. 1 paper, irrigated with *n*-butanol-water-ammonia, S.G. 0·880 (20 : 19 : 1, by vol); or (c) silica gel (Kieselgel G) thin-layer plates irrigated with benzyl alcohol-chloroform-methanol-water-15 N ammonia (30 : 30 : 30 : 6 : 2, by vol). The di-DNP derivative of *meso*-diaminopimelic acid travels slower than that of the other two isomers in any of these systems. Thus a combination of chromatography of free amino-acid and DNP derivatives enables all three isomers to be separated and, if desired, estimated by standard methods (III.C.4.c, IV.B.2.a).

If *meso*-diaminopimelic acid is present in a mucopeptide, the question arises as to which asymmetric centre is situated in the main peptide chain, between D-glutamic acid and D-alanine. This was determined for *E. coli* and *Bacillus megaterium* by isolation of the mono-DNP derivatives from dinitro-phenylated mucopeptides and measurement of their optical rotations (Diringer and Jusic, 1966; Bricas *et al.*, 1967). Dinitrophenylation increased the optical rotation in a predictable manner, and the mono-DNP-derivative from *B. megaterium* was identical with synthetic mono-DNP-(D)+*meso*-diaminopimelic acid. It was thus possible to identify the dinitrophenylated centre as the D centre, showing that in both organisms the L-centre is in the main peptide chain.

D. Enzymic methods

Table VII summarizes the methods used. It should be emphasized that in all the methods to be described, absolute cleanliness of tubes and all equipment is quite essential; many failures are due to lack of attention to this point.

TABLE VII

Determination of amino-acid isomers by enzymic reactions

Amino-acid	Enzymes	Reference
L-alanine	Glutamate-pyruvate transaminase + lactate dehydrogenase	Pfleiderer, 1963 Ghuysen et al., 1966.
D-alanine	D-amino-acid oxidase + lactate dehydrogenase	Allsop and Work, 1963 Schleifer and Kandler, 1967 Ghuysen et al., 1966.
L-glutamate	L-glutamate dehydrogenase	Bernt and Bergmeyer, 1963.
D-glutamate	Total glutamate-L-glutamate	
L-aspartate	Aspartate decarboxylase	Ikawa, 1964.
D-aspartate	Total aspartate-L-aspartate	Ikawa, 1964
L-ornithine	Ornithine transcarbamylase	Work, 1964.
D-ornithine	D-amino-acid oxidase	Guinard et al., 1969.
Meso-Dap†	Dap decarboxylase	White and Kelly, 1965.
Meso- + LL-Dap	Dap epimerase + Dap decarboxylase	White, Lejeune and Work, 1969; Work, 1963.
LL-Dap	(Meso + LL-Dap)-Meso-Dap	
DD-Dap	Total Dap-(Meso + LL-Dap)	
L-lysine	Lysine decarboxylase	Gale, 1963.

† Dap = Diaminopimelic acid.

1. Isomers of alanine

Although other D and L amino-acids do not interfere with the enzymic determinations of D or L alanine, there is interference by some components of mixtures obtained after hydrolysis of dinitrophenylated peptides or after hydrazinolysis. Therefore, if these estimations are to be combined with end group determinations, the alanine in the hydrolysates must be purified prior to enzymic analysis (Ghuysen et al., 1966). This is done on a column of Dowex 50 or Zeocarb 225-H (× 8, 100–200 mesh) 10–15 cm in length, 0·8 cm diameter. After washing with 4 N HCl (20 ml) the resin is equilibrated with 1·5 N HCl (20 ml). The sample, dissolved in about 1 ml of 1·5 N HC is applied to the column and eluted with 20 ml of 1·5 N HCl

(0·2 ml/min). Fractions of 1 ml are collected. Alanine is eluted in a volume of 4–5 ml, centred at about 12 ml after the beginning of the elution. It can be located by any suitable means. The alanine-containing fractions are lyophilized.

L-*alanine* (Pfleiderer 1963).

Principle: L-alanine is converted to pyruvate by glutamate-pyruvate transaminase and α-oxoglutarate. The pyruvate is reduced to lactic acid by lactic dehydrogenase + NADH$_2$.

$$\text{Pyruvate} + \text{NADH}_2 \xrightarrow{\text{LDH}} \text{L-lactate} + \text{NAD}$$

The disappearance of NADH$_2$ can be followed spectrophotometrically at 340 or 366 nm. The equilibrium of the lactic dehydrogenase reaction is well to the right; however, a quantitative conversion of alanine to pyruvate is not possible because the Km of the transaminase is too high. With excess of both enzymes and NADH$_2$ the rate of the coupled reaction with limited alanine concentrations is strictly proportional to the amount of alanine added. Measurement of the reaction rate permits the determination of L-alanine by use of a standard curve prepared with known concentrations of L-alanine.

Solutions (all to be made up in doubly distilled water):

Phosphate buffer M/15, pH 7·2

0·1 M α-oxoglutarate.

Reduced diphosphopyridine nucleotide-50 mg NADH$_2$-Na$_2$ in 5 ml water.

L-alanine 2 mg/ml.

Lactic dehydrogenase (LDH) (Boehringer; or Sigma (pyruvic kinase free) ca 1 mg protein/ml—dilute the crystalline suspension with 2·2 M ammonium sulphate solution.

Glutamate-pyruvate transaminase (GPT) (Boehringer) ca 10 mg protein/ml. Dilute with 1·6 M ammonium sulphate solution.

Method

Pipette successively into spectrometer cuvettes—

0·2 ml α-oxoglutarate.

0·1–0·2 ml sample or standard L-alanine.

0·06 ml NADH$_2$.

0·01 ml LDH suspension.

Mix, observe optical density at 340 to 366 nm for any slight changes which may occur (E$_1$/min).

Start transaminase reaction by mixing in 0·04 ml GPT suspension. Take readings of decrease in optical density at 1 min intervals for about 5 min (E$_2$/min).

Both transaminase reaction and any reaction before addition of transaminase are linear with time, therefore the values for the rates can be
averaged. The corrected rate of transaminase reaction is $E_2/min-E_1/min =$
E/min. These values are used to prepare a standard curve from the standard
L-alinine solution or to obtain the L-alanine concentration of unknown
samples.

The method could be scaled down by $\frac{1}{4}$ if 1 ml cuvettes are available.

D-*alanine*. D-amino-acid oxidase is used to convert D-alanine to pyruvic
acid. In one method (a), the pyruvate is coupled with quinolyl hydrazine
and the complex is estimated spectrophotometrically at 305 nm (Allsop
and Work, 1963); in other methods pyruvate is reduced to lactate by lactate
dehydrogenase and $NADH_2$, the NAD so formed is estimated (b) spectrophotometrically (Schleifer and Kandler, 1967) or (c) fluorometrically
(Ghuysen et al., 1966).

Method (a) (Allsop and Work, 1963). Hydrolysed walls (150–500 μg) are
mixed with 1μg of FAD (900 μg/ml), 1 μg of catalase (Boehringer) and
50 μl of D-amino-acid oxidase (2 mg/ml) (Koch Light and Co. Ltd, or
Worthington Biochemical Corp. Freehold, N.J.) and the volume is made
up to 0·5 ml with cold, oxygenated 0·05 M sodium pyrophosphate buffer
pH 8·3. The mixture is incubated for 2 h at 37°C, then treated with 15 μl
of 15% (w/v) trichloroacetic acid; 30 mM quinolylhydrazine (10 μl) is
added to form a complex with the pyruvic acid. The samples are incubated
for a further 30 min at 37°C and then diluted with 0·01 N HCl to 5 ml and
the absorption at 305 nm is read against a reagent blank. There is no
interference by other mucopeptide constituents.

Method (b) (Schleifer and Kandler, 1967). Pyrophosphate buffer
(0·1 M, pH 8·1), 1·58 ml, 0·01 ml D-amino-acid oxidase (Boehringer),
0·01 ml catalase, and 0·1 ml of neutral hydrolysate are mixed and incubated 2·5 h at 37°C, a fine stream of oxygen is bubbled in for a few minutes
three times during incubation. 1 ml of the reaction mixture is mixed with
0·06 ml of $NADH_2$ (10 mg/ml of water) and 1·93 ml of triethanolamine-
EDTA buffer (pH 7·6) and after 3 min the absorption is read at 340 nm.
Lactic dehydrogenase (0·01 ml) is added, and after 3 min the absorption is
read again.

Method (c) Microestimation (Ghuysen et al., 1966). Here $NADH_2$ is
estimated fluorometrically (Lowry and Passonneau, 1963). Triplicate
aliquots containing 0·5–1·5 millimicromoles of D-alanine are dried in 3 ml
tubes. Forty microlitres of 0·1 M pyrophosphate, pH 8·3, containing per
ml, 5 μl catalase and 5 μl D-amino-acid oxidase (17 mg/ml), is added.
Tubes are incubated at 37°C for 3 h. Stock $NADH_2$ (2·5 mg/ml in 0·1 M
Na_2CO_3) is diluted 1 : 80 just before use with 0·05 M potassium phosphate,
pH 6·8 : 1 ml is added to each tube, which is then mixed and read in the

fluorometer. Sensitivity is adjusted so the $NADH_2$ solutions give near maximum deflection of the galvanometer. Lactate dehydrogenase (crystalline suspension from rabbit muscle, type 2 10 mg/ml, Sigma Chemical Co. St. Louis) is diluted 1 : 5 with 0·05 M potassium phosphate (pH 6·8) and 5 μl is added. The tubes are incubated at 37°C for 30 min, and again read in the fluorometer. Decrease in fluorescence, due to oxidation of $NADH_2$ to NAD, is proportional to the amount of D-alanine present. Controls consist of a blank, at least two levels of D-alanine and a whole set of tubes containing excess pyruvate.

Method (d). Microestimation (Guinand et al., 1969). Here D-alanine is estimated by finding the decrease caused by D-amino-acid oxidase in the total free amino-acid. This is carried out on enzymic digests by the fluorodinitrobenzene technique for determination of free amino groups (IV.B.2a)

2. Isomers of glutamic acid

D-Glutamic acid cannot be estimated directly; instead L-glutamic dehydrogenase + NAD is used to estimate any L-glutamic acid present by spectrophotometric measurement of $NADH_2$ formed (Bernt and Bergmeyer, 1963). The value so obtained for L-glutamate is subtracted from the figure for total glutamate measured by other means (Schleifer and Kandler, 1967).

Glutamic acid is first isolated from the hydrolysate by ion exchange or paper chromatography (isopropanol-acetic acid-water 75 : 10; 15 by vol). L-Glutamic acid is assayed in a sample (0·2 ml) containing 0·01–0·1 μmole/ml in 2·5 ml glycine-hydrazine buffer pH 9·0 (0·5 M glycine, 0·4 M hydrazine)* and 0·2 ml NAD (20 mg/ml water). The optical density is read at 340 nm (E_1) and 0·05 ml of glutamic dehydrogenase (Boehringer, Mannheim, 10 mg/ml) is added. The mixture is stood 50 min at room temperature and optical density E_2 is determined. At the same time a blank without enzyme is set up, and a control tube containing sample + 0·1 ml of L-glutamic acid (2 μ mole/ml). Change in optical density (E_2–E_1) for blank is subtracted from that of sample and the concentration of L-glutamic acid is calculated from a standard curve prepared from 0·05–0·2 μm of L-glutamic acid.

3. Isomers of aspartic acid

D-Aspartic is also determined by a difference method, the principle being that L-aspartate is converted by L-aspartate decarboxylase into alanine,

_____

* Dissolve 3·75 g glycine and 5·50 g 24% hydrazine hydrate in doubly distilled water, adjust to pH 9 with ca. 14·8 ml N H_2SO_4, make up to 100 ml.

which is then estimated chromatographically (Ikawa, 1964). As source of aspartate decarboxylase an acetone-dried preparation of *Nocardia globerula* (NCIB No. 8852) (Crawford, 1958) can be used. In the original method (Ikawa, 1964), alanine was estimated by paper chromatography, but if an automated amino-acid analyser is available, it could be used. In any case, the aspartate to be examined must first be separated from alanine by column chromatography on Dowex I developed with 0·5 M acetic acid. The aspartate-containing fractions are lyophilized and dissolved in water. Aliquots (0·1 ml) containing 0·1–0·5 mg aspartate are mixed with 0·2 ml of M acetate buffer (pH 5·0) containing (in suspension) 0·5 mg of aspartate decarboxylase, and 2·5 μg of pyridoxal phosphate. The mixtures are covered and shaken for 16 h at 37°C. The contents of the tubes are deproteinized with an equal volume of ethanol and centrifuged, and alanine is estimated in the supernatant. Controls and blanks are as in L-glutamate estimation.

4. Isomers of diaminopimelic acid

Two enzymes could be used to distinguish between the isomers of diaminopimelic acid; diaminopimelate epimerase catalyses the interconversion of *meso*- and LL-isomers (White, Lejeune and Work, 1969), while diaminopimelate decarboxylase attacks only the *meso* form, which is converted quantitatively to L-lysine (Work, 1963). The DD-isomer is not attacked by either enzyme. A crude enzyme preparation obtained from a wild-type *E. coli* contains both enzymes and converts both *meso* and LL-isomers to L-lysine, while a purified decarboxylase attacks only the *meso*-isomer. Therefore a combination of paper chromatography (to distinguish the LL-isomer from the other two, Table IV) and enzymic reactions will enable all three isomers to be identified and estimated, using colorimetric estimations by acidic ninhydrin for determination of diaminopimelate and lysine.

$$\text{Total Dap} = Meso + \text{LL} + \text{DD}$$

Residual Dap from crude enzyme = DD.
Residual Dap from purified decarboxylase = DD + LL.

Total Dap-DD　　　　　= *Meso* + LL
Total Dap-(DD + LL) = *Meso*
(DD + LL)-DD　　　　= LL.

Diaminopimelic acid is examined in the neutral amino-acid fraction of the cell wall hydrolysate, prepared by any suitable means, such as chromatography or electrophoresis. Crude enzyme is supplied by *E. coli* A.T.C.C. No. 9637 grown in an aerated glucose-salts medium and harvested just before the end of logarithmic growth; the organisms are washed, then rendered permeable to substrate by acetone-drying, or are broken (Work, 1963). The dried organisms (10 mg) or extract (about 5 mg of protein) in 0·3 ml of 0·25 M phosphate buffer, pH 7·0, containing one drop of chloroform, plus 0·1 ml of pyridoxal phosphate (1 mg/ml) are incubated at 37°C for 18 h with the diaminopimelic acid solution under test (0·3 ml containing about 0·3 mg). For qualitative examination to distinguish between meso- and DD-diaminopimelate (not separated by paper chromatography, Table IV), samples (0·2 ml) are taken at the beginning and end of incubation, deproteinized with 0·4 ml of ethanol and 0·1–0·2 ml of supernatant solution are examined by paper chromatography. The presence of DD-diaminopimelic acid is suggested when the spot in the position of meso- and DD-isomers has not all disappeared: all meso-isomer should have been converted to lysine, and LL-diaminopimelic acid, if present in the original sample, should also have disappeared. Controls should be included to ensure that meso- and LL-isomers are completely removed by incubation with the enzyme preparation.

For quantitative estimation of meso- plus LL- and of DD-diaminopimelic acid, tubes may be set up containing (a) test diaminopimelate (2 μmole), in duplicate; (b) standard meso- plus LL-diaminopimelate, as obtained by fermentation (2 μmole); (c) L-lysine (2 μmole). Each tube contains 0·25 M phosphate buffer, pH 7·0 (to give 1·0 ml total vol), 0·1 ml of pyridoxal phosphate (1 mg/ml) and acetone-dried organisms (10 mg) or extract (about 5 mg of protein), with one drop of chloroform. Acetic acid equal in volume to the total contents of each of the tubes is added to one of the tubes (a) before the enzyme, and to each of the other tubes after overnight incubation at 37°C. After centrifuging, diaminopimelate is estimated in the supernatant liquid (0·2 ml) as described on p. 404 using one of the methods insensitive to lysine. Provided no diaminopimelate remains in tube (b) after incubation, and none is detected in tube (c), then any found in tube (a) after incubation is the DD-isomer. The initial amount in tube (a) before incubation is the total diaminopimelic acid. The initial amount minus the final amount in tube (a) is meso-, or meso- plus LL-isomers.

If it is desired to estimate LL- and meso-isomers separately, a partly-purified decarboxylase preparation (White and Kelly, 1965) which is free from epimerase must be used to differentiate between them. Procedure is as above, except that 2,3-dimercaptopropan-l-ol (1 μmole) is added to each tube as an activator, and the dried organisms or extract are replaced by 0·1 ml of enzyme solution (purified as far as precipitation with ammonium sulphate) containing about 0·05 unit. Tube (b) is prepared containing only meso-diaminopimelate (2 μmoles). The results are calculated on the supposition that the diaminopimelate remaining after incubation is LL- plus DD-isomers.

As an additional check on these quantitative tests, the lysine produced can be estimated by a method in which lysine can be measured in the presence of diaminopimelate (Shimura and Vogel, 1966). Samples (0·4 ml) are taken (prior to addition of acetic acid) from tube (a) before and after incubation, and from (b) and (c) after incubation, then quickly mixed with 0·1 ml of 6 N HCl and centrifuged. To the supernatant liquid is added 0·5 ml of 15% (w/v) ninhydrin in 2-methoxyethanol. The tubes are covered, heated for 60 min in boiling water and cooled to room temperature. Orthophosphoric acid (15 M; 4 ml) is added and mixed in well. Absorbancy is read at 515 nm against a blank in which diaminopimelate (0·8 µmole) was initially omitted from the sample but was added after the HCl. Since diaminopimelate as well as lysine gives some colour in this reaction, and since for each mole of lysine formed (by action of the decarboxylase) one mole of diaminopimelate disappears, lysine standards are provided by partly replacing diaminopimelate in the blank by L-lysine (x µmole, where x is 0·1 to 0·8); the amount of diaminopimelate present with the lysine is 0·8—x µmole.

5. Isomers of lysine

L-Lysine can be estimated by L-lysine decarboxylase (Gale, 1963). The source of enzyme is an acetone powder of Bacterium cadaveris (NCIB 6578) grown for 30 h at 25°C in a medium containing 2% casein hydrolysate and 2% glucose. The usual method of carrying out the estimation in a Warburg apparatus is probably on too big a scale (about 0·5 mg lysine is needed). It is therefore suggested that the disappearance of lysine is followed colorimetrically by the method of Shimura and Vogel (1966) described above or by the method for estimating diaminopimelic acid (Work, 1957) described on p. 404, using a heating time of 60 min. The reaction with acetone powder (4 mg/ml) is carried out in 0·2 M phosphate buffer pH 6·0. Suitable controls should be included with water blank and with a standard L-lysine solution, both on its own and added to the cell wall hydrolysate.

Any lysine remaining after overnight incubation will be D-lysine, providing L-lysine standards have been completely decarboxylated.

6. Isomers of ornithine

L-Ornithine, in the presence of carbamyl phosphate is specifically converted to citrulline by ornithine transcarbamylase. Measurement of initial and residual ornithine, and of citrulline formed, gives a measure of total D- and L-ornithine respectively (Work, 1964). D-Ornithine may be directly estimated by D-amino-acid oxidase (Guinand et al., 1969).

Reagents. Ornithine transcarbamylase: the preparation is made from *Streptococcus faecalis* N.C.T.C. 6782 grown for 16 h at 30°C in the following medium (Knivett, 1954): L-arginine HCl, 1·5%; yeast autolysate, 10%; tryptic digest of casein, 0·5%; K_2HPO_4, 0·1%; $MgSO_4.7H_2O$, 0·05%; $MnSO_4.4H_2O$, 0·01%; glucose, 1% (sterilized separately). Harvested cells are washed twice with water, resuspended to give approx. 20 mg dry wt/ml, and broken in the Mickle disintegrator with an equal volume of glass beads at 0°C. The cell debris is removed by centrifuging at 20,000 g for 20 min, the supernatant is dialysed overnight against 0·02 M phosphate.

Ninhydrin reagent for ornithine estimation (prepared fresh daily): 250 mg ninhydrin and 37·6 mg of hydrindantin are dissolved by heating in a mixture of 4·0 ml of 6·0 M H_3PO_4 and 6·0 ml of acetic acid.

Diacetyl reagent for citrulline estimation: 25 ml of 1 mM citrulline, 37·5 ml arsenious oxide (10% w/v in conc HCl), 20 ml diacetyl monoxime (1% w/v in 5% v/v acetic acid).

Procedure. The enzymic reaction mixture contains (in a total volume of 1·7 ml) 0·3 ml of ornithine transcarbamylase (35 mg protein/ml), 0·3 ml of 0·1 M tris buffer pH 8·4, 1·3 μmoles of ornithine and, added last, 16 mg of lithium carbamyl phosphate (Spector et al., 1957) dissolved in 0·5 ml water (stored frozen). After incubation for 5 min at 37°C, 0·17 ml of 30% (w/v) trichloroacetic acid is added and the precipitate is removed by centrifugation. Residual ornithine (D-isomer) is estimated on 0·1–0·2 ml of supernatant as follows (Ratner, 1962): Sample and water to 1 ml are mixed with ninhydrin reagent (1·0 ml) and acetic acid (1·5 ml); the mixture is covered and heated for 30 min at 100°C, then cooled at 25°C for 10 min. Acetic acid is added to bring the volume to 6·0 ml and absorbancy is read at 515 nm against a reagent blank.

Citrulline is estimated by mixing 0·2 ml of supernatant from the enzymic reaction mixture with 2 ml (from burette) of diacetyl reagent (Knivett, 1954) and the mixture is stood overnight at room temperature and then heated 15 min, with frequent shaking, in a boiling water. After cooling, the volume is made up to 5 ml with water and the absorbancy is read at 492 nm. Identical tubes to which trichloroacetic acid is added before carbamyl phosphate provide figures for the starting amounts of ornithine. Ornithine and citrulline standards (0·02–0·1 μmole) are also set up in similar tubes to compensate for the effects of constituents of the reaction mixtures on the development of colour.

D-*Ornithine.* May be estimated by D-amino-acid oxidase, by measuring the loss in free amino-acid following incubation with the enzyme. This is carried out by the fluorodinitrobenzene technique for determination of free amino groups (IV.B.2.a) (Guinand *et al.*, 1969).

IV. MURALYTIC ENZYMES

A. Lysis

Many types of enzymes are known which attack mucopeptides. Most of these enzymes induce some degree of lysis of the cell wall (muralysis) through solubilization of part or all of the mucopeptide molecule. Their action can easily be observed by measurement of turbidity of suspensions of cell walls or even of acetone powders of whole bacteria at frequent intervals (Work, 1959). The reaction is most conveniently carried out at room temperature, and enzyme concentrations should be adjusted so that a measurable fall in optical density occurs within, at the most, 5 min; but the change should not occur so rapidly that the initial fall exceeds 10% per min. Care should be taken to have the right pH, ionic strength and necessary cations present in the reaction mixture, as many muralytic enzymes are very dependent on the correct ionic conditions for full activity. Controls should include substrate suspended in buffer without enzyme. The rate of reaction is obtained from the initial slope of the lysis curves. If digestion is to be continued over a long period, a drop of chloroform should be added to prevent bacterial contamination.

B. Linkage split

Apart from lysozymes from various sources, muralytic enzymes are usually obtained from microbes and can be either extra-cellular or intra-cellular. A variety of bonds in the mucopeptide molecule, both glycosidic and peptidic, are attacked; as a rule each enzyme attacks only one specific linkage. Lists of such enzymes and their specificities, where known, are given by Strominger and Ghuysen (1967), and Ghuysen *et al.* (1966). The latter publication contains detailed instructions for micro-determination of the point of attack. Some of the techniques which will help to characterize an enzyme as a glycosidase or a peptidase are described here.

1. *Detection of splitting of glycosidic bonds*

A glycosidase produces soluble products containing reducing groups belonging either to N-acetylglucosamine or N-acetylmuramic acid. Presence of reducing power or free N-acetylhexosamines in the soluble digest will therefore characterize an enzyme as a glycosidase.

(a) *Reducing power* (Thompson and Shockman, 1968).
   Reagents: Potassium ferricyanide, 0·05% (w/v).
   Carbonate-cyanide reagent-5·3 g $Na_2CO_3 + 0·65$ g KCN in 1 litre water.

Colour reagent, made from the following stock solutions which are all made up in 0.05 N H_2SO_4:

(a) Ferric ammonium sulphate 15 g/l.
(b) Dupanol (sodium dodecyl sulphate) 3 g/l.
(c) Carbowax 20 M (polyethylene glycol) (Union Carbide Corp) 10 g/l.
(d) 0.05 N H_2SO_4.

Equal volumes of (a), (b), (c) and (d) are mixed. This reagent has an appreciable blank due to Carbowax; it can be reduced by addition of 0.1% $KMnO_4$ (1·4–1·8 ml/g of Carbowax) until the blank is no higher than that given by a mixture of (a) : (b) : (d) in ratios of 1 : 1 : 2.

Procedure. Solution or suspension of sample, plus water to 2 ml, is mixed with 2 ml ferricyanide and 2 ml carbonate-cyanide reagent. The mixture is heated in boiling water for 15 min, then cooled for 5 min in tap water. 0.1 N H_2SO_4 (5 ml) is added, and after mixing the solution is centrifuged at 2000 g for 10 min. Supernatant (5 ml) is then mixed with 2 ml of colour reagent and stood 15 min. Absorbancy is read at 700 nm. The colour is stable, and a maximum of 40 μg of glucose can be measured.

The estimation can be scaled down by 1/10. All glassware must be scrupulously clean. New Teflon-lined screw caps, if used, must be heated for 10 min at 100°C in a mixture of equal volumes of ferricyanide and cyanidecarbonate reagents, and then washed.

Since the reducing power of different sugars is not identical, no absolute measure of sugar in solution is obtainable. However, the method is suitable for following changes in reducing power, which can be expressed in terms of a glucose standard.

(b) N-*acetyl-amino sugars.* The Morgan–Elson reaction can be used to follow the appearance of soluble fragments containing end N-acetyl-amino sugars (Reissig *et al.*, 1955). To the sample in a volume of 0·5 ml in a 13 × 100 mm test-tube, is added 0·1 ml of potassium tetraborate* solution, final pH 9·1. Heat the solution 3–30 min in boiling water and cool in tap water. Add 3 ml of colour reagent (made as on p. 393, but diluted 1 : 8 with acetic acid immediately before use); mix and immediately place in a water-bath at 37°C. After exactly 20 min, cool the tubes in tap water and read without delay at 585 nm. This method was developed for free N-acetyl-hexosamines: higher colour values are obtained with enzymic digests of mucopeptides by using a pH of 10·8 (Na_2CO_3, Aminoff *et al.*, 1952) for the initial heating step (Perkins, 1960). In either case, the optimal time at 100°C will have to be determined.

* Prepare tetraborate salt by adding the calculated amount of KOH to a solution of H_3BO_3, concentrating and recovering the crystals. To make the reagent, prepare a solution 0·8 M in borate and adjust pH to 9·1 with KOH.

Microdetermination of N-*acetylamino sugars* (Ghuysen *et al.*, 1966). Sample containing the equivalent of 1–5 millimicromoles of N-acetyl-glucosamine in 20 μl of 1% potassium tetraborate is heated for 30 min in a boiling water bath. After cooling, freshly-made (p. 000) colour reagent (90 μl) is added, the mixture is incubated for 20 min at 37°C and extinction is read at 585 nm. Molar extinction coefficients are as follows: N-acetyl glucosamine (AG) 14,000; N-acetylmuramic acid (AM) 13,500; dimer AG-AM 9500; tetramer, 1000; dimer AM-AG 1600. The presence of peptides amidically linked to N-acetylmuramic acid does not prevent colour development, but the use of Na_2CO_3 in the initial step may improve colour of dimers and tetramers (see above).

Free N-acetylglucosamine and N-acetylmuramic acid can best be determined by a modification of this procedure, with the initial heating time at 100°C decreased to 7 min; the extinction coefficients are then increased to 20,000 and 19,000 respectively.

(c) *Characterization of glycosidic linkage split.* When fragments contain oligosaccharides of chain length six or less, the nature of the linkage split by the enzyme can be determined by identification of the amino sugar hexitol produced by sodium borohydride reduction of the oligosaccharide (Salton and Ghuysen, 1960; Ghuysen *et al.*, 1966; Ward and Perkins, 1968) A sample containing 100 mμmoles of total hexosamine is dissolved in 30 μl of fresh unbuffered 0·1 M $NaBH_4$ and held for 3 h at room temperature. A control is an identical sample, also in 30 μl of $NaBH_4$, but here the $NaBH_4$ has been previously destroyed by acidification with acetic acid to pH 5 followed by neutralization. Product and control are acidified with 15 μl conc HCl and hydrolyzed in a sealed tube for 3 h at 95°C. After removal of HCl by evaporation, residues are chromatographed in two dimensions in either freshly prepared butanol-acetic acid-water (3 : 1 : 1, by vol) or 2,6-lutidine-water (4 : 1, by vol) and then pyridine water (4 : 1, by vol). When spots are revealed with ninydrin, the hexosamine which had a free reducing group in the oligosaccharide (e.g. muramic acid in AG-AM) will be seen to have decreased in intensity, as compared with the control, while a faint spot of the corresponding hexoseaminicitol (in the example, muramicitol) will have appeared. The hexoseaminicitol can be better visualized by spraying with periodate-benzidine reagents. Thus, the enzyme producing AG-AM as a split product can be established as an acetyl-muramidase.

2. *Detection of splitting of peptide bonds*

If an enzyme hydrolyses peptide bonds, the product will show a higher level of free amino groups (total and soluble) than the original substrate.

If the new terminal amino acids are identified (by end amino group and end carboxyl group assays) the site of action of the enzyme can be located.

(a) *Determination of free amino groups in soluble digestion products* (Ghuysen *et al.*, 1966). Residual insoluble material is removed from the digestion mixture by centrifugation. Samples, containing 5–25 millimicromoles of amino groups dissolved in 20 μl of 1% potassium perborate, are mixed with 2 μl of 1-fluoro-2,4,-dinitrobenzene (130 μl in 10 ml ethanol) and immediately heated for 30 min at 60°C. 2 N HCl (80 μl) is added and extinction is read at 420 nm. On a slightly larger scale, 50–100 millimicromoles in 2·2 ml perborate is treated with 0·1 ml of 1% fluorodinitrobenzene. The molar extinction coefficient of DNP-amino-acids is 5200.

(b) *Characterization and determination of* N-*terminal and free amino-acids in soluble digestion products* (Ghuysen *et al.*, 1966, 1968). Soluble product (0·1 μ mole) is dissolved in 50 μl of water and mixed with 8 μl of triethylamine (10% v/v) in ethanol, and 25 μl of 0·1 M fluorodinitrobenzene in ethanol. The mixture is heated at 60°C for 30 min, evaporated to dryness and dissolved in 50 μl of 6 N HCl. Free dinitrophenyl (DNP) amino-acids, if present, are removed by three extractions with ether (100 μl). Residual ether is removed from the aqueous phase by evaporation at 60°C, the tubes are sealed and the DNP-peptide is hydrolysed for 18 h at 100°C. DNP derivatives of N-terminal amino-acids (except those of diamino-acids) are extracted with ether three times, leaving the mono-DNP derivatives of diamino-acids in the aqueous phase. Ether extracts are evaporated at 37°C and dried *in vacuo*, redissolved in 0·05 M NH₃ (20 μl) and aliquots are chromatographed on thin-layer plates of silica gel, using markers of known DNO-amino-acids. Sequential development in the same direction with a basic and then an acidic solvent separates all DNP-derivatives present. Plates are developed first in *n*-butanol-1% (w/v) ammonia 1 : 1 (upper phase) at room temperature, then are thoroughly dried in a stream of cold air. The second solvent, chloroform-methanol-acetic acid 85 : 14 : 1-single phase) is run at 2°C in the same direction.

The mono-DNP derivatives of diamino-acids remaining in the aqueous HCl phase are extracted with *n*-butanol (100 μl) twice, and extracts are dried *in vacuo*: the mixtures are chromatographed on thin layer chromatograms (Keiselgel G) in benzyl alcohol-chloroform-methanol-water-conc ammonia (30 : 30: 30 : 6 : 2, by vol).

After drying, yellow spots on the plates are transferred separately to 1 ml tubes, using a suction device, and the DNP derivatives are eluted from the gel by vigorous mixing (twice for 10 sec in a Vortex mixer) with 0·01 M ammonia-methanol 1 : 1 (200 μl). After clarification by centrifugation, the extinction of the eluates is measured at 360 nm. Molar extinction coefficients are about 15,000 and 25,000 for unhydrolysed mono- and bis-DNP derivatives respectively, and are reduced by 10–20% by hydrolysis. The whole procedure should be carried out in the dark or with tungsten lamp illumination as DNP derivatives are photosensitive. Controls consist of two standard mixtures of cell-wall amino-acids, one of which is hydrolysed after dinitrophenylation, and also a sample, of known concentration, of mono-DNP ω-derivative of lysine or ornithine, or of mono-DNP-diaminopimelic acid, which is also hydrolysed.

(c) *Characterization and determination of carboxy-terminal amino-acids in walls or digestion products* (Salton, (1961). The method is based on the fact that carboxy-terminal amino-acids of peptides are liberated by treatment with anhydrous hydrazine. All other amino-acids in peptide linkage are converted to the hydrazides which may be condensed with benzaldehyde to form water-insoluble dibenzal derivatives. The carboxy-terminal amino-acids remaining in solution are converted to the DNP-derivatives and estimated. Recovery of control amino-acids is usually only 60–80% of the theoretical, and therefore a correction should be applied for destruction.

Hydrazinolysis is carried out, using 0·2 μmole of rigorously dried soluble digestion product (Ghuysen *et al.*, 1968), by heating with 50 μl of freshly distilled hydrazine (prepared according to Smith and Howard, 1944). Extreme care should be taken to avoid contact with water. Tubes are sealed and heated 24 h at 80°C. Hydrazine is removed by evaporation *in vacuo* over H_2SO_4. The residue is dissolved in 150 μl of water and 75 μl of freshly distilled benzaldehyde is added, the tubes are stoppered and shaken vigorously for 1 h using a rotary Evapo-Mix (Buchler Instruments Inc. Fort Lee, N.J.). The solids are removed by centrifugation, and the aqueous phase is treated again with benzaldehyde. The aqueous phase is extracted with ether and 75 μl is examined for amino-acids by a convenient method, e.g. by dinitrophenylation, identification and estimation of DNP-amino-acids as previously described. The optical configuration of the liberated amino-acid could also be determined on the DNP derivative if required (p. 411).

The site of action of a peptidase may be confirmed by determination of the configuration of the terminal alanine. For example, a peptidase splitting interpeptide cross links may liberate C-terminal D-alanine, while an acetyl-muramyl-L-alanine amidase type enzyme will liberate a peptide containing an N-terminal amino-acid, frequently L-alanine. The configuration of the alanine in either case could be determined directly by measurements on DNP-alanine (molar optical rotations or optical rotary dispersions, p. 411). Alternatively, enzymic methods (p. 400) could be used, after ion exchange separation of alanine from the reaction mixtures to avoid inhibition of the enzymes by substances in the reaction mixtures. Change in C-terminal D-alanine is determined by comparing the amount of D-alanine liberated by hydrazinolysis of the starting material and the digestion product. Liberation of N-terminal L-alanine is also recognized by comparing the L-alanine content of the digestion product when hydrolysed with and without prior dinitrophenylation.

C. Fractionation of soluble products

The extent of fragmentation of mucopeptides by lytic enzymes varies considerably with the species of organism and the type of enzyme used (Strominger and Ghuysen, 1967). The soluble digestion products ("soluble mucopeptide", p. 408) may be separated into fractions of high and low molecular weights by dialysis or gel filtration. The low molecular weight fraction from lysozyme digests can be separated by paper chromatography e.g. iso-butyric acid-ammonia (0·5 N) (5 : 3 by vol), Ghuysen et al. (1967) or butanol-acetic acid-water (4 : 1 : 5, by vol), Peltzer et al. (1963). Four or five main peptide-containing fragments have been isolated and characterized from several Gram-negative species (Martin, 1966). Small fragments from the action of various enzymes on more complex mucopeptides from Gram-positive species have been isolated by ion exchange chromatography (Mirelman and Sharon, 1966; Hughes, 1965) or by using Sephadex G 25 or G 50 eluted with water, lithium chloride, or volatile eluants; peaks being identified by ninhydrin or by a reaction for N-acetylhexosamine (Bricas et al., 1967; Ghuysen et al., 1968; Hughes, 1968), or by combinations of these techniques.

The sequential use of muralytic enzymes of known specificity is of great value in fragmenting mucopeptides for detailed structural studies, but the subject is too specialized to be dealt with here; it is described by Strominger and Ghuysen (1967). Mention should also be made of the use of partial acid hydrolysis (in 4 N HCl at 100°C, optimal time usually between 0·5–2 h) followed by separation by 2-dimensional paper chromatography of the split products to produce characteristic patterns of peptide spots ("finger prints"). This technique was successfully used by Kandler and co-workers as steps in elucidating structures of a great variety of mucopeptides and their soluble digestion products (Kandler, 1967). On a larger scale, peptide fragments from partial acid hydrolysates may be separated on Sephadex G 15 in 0·1 M pyridine-acetic acid buffer at pH 5·1, followed by ion exchange chromatography on Dowex 50 (Hughes, 1958). A different type of fragmentation is that obtained from the β-elimination of the D-lactyl residue from muramic acid by alkali treatment (Tipper, 1968). This produces D-lactyl peptides from soluble mucopeptide preparations (Ghuysen et al., 1968).

414   ELIZABETH WORK

Small fragments of mucopeptides, produced by any of the techniques described here, are amenable to detailed analysis. Thus, the way is open to the elucidation of chemical structure and molecular architecture of the mucopeptides, those essential, unique, complex and diverse polymers of bacterial cell walls.

REFERENCES

Allsop, J., and Work, E. (1963). *Biochem. J.*, **87**, 512–519.
Aminoff, D., Morgan, W. T. J., and Watkins, W. M. (1952). *Biochem. J.*, **51**, 379–389.
Anderson, N. G. (1967). *Methods of Biochemical Analysis*, **15**, 272–310.
Anderson, F. B., and Millbank, J. W. (1966). *Biochem. J.*, **99**, 682–687.
Applegarth, D. A. (1967). *Arch. biochem. biophys.*, **120**, 471–478.
Archibald, A. R., and Baddiley, J. (1966). *Adv. in Carbohydrate Chemistry*, **21**, 323–376.
Archibald, A. R., Baddiley, J., and Buchanan, J. G. (1961). *Biochem. J.*, **81**, 124–134.
Archibald, A. R., Baddiley, J., and Button, D. (1968). *Biochem. J.*, **110**, 543–557.
Armstrong, J. J., Baddiley, J., Buchanan, J. G., Carss, B., and Greenberg, G. R. (1958). *J. Chem. Soc.*, 4344–4354.
Baddiley, J. (1968). *Proc. Roy. Soc. B.*, **170**, 331–348.
Barkulis, S. S., and Jones, M. F. (1957). *J. Bact.*, **74**, 207–216.
Bernt, E., and Bergmeyer, H. U. (1963). *Methods of Enzymatic Analysis*, pp. 384–386 (Ed. H. U. Bergmeyer). Academic Press, N.Y., and Verlag Chimie, Weinheim.
Bleiweis, A. S., Karakawa, W. W., and Krause, R. M. (1964). *J. Bact.*, **88**, 1198–1200.
Bricas, E., Ghuysen, J. M., and Dezélée, P. (1957). *Biochemistry, N.Y.*, **6**, 2598–2607.
Brundish, D. E., and Baddiley, J. (1968). *Carbohyd. Res.*, **8**, 308–316.
Cessi, C., and Piliego, F. (1960). *Biochem. J.*, **77**, 508–510.
Chattaway, F. W., Holmes, M. R., and Barlow, A. J. E. (1968). *J. gen. Microbiol.*, **51**, 367–376.
Conway, E. J. (1942). *Microdiffusion Analysis and Volumetric Error*, 5th Ed. p. 258. Crosby, Lockwood and Son Ltd., London.
Crawford, L. V. (1958). *Biochem. J.*, **68**, 221–225.
Crook, E. M., and Johnston, I. R. (1962). *Biochem. J.*, **83**, 325–331.
Crumpton, M. J. (1959). *Biochem. J.*, **72**, 479–486.
Cummins, C. S., and Harris, H. (1956). *J. gen. Microbiol.*, **14**, 583–600.
Cummins, C. S., and Harris, H. (1958). *J. gen. Microbiol.*, **18**, 173–189.
Dennis, D. (1962). *Methods in Enzymology*, Vol. 5, p. 430 (Ed. S. P. Colowick and N. O. Kaplan). Academic Press, New York.
Diringer, H., and Jusic, D. (1966). *Z. Naturforsch.*, 21b, 603.
Dische, Z. (1947). *J. biol. Chem.*, **167**, 189–198.
Dubois, M., Gilles, K. A., Hamilton, J. K., Rebers, P. A., and Smith, F. (1956). *Analyt. Chem.*, **28**, 350–356.
Ellwood, D. C., and Tempest, D. W. (1969) *Biochem. J.*, **111**, 1–5.
Ellwood, D. C., Keleman, M. V., and Baddiley, J. (1963). *Biochem. J.*, **86**, 213–225
Esser, K., (1965). *J. Chromatog.*, **18**, 414–416.
Galanos, C., Lüderitz, O., and Westphall, O. (1969). *Europ. J. Biochem.*, **9**, 245–249.

Gale, E. F., (1963). *Methods of Enzymatic Analysis* (Ed. H. U. Bergmeyer), pp. 373–377. Academic Press, New York, and Verlag Chemie, Weinham, Germany.

Ghuysen, J. M., Bricas, E., Lache, M., and Leyh-Bouille, M. (1968). *Biochemistry, N.Y.*, **7**, 1450–1460.

Ghuysen, J. M., Bricas, E., Leyh-Bouille, M., Lache, M. and Shockman, G. D. (1967). *Biochemistry, N.Y.*, **6**, 2607–2619.

Ghuysen, J. M., Tipper, D. J., and Strominger, J. L. (1965). *Biochemistry, N.Y.*, **4**, 474–485.

Ghuysen, J. M., Tipper, D. J., and Strominger, J. L. (1966). In *Methods in Enzymology*, Vol. 8, pp. 685–699 (Ed. E. F. Neufield and V. Ginsberg). Academic Press, New York.

Gilvarg, C. (1958). *J. Biol. Chem.*, **233**, 1501–1504.

Glaser, L., and Burger, M. M. (1964). *J. Biol. Chem.*, **239**, 3187–3191.

Graham, R. K., and May, J. W. (1965). *J. gen. Microbiol.*, **41**, 243–249.

Gregory, N. L. (1968). *J. Chromatog.*, **36**, 342–343.

Grisaro, V., Sharon, N., and Barkai-Golan, R. (1968). *J. gen. Microbiol.*, **51**, 145–150.

Grov, A., and Rude, S. (1967). *Acta. path. microbiol Scand.*, **71**, 409–421.

Guinand, M., Ghuysen, J. M., Schleifer, K. H., and Kandler, O. (1969). *Biochemistry*, **8**, 200–207.

Haukenes, G. (1962). *Acta path. microbiol. Scand.*, **55**, 463–474.

Hay, J. B., Archibald, A. R., and Baddiley, J. (1965). *Biochem. J.*, **97**, 723–730.

Hellerquist, C. G., Lindberg, B., Svensson, S., Holme, T., and Lindberg, A. A. (1969). *Carbohydrate research*, **9**, 237–241.

Hestrin, S. (1949). *J. biol. Chem.*, **180**, 249–261.

Heymann, H., Manniello, J. M., and Barkulis, S. S. (1964). *J. biol. Chem.*, **239**, 2981–2985.

Heyworth, R., Perkins, H. R., and Walker, P. G. (1961). *Nature, Lond.*, **190**, 261–262.

Hoare, D. S., and Work, E. (1955). *Biochem. J.*, **61**, 562–568.

Hoare, D. S., and Work, E. (1957). *Biochem. J.*, **65**, 441–447.

Höcht, H., Martin, H. H., and Kandler, O. (1965). *Z. Pflanzen-physiol.*, **53**, 39–57.

Huff, E., Oxley, H., and Silverman, C. S. (1964). *J. Bact.*, **88**, 1155–1162.

Hughes, R. C. (1965). *Biochem. J.*, **96**, 700–709.

Hughes, R. C. (1968). *Biochem. J.*, **106**, 41–48, 49–59.

Hughes, R. C., and Tanner, P. J. (1968). *Biochem. biophys. res. commun.*, **33**, 22–28.

Humphrey, B., and Vincent, J. M. (1962). *J. gen. Microbiol.*, **29**, 557–561.

Ikawa, M. (1964). *Biochemistry, N.Y.*, **3**, 594–597.

Ingram, V. M., and Salton, M. R. J. (1957). *Biochim. biophys Acta*, **24**, 9–14.

Janczura, E., Perkins, H. R., and Rogers, H. J. (1961). *Biochem. J.*, **80**, 82–93.

Kandler, O. (1967). *Ztbl. Bakt. Abt. Orig.*, **205**, 197–209.

Knivett, V. A. (1954). *Biochem. J.*, **56**, 602–610.

Knox, K. W. (1966). *Biochem. J.*, **100**, 73–78.

Knox, K. W., and Hall, E. (1965). *Biochem. J.*, **96**, 302–309.

Knox, K. W., Vesk, M., and Work, E. (1966). *J. Bact.*, **92**, 1206–1217.

Korn, E. D., and Northcote, D. H. (1960). *Biochem. J.*, **75**, 12–17.

Krause, R. M., and McCarty, M. (1961). *J. exp. Med.*, **114**, 127–140.

Leutgeb, W., Maass, D., and Weidel, W. (1963). *Z. Naturf.*, **18 (b)**, 1062–1064.

Levvy, G. A., and Conchie, J. (1966). In *Methods in Enzymology*, VIII, pp. 575–580 (Ed. E. F. Newfield and V. Ginsberg). Academic Press, New York.

Lüderitz, O., Simmons, D. A. R., Westphal, O., and Strominger, J. L., (1964). *Analyt. Biochem.* 9, 263–271.

Ludowieg, L., and Dorfman, A. (1960). *Biochim. biophys. Acta*, 38, 212–218.

Lowry, O. H., and Passonneau, J. V. (1963). In *Methods in Enzymology*, Vol. 6, p. 796 (Ed. S. P. Colowick and N. O. Kaplan). Academic Press, New York.

Mandelstam, J. (1962). *Biochem. J.*, 84, 294–299.

Mandelstam, J., and Rogers, H. J. (1959). *Biochem. J.*, 72, 654–662.

Manning, J. M., and Moore, S. (1968). *J. biol. Chem.*, 243, 5591–5597.

Martin, H. H. (1966). *Ann. Rev. Microbiol.*, 35, 457–484.

Mergenhagen, S. E., Bladen, H. A., and Hsu, K. C. (1966). *Ann. N.Y. Acad. Sci.*, 133, 279–291.

Merkenschlager, M., Schlossman, K., and Kurz, W. (1957). *Biochem. Z.*, 329, 332–340.

Mickel, H. (1948). *Jl. R. microscop. Soc.*, 68, 10–12.

Milner, H. W., Lawrence, N. S., and French, C. S. (1950). *Science N.Y.*, 111, 633–634.

Mirelman, D., and Sharon, N. (1966). *Biochem. biophys. Res. Commun.*, 24, 237–243.

Northcote, D. H., and Horne, R. W. (1952). *Biochem. J.*, 51, 232–236.

Northcote, D. H., Goulding, K. J., and Horne, R. W. (1960). *Biochem. J.*, 77, 503–508.

Novaes-Ledieu, M., Jiménez-Martínez, A., and Villanueva, J. R. (1967). *J. gen. Microbiol.*, 47, 237–245.

Osborn, M. J. (1963). *Proc. nat. Acad. Sci. Wash.*, 50, 499–506.

Partridge, S. M. (1948). *Biochem. J.*, 42, 238–248.

Partridge, S. M. (1949). *Nature, Lond.*, 164, 443.

Pelzer, H., Maass, D., and Weidel, W. (1963). *Naturwiss*, 50, 722–723.

Perkins, H. R. (1960). *Biochem. J.*, 74, 182–192.

Perkins, H. R. (1963). *Biochem. J.*, 86, 475–483.

Perkins, H. R. (1965a). *Nature, Lond.*, 208, 872–873.

Perkins, H. R. (1965b). *Biochem. J.*, 95, 876–882.

Perkins, H. R. (1965c). *Biochem. J.*, 97, 3C–5C.

Perkins, H. R. (1967). *Proc. Roy. Soc. B.*, 176, 443–445.

Perkins, H. R. (1969). *Biochem. J.*, 115, 797–805.

Perkins, H. R., and Cummins, C. S. (1964). *Nature, Lond.*, 201, 1105–1107.

Pfleiderer, G. (1963). In *Methods of Enzymatic Analysis*, p. 378 (Ed. H. U. Bergmeyer). Academic Press, New York, and Verlag Chemie, Weinheim, Germany.

Pickering, B. T. (1966). *Biochem. J.*, 100, 430–440.

Ratner, S. (1962). In *Methods in Enzymol.*, 5, p. 844, Ed. S. P. Colowick and N. O. Kaplan, Academic Press, N.Y.

Reissig, J. L., Strominger, J. L., and Leloir, L. F. (1955). *J. biol. Chem.*, 217, 217, 959–966.

Rhuland, L. E., Work, E., Denman, R. F., and Hoare, D. S. (1955). *J. Amer. chem. Soc.*, 77, 4844–4846.

Ribi, E., and Hoyer, B. H. (1960). *J. Immunol.*, 85, 314–318.

Ribi, E., Perrine, T., List, R., Brown, W., and Goode, G. (1959). *Proc. Soc. exp. Biol. Med.*, 100, 647–648.

Roberson, B. S., and Schwab, J. H. (1960). *Biochim. biophys. Acta*, 44, 436–444.

Roelofsen, P. A., and Hoette, I. (1951). *Antonie van Leeuwenhoek*, 17, 297–313.

Roseman, S., and Dorfman, A. (1951). *J. biol. Chem.*, **191**, 607.

Salton, M. R. J. (1961). *Biochim. biophys. Acta*, **52**, 329–342.

Salton, M. R. J. (1967). *Ann. Rev. Microbiol.*, **21**, 417–442.

Salton, M. R. J., and Ghuysen, J. M. (1960). *Biochim. biophys. Acta*, **45**, 355–363.

Salton, M. R. J., and Pavlik, J. G. (1960). *Biochim. biophys. Acta*, **39**, 398–407.

Sanderson, A. R., Strominger, J. C., Nathenson, S. G. (1962). *J. biol. Chem.*, **237**, 3603–3613.

Schleifer, K. H., and Kandler, O. (1967). *Arch. Mikrobiol.*, **57**, 335–364.

Schleifer, K. H., Plapp, R., and Kandler, O. (1967). *Biochem. biophys. res. comm.*, **28**, 566–570.

Schmidt, D. I. (1966). *Techniques in Amino Acid Analysis* pp. 112–113. Technicon Instruments Co. Ltd., Chertsey, England.

Schocher, A. J., Bayley, S. T., and Watson, R. W. (1962). *Can. J. Microbiol.*, **8**, 89–98.

Sentandreu, R., and Northcote, D. H. (1968). *Biochem. J.*, **109**, 419–432.

Shafa, F., and Salton, M. R. J. (1960). *J. gen. Microbiol.*, **23**, 137–141.

Shimura, Y., and Vogel, H. J. (1966). *Biochim. biophys. Acta*, **118**, 396–404.

Shockman, G. D. (1962). *Biochim. biophys. Acta.*, **59**, 234–235.

Shockman, G. D. (1965). *Bact. Rev.*, **29**, 345–348.

Shockman, G. D., Kolb, J. J., and Toennies, G. (1957). *Biochim. biophys. Acta*, **24**, 203–204.

Shockman, G. D., Thompson, J. S., and Conover, M. J. (1967). *Biochemistry, N.Y.*, **6**, 1054–1065.

Smith, L. I., and Howard, K. L. (1944). *Org. Syn.*, **24**, 53.

Smith, H., Strange, R. E., and Zwartouw, H. T. (1956). *Nature, Lond.*, **178**, 865–866.

Spector, L., Jones, M. E., and Lipmann, F. (1957). In *Methods in Enzymology*, 3, p. 653 (Eds. S. P. Colowick and N. O. Kaplan). Academic Press, New York.

Stewart-Tull, D. E. S. (1968). *Biochem. J.*, **109**, 13–18.

Stoffyn, P. J., and Jeanloz, R. W. (1954). *Arch. Biochem. Biophys.*, **52**, 373–379.

Strange, R. E., and Kent, L. H. (1959). *Biochem. J.*, **71**, 333–339.

Strominger, J. L., and Ghuysen, J. M. (1963). *Biochem. biophys. res. Comm.*, **12**, 418–424.

Strominger, J. L., and Ghuysen, J. M. (1967). *Science*, **156**, 213–221.

Sutow, A. B., and Welker, N. E. (1967). *J. Bact.*, **93**, 1452–1457.

Swallow, D. L., and Abraham, E. P. (1958). *Biochem. J.*, **70**, 364–373.

Taylor, A., Knox, K. W., and Work, E. (1966). *Biochem. J.*, **99**, 53–61.

Ternberg, J. L., and Hershey, F. B. (1960). *J. lab. clin. Med.*, **56**, 766–776.

Thompson, J. S., and Shockman, G. D. (1968). *Analyt. Biochem.*, **22**, 260–268.

Tinelli, R. (1966). *Bull. Soc. Chim. Biol.*, **48**, 182–185.

Tipper, D. J. (1968). *Biochemistry, N.Y.*, **7**, 1441–1449.

Tipper, D. J., Katz, W., Strominger, J. L., and Ghuysen, J. M. (1967). *Biochem. N.Y.*, **6**, 921–929.

Torii, M., Kabat, E. A., and Bezer, A. E. (1964). *J. exp. Med.*, **120**, 13–29.

Ward, J. B., and Perkins, H. R. (1968). *Biochem. J.*, **106**, 69–76.

Weidel, W., Frank, H., and Leutgeb, W. (1963). *J. gen. Microbiol.*, **30**, 127–130.

Weidel, W., Frank, H., and Martin, H. H. (1960). *J. gen. Microbiol.*, **22**, 158–166.

Wheat, R. W. (1966). In *Methods in Enzymology*, Vol. 8, pp. 60–78 (Ed. E. F. Newfield and V. Ginsberg). Academic Press, N.Y.

White, P. J. (1968). *J. gen. Microbiol.*, **50**, 107–120.

White, P. J., and Kelly, B. (1965). *Biochem. J.*, **96**, 75–84.
White, P. J., de Jeune, B., and Work, E. (1969). *Biochem. J.*, **113**, 589–601.
de Wij, H., and Jolles, P. (1964). *Biochim. biophys. Acta*, **83**, 326–332.
Wolin, M. J., Archibald, A. R., and Baddiley, J. (1966). *Nature, (Lond.)*, 484–486.
Wood, W. H., and Wisseman, C. L. (1967). *J. Bact.*, **93**, 1113–1118.
Work, E. (1957). *Biochem. J.*, **67**, 416–423.
Work, E. (1959). *Ann. Inst. Past.*, **96**, 468–480.
Work, E. (1963). In *Methods in Enzymol*, Vol. 6, p. 624 (Ed. S. P. Colowick and N. O. Kaplan). Academic Press, New York.
Work, E. (1964). *Nature, Lond.*, **201**, 1107–1109.
Work, E. (1967). *Proc. Roy. Soc. B.*, **176**, 446–447.
Work, E., and Griffiths, H. (1968). *J. Bact.*, **95**, 641–657.
Yamaguchi, T. (1965). *J. Bact.*, **89**, 444–453.
Yoshida, A., Heden, C. G., Cedergen, B., and Edebo, L. (1961). *J. biochem. microbiol. Technol. Engng.*, **3**, 151.
Young, F. E., Spizizen, J., and Crawford, I. P. (1963). *J. biol. Chem.*, **238**, 3119–3125.

Author Index

Numbers in *italics* refer to the pages on which references are listed at the end of each chapter.

A

Abbot, A., 169, 170, *170*, *171*
Abraham, E. P., 384, *417*
Abram, D. 155, *162*, 165, 166, 167, 168, 170, *170*, *171*
Abron, H. E., 169, *171*
Ada, G. L., 169, 170, *170*, *171*
Adams, D. A. W., 135, *144*
Adler, J., 147, 151, 156, 159, 160, 161, *162*, *163*
Adye, J., 168, *171*
Afonso, E., 192, *217*
Alcock, Doris, 174, *217*
Allen, R. D., 75, *102*
Allen, R. M., *102*
Allsop, J., 365, 369, 370, 382, 390, 400, 402, *414*
Ambler, R. P., 168, 169, *171*
Aminoff, D., 409, *414*
Anderson, F. B., 374, 375, *414*
Anderson, N. G., 370, *414*
Apirion, D., 333, 355, *360*
Applegarth, D. A., 374, *414*
Archibald, A. R., 376, 384, 385, 388, 389, 393, *414*, *415*, *418*
Armstrong, J. B., 159, 161, *162*
Armstrong, J. J., 376, 388, *414*
Armstrong, R. L., 348, *359*
Asakura, S., 168, 169, 170, *171*, *172*
Attardi, G., 351, 355, *359*
Avery, R. J., 350, 351, 352, 354, 355, *359*
Avrameas, S., 250, 251, *254*

B

Bachrach, H. L., 337, *360*
Baddiley, J. 376, 380, 384, 385, 388, 389, 392, 393, *414*, *415*, *418*
Baerer, R., 231, *246*
Baillie, Ann, 200, 210, *218*, 241, 242, *247*
Barer, R., 75, 100, *102*

Barkai-Golan, R., 374, *415*
Barkulis, S. S., 368, 369, 373, 376, *414*, *415*
Barlow, A. J. E., 374, 375, *414*
Barr, Mollie, 215, *218*, 261, 262, 279, *279*
Barron, A. L. E., 76, 77, *102*
Batty, I., 222, 231, 232, 241, 242, *246*, *247*, 268, *279*
Batty, Irene, *218*
Bautz, E. K. F., 332, *359*
Bayley, S. T., 377, *417*
Behring, E. A., von, 255, *279*
Bekhor, I., 342, *359*
Bendich, A. J., 328, *328*
Bennett, A. H., *102*
Bergmeyer, H. U., 400, 403, *414*
Bernhard, W., 250, *254*
Berns, K. I., 302, *309*
Bernt, E., 400, 403, *414*
Bettelheim, K. A., 195, *217*
Betz, J. V., 165, 166, *171*
Bezer, A. E., 387, *417*
Bisset, R. A., 161, *162*
Bissett, K. A., 174, *217*
Bladen, H. A., 378, *416*
Bleiweis, A. S., 365, *414*
Bode, H. R., 302, 305, 306, 307, *309*
Bodily, H. L., 228, *247*
Boezi, J. A., 348, *359*
Bogdanik, T., 192, *217*
Bogdanikowa, Beata, 192, *217*
Bolton, E. T., 315, 318, 320, 328, *329*, *329*, 332, 336, 339, 340, 341, 348, 349, 359, *359*, *360*
Bonner, J., 342, *359*
Bordet, J., 202, *217*
Borek, F., 240, 241, *246*
Bormann, E. K., 222, *246*
Boyden, S. V., 200, 201, 202, *217*
Bradley, S. G., 328, *329*
Braun, R. E., 166, *171*
Brenner, D. J., 328, *328*
Brenner, S., 154, *162*

Bricas, E., 376, 377, 398, 399, 411, 412, 413, *414*, *415*
Brighton, W. D., 227, *246*
British Standard Specification, 95, *102*
Britten, R. J., 321, *328*, 332, 339, 341, *359*
Brokaw, C. J., 148, *162*
Brooks, J. B., 222, *246*
Brooks, J. I., 160, *163*
Brown, E. R., 159, *162*
Brown, R., 265, *280*
Brown, W., 366, *416*
Brundish, D. E., 389, *414*
Bryner, J. H., 155, *163*, 165, 166, *172*
Buchanan, J. G., 376, 388, *414*
Bui, J., 169, *172*
Bujard, H., 306, 307, *309*
Burge, R. E., *171*
Burger, M. M., 380, *415*
Burkhalter, J. H., 232, *247*
Burkholder, P. M., 204, *217*
Burnet, F. M., 208, *217*
Burton, K., 304, *309*, 345, *359*
Button, D., 385, 389, *414*

C

Cairns, J., 302, 305, 308, *309*
Caldwell, P. J., 302, *309*
Campbell, I., 196, *217*
Carske, T. R., 222, *246*
Carss, B., 376, 388, *414*
Carver, D. H., 210, *218*
Casals, J., 208, *217*
Castro, E., 302, *309*
Causley, D., 148, *162*
Cedergen, B., 369, *418*
Cessi, C., 394, *414*
Chadwick, C. S., 221, 225, *246*
Chakravarti, 212
Champness, J. N., *171*
Chanock, R. M., 328, *329*
Chase, M. W., 226, 227, *246*
Chattaway, F. W., 374, 375, *414*
Chatterjee, S. N., 166, *171*
Chen, C., 302, *309*
Chen, P. S., Jr., 345, *359*
Cherry, W. B., 159, *162*, 222, 237, *246*
Clark, G. L., *102*
Clark, S. P., 232, *246*

Clarke, H. G. M., 193, *217*
Cleverdon, R. C., 302, *309*
Clowes, R. C., 150, *162*
Cocking, 344
Coetznee, J. N., 161, *162*
Coghlan, Joyce, D., 198, *218*
Cohen, 326, *329*
Cohen–Bazire, G., 165, *171*
Cole, R. M., 328, *329*
Colowick, S. P., 407, *416*
Conchie, J., 387, *416*
Conover, M. J., 368, *417*
Conway, E. J., 393, *414*
Cooke, R. A., 211, *217*
Coombs, R. R. A., 197, 199, 200, *217*, *218*
Coons, A. H., 219, 223, 227, 233, *246*, *247*
Cooper, G. N., 195, *217*
Cooper, P. D., *217*
Coslett, V. E., 100, *102*
Cowie, D. B., 315, *328*
Crawford, I. P., 379, *418*
Crawford, L. V., 404, *414*
Creech, H. J., 219, 223, *246*
Crook, E. M., 363, *414*
Crumpton, M. J., 381, *414*
Culling, C. F. A., 105, *134*
Cummins, C. S., 363, 368, 383, 384, 396, *414*, *416*
Curtain, C. C., 227, *246*

D

Dahl, M. M., 151, 160, 161, *162*
Daniel, Mary, R., 199, *217*
Danysz, J., 263, *279*
Darken, M. A., 136, *144*
Das, J., 166, *171*
Davenport, D., 148, *162*
Davis, B. D., 337, *360*
Deacon, W. E., *217*
Deflandre, G., *103*
De Klerk, H. C., 161, *162*
De Ley, J., 302, 306, 308, *309*, 312, 326, 328, *328*, *329*
Demerec, M., 305, *309*
Dempster, G., 195, *217*
Denhardt, D. J., 324, *328*

Denman, R. F., 382, *416*
Dennis, D., 396, *414*
Dennis, E. S., 302, *309*
De Robertis, E., 147, 161, *162*, 166, *171*
Dezéleé, P., 376, 398, 399, 413, *414*
Diringer, H., 399 ,*414*
Dische, Z., 381, *414*
Doetsch, R. N., 168, *172*
Dohadwalla, 212
Doi, R. H., 328, *329*, 359, *359*
Donah, E. J., 241, *247*
Dorfman, A., 387, 391, *416*, *417*
Doty, P., 326, 328, *329*
Downs, C. M., 223, *247*
Drozd, Jadwiga, 192, *217*
Dubin, D. T., 332, 333, *360*
Dubinska, Lidia, 192, *217*
Dubnau, D., 355, 359, *359*, *360*
Dubois, M., 390, *414*
Duguid, J. P., 195, 196, *217*
Duncan, J. F., 328, *329*
Dutta, 212

E

Eastman Kodak, Co., 85, *103*
Easton, J. M., 242, *246*
Edebo, L., 369, *418*
Edmunds, P. N., 195, *217*
Edwards, P. R., 150, 160, *162*, 170, *171*
Eguchi, G., 168, 169, *171*
Ehringhaus, A., *103*
Ehrlich, P., 256, *279*
Elek, S. D., 170, *171*
Elliott, G. F., 168, *172*
Ellis, E. C., 226, *247*
Ellwood, D. C., 362, 380, 385, 388, 389, 392, *414*
Enomoto, M., 150, 156, 157, 158, 159, *162*, 169, *171*
Erlander, S., 168, *171*
Esser, K., 302, *309*, 382, 386, *414*
Evaland, W. C., 223, *246*
Ewing, W. H., 160, *162*

F

Falcone, V. H., *217*
Falkow, S., 328, *329*
Farquhar, M. N., 168, 169, *171*

Farrell, I. D., 198, *218*
Filshie, B. K., 165, *172*
Florey, Lord, 238, 239, *246*
Follett, E. A. C., 166, *171*
Forslind, B., *171*, *172*
Foster, J. F., 168, *171*, *172*
Fox-Carter, E., 315, *329*
Franchi, C. M., 166, *171*
Francon, M., *103*
Frank, H., 363, 377, *417*
Frankel, R. W., 156, 160, *162*
Freeman, J. A., 242, *246*
Freeman, T., 193, *217*
French, A. McK., 136, 140, *144*
French, C. S., 336, *360*, 365, *416*
Freund, J., 215, *217*
Friedman, S., 326, 328, *328*, *329*
Friesen, J. D., 347, *360*
Fukumi, H., 169, *172*
Fulthorpe, A. J., 185, 187, 188, 189, 201, *217*, *218*, 264, 265, 275, *279*
Furness, G., 150, *162*

G

Gaertner, F. H., 168, 170, *171*, *172*
Galanos, C., 378, *414*
Gale, E. F., 400, 406, *415*
Gause, H., *103*
Gay, F. P., 202, *217*
Gerloff, R. K., 328, *329*
Ghuysen, J. M., 376, 377, 379, 393, 398, 399, 400, 402, 403, 407, 408, 410, 411, 412, 413, *414*, *415*, *417*
Giacomoni, D., 350, *360*
Gilleś, K. A., 390, *414*
Gillespié, D., 324, 325, *329*, 332, 339, 342, 346, 347, *360*
Gillies, R. R., 196, *217*
Gilvarg, C., 397, *415*
Glaser, L., 380, *415*
Glavert, A. M., 155, *162*, 165, 166, 168, 169, *171*, 242, *246*
Glavert, R. H., 242, *246*
Glenny, A. T., 215, *218*, 261, 262, 268, 279, *279*
Goldberg, B., 242, *246*
Goldman, M., 222, *246*
Goldstein, G., 226, 227, *246*
Goldwasser, 204

Goode, G., 366, *416*
Gooder, H., 161, *162*
Goodman, H. M., 328, *329*, 350, *360*
Gordon, J., 166, *171*
Gordon, M. A., 159, *162*, 222, *246*
Goudie, R. B., 214, *218*
Goulding, K. J., 374, *416*
Grabar, P., 182, 191, *218*
Graham, R. K., 377, *415*
Granick, S., 237, 238, *246*
Gray, J., 148, *162*
Green, D. M., 328, *329*
Green, H., 242, *246*
Greenberg, G. R., 376, 388, *414*
Gregory, N. L., 388, *415*
Griffiths, H., 369, 372, *418*
Grimstone, A. V., 154, *162*
Grisaro, V., 374, *415*
Grov, A., 369, *415*
Guest, M., 315, *329*
Guild, W. R., 302, *309*
Guinand, M., 400, 403, 407, 408, *415*
Gunsalus, I. C., 161, *163*
Gurner, B. W., 199, *217*
Guruswamy, S., 96, *102*

H

Hall, B. D., 328, *329*, 332, 347, *359*, *360*
Hall, E., 369, 376, *415*
Hall, G. T., 227, *246*
Hamilton, J. K., 390, *414*
Hanson, J., 154, *163*, 166, 168, *171*, *172*
Hanson, P. A., 227, *246*
Hardy, R., 242, *247*
Hardy, R. D., 200, *218*
Harris, A., *217*
Harris, H., 363, 368, 384, *414*
Harris, J. O., 149, *162*
Harter, J. G., 227, *247*
Haukenes, G., 376, *415*
Hawkins, D. A., 151, *163*
Hay, J. B., 393, *415*
Hayashi, M., 332, 347, *360*
Heberlein, G., 312, 328, *329*
Hedén, C. G., 369, *418*
Heidelberger, 213
Heimer, G. V., 226, 231, 232, *246*, *247*

Hellerquist, C. G., 378, *415*
Henle, G., 242, *246*
Henle, W., 242, *246*
Herbert, W. J., 136, 140, *144*
Heremans, J. F., 192, *218*
Hershey, F. B., 397, *417*
Hestrin, S., 392, *415*
Heumann, W., *172*
Heymann, H., 376, *415*
Heyworth, R., 381, *415*
Higa, 336, 348, *360*
Highman, W., 170, *171*
Hinshelwood, C., 302, *309*
Hoare, D. S., 365, 382, 399, *415*, *416*
Höcht, H., 363, *415*
Hoeniger, J. F. M., 155, 161, *162*, 165, *171*
Hoette, I., 375, *416*
Holme, T., 378, *415*
Holmes, M. R., 374, 375, *414*
Holt, L. B., *218*
Hopwood, D., 112, *134*, 305, *309*
Horne, C. H., 214, *218*
Horne, R. W., 154, 155, *162*, 165, 166, 168, 169, *171*, 374, *416*
Horsjai, J., 223, *246*
Hotani, H., 169, 170, *172*
Hotchkiss, R. D., 124, *134*
Hough, H. B., 215, *217*
Houwink, A. L., 165, *171*
Howard, K. L., 412, *417*
Hoyer, B. H., 315, 322, 328, *328*, *329*, 339, *360*, 369, 370, *416*
Hsu, K. C., 239, 240, 241, *247*, 378, *416*
Huang, P. C., 351, 355, *359*
Huff, E., 365, 369, *415*
Hughes, D. E., 336, *360*
Hughes, R. C., 376, 377, 394, 413, *415*
Humphrey, B., 363, *415*
Humphries, D. W., 100, *102*
Hunt, W. B., Jr., 227, *246*
Hybner, C. J., 328, *329*

I

Igarashi, R. T., 328, *329*, 359, *359*
Iino, T., 152, 154, 156, 157, 158, 159, *162*, *163*, 168, 169, 170, *171*, *172*

Ikawa, M., 400, 404, *415*
Ikeda, Y., 328, *329*, 359, *360*
Ingram, V. M., 390, *415*
Inman, R. B., 307, *309*
Inouye, Y., 228, *247*
Iterson, W., van, 155, *162*, 165, 166, *171*

J

Jacherts, D., 306, *309*
Jacobs, J. W., 207, *218*
Janczura, E., 376, 379, *415*
Jeanloz, R. W., 382, *417*
Jeune, B., de, 400, 404, *418*
Jimenez–Martinez, A., 364, *416*
Johnson, B. K., *103*
Johnston, I. R., 363, *414*
Jolles, P., 363, *418*
Jones, D. M., *218*
Jones, H. E., 160, *162*
Jones, L. A., 328, *329*
Jones, M. F., 368, 369, 373, 407, *414*, *417*
Jones, R. N., 219, 223, *246*
Jones, W. L., 222, *246*
Joseph, S., 75, *102*
Joys, T. M., 156, 160, *162*
Jupnik, H., *102*
Jusic, D., 399, *414*

K

Kabat, E. A., *218*, 387, *417*
Kabat, S., 351, 355, *359*
Kandler, O., 363, 377, 379, 382, 384, 400, 402, 403, 407, 408, 413, *415*, *417*
Kaplan, A. S., 265, *279*
Kaplan, M. H., 219, 233, *246*
Kaplan, N. O., 407, *416*
Karakawa, W. W., 222, *246*, 365, *414*
Katz, W., 398, *417*
Kawarai, Y., 250, 254, *254*
Keeler, R. F., 155, *163*, 165, 166, *172*
Keleman, M. V., 380, 385, 388, 389, 392, *414*
Kellog, D. S., Jr., 228, *247*
Kelly, B., 400, 405, *418*
Kelus, A. S., 199, *217*
Kennell, D., 333, *360*
Kenti, L. H., 382, *417*

Kerr, W. R., 197, 198, *218*
Kerridge, D., 155, 161, *162*, 165, 166, 168, 169, *171*
Keynan, A., 336, 348, *360*
Kiho, Y., 337, *360*
Kingsbury, D. T., 328, *329*
Kingsley-Smith, B. V., 170, *171*
Kirby, K. S., 315, *329*, 333, *360*
Kitasato, S., 255, *279*
Kleinschmidt, A. K., 306, *309*
Kline, R. M., 149, *162*
Klug, A., 154, *162*
Knivett, V. A., 407, *415*
Knox, K. W., 369, 376, 378, 380, *415*, *417*
Kobayashi, T., 168, 169, *171*
Koffler, H., 155, *162*, 165, 166, 167, 168, 169, 170, *170*, *171*, *172*
Kohler, R. E., 337, *360*
Kohn, J., 240, *246*
Kohne, D. E., 321, *328*
Kolb, J. J., 365, *417*
Korn, E. D., 374, *415*
Kotoulas, A., 333, *360*
Krause, R. M., 365, 376, *414*, *415*
Kuenen, R., 302, *309*
Kung, G., 342, *359*
Kurz, W., 365, *416*
Kustner, H., 211, *218*

L

Labaw, L. W., *171*
Lacey, B. W., 161, *162*
Lache, M., 377, 411, 412, 413, *415*
Lamore, Lewis, *103*
Lang, D., 306, 307, *309*
Langeron, M., *103*
Laurell, C. B., 192, *218*
Lawrence, N. S., 365, *416*
Lawson, D. F., *103*
Lea, D. J., 232, *247*
Lederberg, J., 150, 156, 159, 160, *162*, *163*, 169, 170, *171*
Ledingham, J. C. G., 196, *218*
Leduc, E. H., 250, *254*
Ledwell, D. M., 232, *246*
Lehman, I. R., 326, 327, *329*
Leifson, E., 151, 152, *163*
Leloir, L. F., 409, *416*

Lennette, E. H., 222, *246*
Lennox, E. S., 265, *279*
Leutgeb, W., 363, 369, 377, *415*, *417*
Levine, J., 328, *329*
Levinthal, C., 336, 348, *360*
Levvy, G. A., 387, *416*
Lewis, J. J., 222, *246*
Leyh-Bouille, M., 377, 411, 412, 413, *415*
Lim, F., 168, *172*
Linberg, B., 378, *415*
Lindberg, A. A., 378, *415*
Longhurst, R. S., *103*
Lipmann, F., 407, *417*
List, R., 366, *416*
Loeffler, H., 242, *246*
Löffler, F., 151, *163*
London, J., 165, *171*
Lowry, O. H., 402, *416*
Lowry, J., 154, *163*, 166, 168, *171*, *172*
Lüderitz, O., 378, 395, *414*, *416*
Ludowieg, L., 391, *416*
Lumsden, W. H. R., 136, 140, *144*

M

Maass, D., 369, 377, 413, *415*, *416*
McCapra, J., 161, *163*
McCarthy, B. J., 315, 318, 320, 323, 325, 328, *328*, *329*, 332, 336, 339, 340, 341, 348, 249, 359, *359*, *360*
McCarty, M., 376, *415*
McClean, J. D., 241, 242, *247*
McConkey, E. H., 332, 333, *360*
McDevitt, H. O., 227, *247*
McDonough, M. W., 168, 169, 170, *172*
McEntergert, M. G., 221, 225, *246*
McFarland, C. R., 222, *246*
McGaughey, W. J., 198, *218*
McGee, Z. A., 328, *329*
MacHattie, L. A., 306, *309*
McLaren, A., 328, *329*
Maitland, H. B., 161, *163*
Mäkelä, H., 159, *163*
Mallett, G. E., 168, 170, *171*, *172*
Mandelstam, J., 377, 396, *416*
Mangiarotti, G., 333, 355, *360*
Manniello, J. M., 376, *415*

Manning, J. M., 398, *416*
Marcus, P. I., 210, *218*
Marigault, P., *103*
Marmur, J., 314, 316, 326, 328, *329*, 333, 355, 359, *359*, *360*
Marshall, J. D., 223, *246*
Marsland, D. A., 151, *163*
Martin, H. H., 363, 377, 413, *415*, *416*, *417*
Martin, L. C., *103*
Martin, M. A., 328, *328*
Martinez, R. J., 166, *172*
Marx, R., *172*
Massie, H. R., 302, *309*
Maung, R. T., 174, *218*
May, J. W., 377, *415*
Mayer, M. M., 202, 207, *218*
Meinké, W. J., 328, *329*
Mekler, L. B., 199, *218*
Mergenhagen, S. E., 378, *416*
Merkerschlarger, M., 365, *416*
Meselson, M., 326, *329*
Metcalf, J. J., 223, *247*
Metzner, P., 149, *163*
Meynell, E. W., 156, *163*
Meynell, G. G., 161, *162*
Michaelis, L., 237, *247*
Mickel, H., 364, *416*
Midgley, J. E. M., 336, 337, 338, 348, 349, 350, 351, 352, 354, 355, *359*, *360*
Midura, T. F., 228, *247*
Miles, A. A., 150, *163*
Millbank, J. W., 374, 375, *414*
Millman, B. M., 168, *172*
Milner, H. W., 336, *360*, 365, *416*
Mirelman, D., 413, *416*
Mitani, M., 152, 154, 156, 158, *162*
Mitchen, J. R., 165, 166, 167, *170*, *172*
Mityushin, V. M., 199, *218*
Moody, M. D., 159, *162*, 222, 226, 237, *246*, *247*
Moore, R. L., 359, *360*
Moore, S., 398, *416*
Morell, P., 355, 359, *359*, *360*
Morgan, C., 239, 240, 241, *247*
Morgan, W. T. J., 409, *414*
Morowitz, H. J., 302, 305, 306, 307, *309*
Morrison, R. B., 146, 161, *163*
Mosley, V. M., *171*

N

Nairn, N. C., 221, 225, *246*
Nairn, R. C., 232, *247*
Nakane, P. K., 242, *247*, 250, 251, 254, *254*
Nathenson, S. G., 376, 380, 387, 389, *417*
Nauman, R. K., 166, *172*
Neeham, C., *103*
Nelson, R. A., 207, *218*
Nichols, T. H., 136, 142, *144*
Nivogi, S. K., 351, *360*
Norris, J. R., 185, *218*
Northcote, D. H., 374, *415*, *416*, *417*
Nossal, G. J. V., 169, 170, *170*, *171*
Novaes–Ledieu, M., 364, *416*
Nygaard, A. P., 332, 347, *360*

O

Oakley, C. L., 176, 185, 187, 188, 189, 208, 214, 216, *218*, 262, *280*
Ogiuti, K., 146, 161, *163*
Oishi, M., 355, *360*
Ørskov, F., 159, *163*
Ørskov, I., 159, *163*
Osborn, M. J., 381, *416*
Osechinsky, I. V., 199, *218*
Osterberg, H., *102*
Oxley, H., 365, 369, *415*

P

Panse, 212
Pappenheimer, A. M., 265, *280*
Pappenheimer, A. M., Jr., 180, *218*
Park, I. W., 302, 306, 308, *309*, 328, *328*, *329*
Park, R. W., 160, *162*
Partridge, S. M., 381, *416*
Passonneau, J. V., 402, *416*
Paton, A. M., 136, 138, 140, 142, *144*, 228, *247*
Patton, R. F., 136, 142, *144*
Pavlik, J. G., 378, *417*
Payne, B. O., *103*
Payne, D. J. H., 197, 198, *218*
Peacock, D. B., 207, *218*
Pearce, A. G. E., 221, *247*, 251, *254*
Pease, P., 161, *162*

Peluffo, C. A., 147, 161, *162*
Pelzer, H., 413, *416*
Peris, J. Aguila, *103*
Perkins, H. R., 375, 376, 379, 381, 383, 384, 396, 399, 409, 410, *415*, *416*, *417*
Perrine, T., 366, *416*
Peters, J. H., 227, *247*
Petrali, J. P., 241, *247*
Petrov, R. V., 199, *218*
Pfleiderer, G., 400, 401, *416*
Picard, J. J., 192, *218*
Pickering, B. T., 363, *416*
Pierce, G. B., Jr., 242, *247*, 250, 251, *254*
Pigott, G. H., 336, 349, 350, 354, 355, *359*, *360*
Pijper, A., 147, 149, 153, 161, *163*
Piliego, F., 394, *414*
Pisani, T. M., 215, *217*
Plapp, R., 384, *417*
Polatnick, J., 337, *360*
Pollard, L. W., 227, *247*
Pope, C. G., 275, *280*
Porter, R. R., 224, 240, *247*
Porterfield, J. S., 209, 210, *218*
Prausnitz, C., 211, *218*
Preer, R. J., Jr., 189, *218*
Preston, N. W., 147, 161, *163*, *218*
Pye, J., 169, 170, *170*, *171*

Q

Quadling, C., 161, *163*
Quaife, R. A., 198, *218*

R

Ratner, S., 407, *416*
Rebers, P. A., 390, *414*
Reese, M. W., 168, 169, *171*
Reich, P. R., 328, *329*
Reissig, J. L., 409, *416*
Rhuland, L. E., 382, *416*
Ribi, E., 366, 369, 370, *416*
Rice, C. E., 204, *218*
Rich, A., 328, *329*, 331, 337, 350, *360*
Richards, O. W., *102*
Richter, A., 150, *163*
Rifkind, R. A., 239, 240, 241, *247*
Riggs, J. L., 223, *247*

Rinker, J. N., 165, 168, 169, *171*, *172*
Ris, H., 156, *163*
Ritchie, A. E., 155, *163*, 165, 166, *172*
Ritter, D. B., 328, *329*
Roberson, B. S., 369, *416*
Roberts, F., 148, *163*
Roberts, F. F., Jr., 168, *172*
Roberts, R. B., 339, *360*
Robertson, L., 197, 198, *218*
Robinson, E. S., 180, *218*
Roelofsen, P. A., 375, *416*
Rogers, G. E., 165, *172*
Rogers, H. J., 376, 379, 396, *415*, *416*
Rogerts, G. E., 242, *246*
Rogul, M., 328, *329*
Romeis, B., *103*
Ron, E. Z., 337, *360*
Rose, J. A., 328, *329*
Roseman, S., 387, *417*
Ross, H. E., 261, 262, *279*
Ross, K. F. A., 65, 67, 73, 75, 100, *102*
Roux, E., 255, *280*
Rowley, D., 150, *162*
Rude, S., 369, *415*
Russell, D., 306, 307, *309*
Ruthmann, A., *103*

S

Sacks, T. G., 161, *162*
Saito, H., 328, *329*, 359, *360*
Sala, F., 170, *172*
Salton, M. R. J., 361, 369, 378, 390, 410, 412, *415*, *417*
Sanderson, A. R., 376, 380, 387, 389, *417*
Sanderson, K. E., 305, *309*
Schade, S. Z., 156, *163*
Schaechter, M., 337, *360*
Schick, A. F., 239, 240, *247*
Schildkraut, C. L., 326, 328, *329*
Schliefer, K. H., 377, 379, 382, 384, 400, 402, 403, 407, 408, *415*, *417*
Schlessinger, D., 333, 355, *360*
Schlossman, K., 365, *416*
Schmidt, D. I., 383, *417*
Schmidt, N. J., 222, *246*
Schocher, A. J., 377, *417*
Schwab, J. H., 369, *416*

Scolari, L., 192, *218*
Scott, A., 146, *163*
Scott, G. B., 250, *254*
Seaman, E., 328, *329*
Seguy, E., *103*
Seidler, R. J., 166, *172*
Selwyn, E. W., H., *103*
Sentandreu, R., 374, *417*
Sequeira, P. J. L., *218*
Seward, R. J., 223, *247*
Shafa, F., 369, *417*
Sharon, N., 374, 413, *415*, *416*
Shelokov, A., 210, *218*
Sherris, J. C., 147, 161, *163*
Shilo, M., 166, *171*
Shimura, Y., 406, *417*
Shockman, G. D., 362, 365, 368, 408, 413, *415*, *417*
Shoesmith, J. G., 147, 161, *163*
Silengo, L., 333, 355, *360*
Silverman, C. S., 365, 369, *415*
Silverstein, A. M., 240, 241, *246*
Simmons, D. A. R., 395, *416*
Singer, S. J., 237, 239, 240, 241, 242, *247*
Skaar, P. D., 150, *163*
Slater, P. N., 60, *102*
Slizys, I. S., 226, 227, *246*
Smetna, R., 223, *246*
Smith, A. G., 304, *309*
Smith, C. W., 223, *246*
Smith, F., 390, *414*
Smith, H., 377, *417*
Smith, I., 355, 359, *359*, *360*
Smith, Isabel W., 195, *217*
Smith, L. I., 412, *417*
Smith, R. W., 169, 170, *172*
Smith, S. M., 169, *172*
Sokolski, W. T., 160, *163*
Somerson, N. L., 328, *329*
Sommer, H. E., 215, *217*
Sorkin, E., 201, *217*, *218*
Spalding, B. H., 227, *246*
Spasojević, Vera, *218*
Spector, L., 407, *417*
Spiegelman, S., 324, 325, 328, *329*, 332, 339, 342, 346, 347, 348, 350, 355, *360*
Spizizen, J., 379, *418*
Spurlock, B. O., 242, *246*
Sri, Ram, J., 240, *247*

Stahl, F. W., 326, *329*
Stanier, R. Y., 161, *163*
Stapert, E. M., 160, *163*
Starr, M. P., 166, *172*
Stavitsky, A. B., 264, *280*
Steele, A. S. V., 200, *218*
Stenesh, J., 168, 169, *172*
Sternberger, L. A., 241, 242, 244, *247*
Stevens, M. F., 261, 262, 275, *279*, *280*
Stewart-Tull, D. E. S., 394, *417*
Stiles, K. A., 151, *163*
Stocker, B. A. D., 150, 156, 159, 160, 161, *162*, *163*, 169, *172*
Stoffyn, P. J., 382, *417*
Strange, R. E., 377, 382, *417*
Striker, G. E., 241, *247*
Strominger, J. C., 376, 380, 387, 389, *417*
Strominger, J. L., 376, 379, 393, 395, 398, 400, 402, 408, 409, 410, 411, 413, *415*, *416*, *417*
Sueoka, N., 355, *360*
Sullivan, Ann, 169, 170, *172*
Sunakawa, S., 169, *172*
Sutow, A. B., 369, 378, *417*
Suzuki, H., 169, 170, *172*
Svensson, S., 378, *415*
Swallow, D. L., 384, *417*
Swanbeck, G., *171*, *172*
Swift, M. E., 136, *144*

T

Takahashi, H., 328, *329*, 359, *360*
Tanner, P. J., 376, *415*
Taude, S. S., 240, *247*
Taylor, A., 380, *417*
Taylor, C. D. E., 232, *246*, *247*
Taylor, Joan, 195, *217*
Tempest, D. W., 362, *414*
Templeton, B., 160, 161, *162*
Ternberg, J. L., 397, *417*
Tewfik, E., 328, *329*
Thewaini Ali, A. J., 176, *218*
Thomas, C. A., 302, 306, *309*
Thomas, C. A., Jr, 348, *360*
Thompson, J. S., 368, 408, *417*
Thompson, K. J., 215, *217*
Thomson, R. O., 200, 210, *218*, 241, 242, *247*

Tijtgat, R., 302, *309*, 312, 328, *328*, *329*
Tinelli, R., 382, *417*
Tipper, D. J., 376, 379, 393, 395, 398, 398, 400, 402, 408, 410, 411, 413, *415*, *417*
Tiselius, A., 166, 168, *172*
Toennies, G., 365, *417*
Tokamuru, T., 224, *247*
Tomlinson, A. J. H., 232, *246*, *247*
Toribara, T. Y., 345, *359*
Torii, M., 387, *417*
Toutelotte, M. E., 302, *309*

U

Uchida, H., 169, *172*
Updyke, E. L., 226, *247*
Uriel, J., 250, *254*

V

Van Ermengem, J., 302, *309*, 328, *328*
Vatter, A. E., 155, *162*, 165, 166, 167, 170, *170*, *171*
Vegotsky, A., 168, *172*
Vendrely, R., 308, *309*
Vesk, M., 378, *415*
Villanueva, J. R., 364, *416*
Vincent, J. M., 363, *415*
Vinograd, J., 326, *329*. .
Vogel, H. J., 406, *417*
Vogel, J., 210, *218*
Von Mayersbach, H., 226, *247*

W

Wagner, M., 241, *247*
Wakabayashi, K., 169, 170, *172*
Wake, R. G., 302, *309*
Walker, P. D., 200, 210, *218*, 222, 231, 232, 241, 242, *246*, *247*
Walker, P. G., 381, *415*
Walker, P. M. B., 328, *329*
Ward, J. B., 410, *417*
Warnaar, 325, *329*
Warner, H., 345, *359*
Warrack, G. H., 262, *280*
Watkins, W. M., 409, *414*
Watson, R. W., 377, *417*
Weible, C., 161, *163*

Weibull, C., 165, 166, 168, 169, *172*
Weidel, W., 363, 369, 377, 413, *415, 416*
    417
Weinberg, E. D., 160, *163*
Weissman, S. M., 328, *329*
Welker, N. E., 369, 387, *417*
Westphal, O., 395, *416*
Westphall, O., 378, *414*
Wheat, R. W., 381, *417*
White, G. W., 50, 100, *102*
White, L. A., 228, *247*
White, P. J., 377, 400, 404, 405, *417, 418*
Wicker, R., 250, *254*
Wierzchsowski, K. L., 328, *329*
Wij, H., de, 363, *418*
Wilkinson, A. E., *246*
Wilkinson, P. C., 214, *218*
Williams, C. A., 182, 191, *218*
Williams, R. C., 166, *172*
Wilson, G. S., 150, *163*
Wilson, H. M., 136, 142, *144*
Wisseman, C. L., 369, *418*
Wittler, R. G., 328, *329*
Woese, C., 302, *309*
Wolff, B., 306, 307, *309*
Wolin, M. J., 376, *418*

Wood, W. H., 369, *418*
Work, E., 365, 369, 370, 372, 376, 378,
    380, 382, 390, 396, 399, 400, 402, 404,
    405, 406, 407, 408, *414, 415, 416, 417,*
    418
Wright, C. A., 148, *162*

Y

Yamaguchi, T., 328, *329*, 363, *418*
Yankofsky, S. A., 332, 348, 350, 355,
    360
Yersin, A., 255, *280*
Yoshida, A., 369, *418*
Young, F. E., 379, *418*
Young, J. Z., 148, *163*
Young, M. R., 231, 232, *247*

Z

Zahhn, R. K., 306, *309*
Zamon, V., 222, *247*
Zanten, E. N., van, 155, *162*, 165, *171*
Zimm, B. H., 302, *309*
Zinder, N. D., 150, 159, *163*
Zwartouw, H. T., 377, *417*

Subject Index

A

Abbe condenser, 53–54
Abbe principle, 28–30
Abbe 60 refractometer, 327
Absorbance measurement, in DNA hybridization, 313
Acacia gum, in microscopy, 75
Acetic acid, for DNA denaturation, 343
Acetic acid bacteria, ^{15}N labelled DNA from, 326
N-Acetyl galactosamine, in teichoic acids, 384
N-Acetyl glucosamine, determination, 408–410
Acetyl groups, in cell wall, 391–392
N-Acetyl hexosamines, determination, 408–410
N-Acetyl muramic acid, determination, 408–410
Achromatic condensers, 54–55
Achromatic eye-piece, 48–49
Achromatic lenses, nature of, 42
Achromatic objectives, 46
Achromobacter sp., DNA hybridization, 312
Acid-fast staining
    fluorescent, 119
    type, 106
    Ziehl–Neelson, 119
Acid phosphatase, antibody labelled with, 251
Actinomycetales,
    lipid extraction from, 363
    staining of acid-fast, 117
^{14}C-Adenine, in hybridization experiments, 325, 335, 336–337
^{3}H-Adenine, in labelled RNA, 336–337
Aerobacter aerogenes, 302
Affinity, of toxin, meaning of term, 261
Agar-gels, for DNA–DNA hybridization, 315–316, 332, 339–341
Agglutinating antisera, production, 213
Agglutination, antigen-antibody reaction tests by, 193–201
    defects of, 196–200

mixed, 199–200
    prozone formation and, 197–198
    tanned red cell, 200–201
    types of, 194–196
    univalent antibody and, 198–199
Agglutination tests, standard sera in, 176
Agglutinins, meaning of term, 193
Agglutinogen,
    meaning of term, 193
    types of, 194–196
Agrobacterium sp., DNA hybridization, 312
Airy disc, 14, 15
Airy pattern, 14, 15, 30
Alanine,
    identification, 382–386
    isomers, determination of, 400–403
    mucopeptide, 382–383
    teichoic acids, in, 384–386
Alanine amide, 386
Albert's stain, preparation and use, 120–121
Albumin-coated membrane filters, in DNA hybridization, 324–325
Alcian Blue stain, preparation and use of, 125–126
Algae, see also under specific names,
    disintegration of, 364
    polymers from, 373–375
Alexin, see Complement.
Alkaline phosphatase, antibody labelled with, 245
Amboceptor, see Antibody.
Amide ammonia, estimation of cell wall, 397–398
Aminco French press, cell disintegration by, 365–368
Amino-acid autoanalyser, in mucopeptide analysis, 383–384, 396
Amino-acid oxidase, determinations us ing, 400–403, 406–408
Amino-acids,
    cell wall hydrolysates, in, 373, 396–397, 398–408
    estimation of, 396–397, 398–408

Amino-acids—*cont.*
  identification of, 382–384, 386, 388
  mucopeptides, in, 382–384, 398–408
  stereo-isomers, 398–408
  teichoic acids, in, 386, 388
4-Aminosalicylate, sodium, 334
ε-(Amino-succinoyl)-lysine, 384
Ammonia, determination, 397–398
Ammonium molybdate, in assay bound
    DNA, 345
^{15}N-Ammonium sulphate, for labelled
    DNA hybrids, 326
Amoeboid movement, 145
Amplitude specimens, in microscopy,
    18–20
Anaesthesia, of experimental animals,
    290, 292
Anastigmatic lens, 44
Angular velocity, in ultracentrifugation,
    327
Animal house room, basic requirements
    for, 284–285
Animals, *see also under specific names*,
  indicators in toxin-antitoxin assays,
    263
  techniques for handling, 282–299
    anaesthesia, 290, 292
    blood withdrawal, 291–292, 294,
      296–297
    depilation, 288
    euthanasia, 297–298
    general considerations, 283–288
    injections, 285–288, 289–290, 291–
      292, 293–294, 295–296
    killing of, 297–298
    working conditions for, 283–285
Annular diaphragm, in microscopy,
  dark field, 75, 80
  phase contrast, 61–62, 68, 70–72
Anthrax bacillus, precautions, 108
Antibacterial sera, horse, 278–279
Antibodies, *see also* Antisera, Immuno-
      globulins, *etc.*,
  antitoxic, 208
  antiviral, 208–210
  cytophilic, 201–202
  labelled,
    enzyme, 250–254
    ferritin, 237–245
    fluorescent, 219–237

nature of, 174–175
passive transfer, 211–212
purification, 181
Antigen—antibody reactions, *see also*
    Precipitation reactions, 174–
    218
  agglutination, 193–201
  changes observable in, 177
  general considerations, 174–177
  precipitation, 177–199, 201–208
Antigenic areas, of bacterial surfaces,
    175
Antigens, *see also under more specific*
    names,
  localization by labelled antibody
    techniques, 219–254
  nature of, 175–176
Antimycoplasmal antisera, production
    of, 216
Antisera, *see also under* Antibodies,
  ferritin labelled, 237–245
    conjugation procedure, 239–241
    results of application of, 242–245
    uses of, 241–242
  fluorescent labelled, 222–230
    conjugation procedure, 222–228
    preparation for conjugation, 222
    staining with, 228–230
  horse immunization for, 278–279
  meaning of term, 176
  production of, 212–216
Antitoxic antibodies, production of,
    208, 215–216
Antitoxic sera, horse, 278–279
Antitoxin—toxin assay, *see under* Toxin.
Antitoxins,
  antibodies to, 208, 215–216
  avidity of, 261–262
  definition of, 257
  flocculation test for, 270–275
  standard sera in assay of, 176
  units of, 259
Antiviral antibodies, 208–210, 216
Antiviral sera, horse, 278–279
Aplanatic condensers, 54
Apochromatic lens and objectives, 42,
    46–47
Arizona arizonae, fimbrial antigen, 196
Ascorbic acid, in bound DNA assay,
    345

Aspartate, determination of stereo-isomers, 400, 403–404
Aspartate decarboxylase, use for aspartate assay, 400, 403–404
Aspartic acid, in mucopeptide, 382–383
Aspergillus nidulans, size of genome, 302
Astigmatism, in microscopy, 43–44
Auramine phenol, fluorescent stain, 119
Auto-agglutination, 196–197
Autolysis, of cell wall, 362–363
Autolytic enzymes, inactivation of, 368
Autoradiography, for assay of DNA per nucleoid, 308
Auxiliary focusing magnifier, *see* Auxiliary telescope.
Auxiliary microscope, *see* Auxiliary telescope.
Auxiliary telescope,
    dark-field illumination, in, 79
    phase contrast microscopy, in, 62, 71–72
Avidity, of serum with toxin, 261
Azide, as bacteriostat, 369
Azotobacter,
    cysts, staining for, 126
    DNA hybridization, 312

B
Bacillus species,
    chemically defined medium for motile, 160
    DNA preparation from, labelled, 333, 335
    motility—phage for, 156
B. anthracis, motility mutants, 159
B. brevis, factors affecting motility of, 161
B. cereus,
    immuno-electrophoresis of cell antigens, 190
    stained with labelled antibodies, 233, 243, 245, 251, 252, 253
B. licheniformis, mucopeptide from, 377
B. megaterium, fluorescence brightening of, 138
B. pumilus,
    flagellar structures, 165, 166, 167
    flagellins of, 169
B. subtilis,
    DNA preparation from, labelled, 335

ferritin labelled antibody staining of, 244
genome size, 302
motility mutants, 160
RNA hybrid competition in, 355, 356
teichoic acids in, 384, 385
B. thuringiensis, crystal protein analysis, 185
Bacteriophage (T2), DNA hybridization of, 332
Ballotini beads, for cell disintegration, 365, 367, 370
Barrel distortion, 45–46
Barrier filters, in microscopy, 87–88
Basal structure, flagellar, 165
Basic Fuchsin, 118
Basidiospores, optical brightening of, 143
Beggiatoales, motility, 145
Benzoate, sodium, in DNA preparation, 335
Binding efficiency,
    DNA, of, 343
    messenger RNA, of, 348–349, 350
Bispheric condenser, in dark-field illumination, 77
Blood withdrawal techniques,
    guinea-pigs, 294
    mice, 290–291
    rabbits, 296–297
    rats, 292
Blood films,
    optical brighteners for examining, 139–141
    stains for, 132–134
Blue-green algae,
    motility of, 145
    mucopeptide of, 373
Body fluids, *see under specific names.*
Borate buffer saline, preparation, 278
Bordetella pertussis, fluorescent antibody identification of, 237
Bovine plasma albumin, as immersion medium, 74, 77
Bovine plasma globulin, as immersion medium, 75
Bovine serum albumin, DNA hybridization in, 325
Braun disintegrator, 365, 367
Breed's stain, for bacteria in milk, 119–120

Broken cell suspensions, cell wall from, 368–372
Brucella spp., prozone formation by, 197
B. abortus, prozone formation, 197
Brucellosis, 197
BSS 3625 micrometer eyepiece graticule, 95–96
Beccal mucosal cells, phase contrast microscopy of, 73
Buoyant density, 327
2-Butoxyethanol, 335

C

Caesium chloride density gradient,
DNA–DNA hybrid detection in, 326–328
DNA–RNA hybrid detection in, 331–332
Calcium gelatin saline, (Cagsal), preparation, 278
Calcofluor White M2R, 137
Calflex interference heat filter, in microscopy, 87
Camera lucida, in micrometry, 96
Capillary—tube method, for motility studies, 147, 150–151
Capsular agglutinogens, (antigens), 194
Capsular antibodies, production of, 213
Capsule stains, preparation and use, 121–123
Carbohydrate, estimation in cell walls, 390
Carbonate buffer, preparation, 248
Carboxyhaemoglobin, as immersion medium, 75
Carboxylated sugars, lipopolysaccharide 381
Carboxy–terminal amino-acids, identification, 412–413
Carcases, disposal of animal, 283
Cardioid condenser, in dark-field illumination, 77
Cassegrain condenser, in dark-field illumination, 77
Cell disintegration, 363–368
Aminco French press, 365, 367–368
Braun disintegrator, 365
DNA preparation, for, 314
glass beads for, 364–365, 367, 370
Hughes press, 364

media for, 364
Mickle disintegrator, 364–365, 367
resuspension of cells for, 366–367
Ribi press, 366, 368
Shockman head for, 365, 370
Waring blender, 364
Virtis homogeniser, 364
Cell storage, for DNA hybridization, 313
Cell wall,
antigens, labelled antisera to, 222
hydrolysates, analysis of, 393–408
lipids, identification, 384
lipopolysaccharides,
extraction, 377–378
hydrolysis of, 379
lipoprotein, extraction, 377–378
mucopeptide,
amino-acids from, 382–384, 396–408
analysis of, 382–384
extraction of, 363, 376–378
hydrolysis, 379
muralytic enzymes and, 376–377, 408
"soluble", fractionation of, 413–414
phospholipid, extraction of, 377–378
polymers,
analysis, 389–393
constituents of, identification, 380–389
hydrolysate analysis, 393–408
hydrolysis of, 378–380
separation of, 373–378
polysaccharide, 377–378, 379
proteins, 377–378, 379
teichoic acids,
constituents of, identification, 384–389
hydrolysis of, 380
extraction of, 375–376
muralytic enzymes and, 376–377
Cell walls,
appearance, 372
autolysis of, 362–363, 368
composition, 373
contamination of, 369
cytoplasm removal from, 368–370
definition of, 361
fluorescence brightening of, 138

Cell walls—*cont.*
    hydrolysis, 378–380
    muralysis, 408
    preparation, 366–367
    staining, 127–128, 138
    whole, analysis of, 389–393
Cellular motility, *see also under* Motility, 146–149
Cellulose, in cell wall, 373–375
Cellulose nitrate filters, as single-stranded DNA support, 326, 342–343
Cellusolve, 334
Centering telescope, *see* Auxiliary telescope.
Cesares—Gil, flagellar stain, 128–130
Chemotaxis, capillary-tube method for, 147
Chick myoblast, phase-contrast microscopy, 74–75
Chitin, in cell wall, 373–375
Chitinase, analytical use, 375
Chitosan sulphate reaction, 375
Chlorate, DNA determination in, 318
Chloroform,
    bacteriostat, as, 369
    DNA preparations, in, 315
    lipid extraction with, 363
Chromatic aberration, 41–43
Chromatic difference, 42–43
Chromobacterium sp., DNA hybridization, 312
Chromosomal DNA, 323
Cilia, motility and, 145
Circle of confusion, in microscopy, 34–35
Cistrons, number per nucleoid, 301, 311, 348
Citrobacter ballerupensis, fimbrial antigen, 196
Clonal motility, 149–151
Clostridium, autoagglutination of, 196
C. botulinum,
    fluorescent labelled, 235
    toxins and indicator effects, 258
C. chauvoei, fluorescent labelled, 233
C. oedematiens, 236, 263
C. perfringens,
    antitoxin, 259, 276, 277
    toxin-antitoxin assay indicators, 263, 264

toxins, 257, 258
C. septicum,
    antitoxin unit,259
    fluorescent labelled, 233, 234
    toxins and indicators, 258, 263
C. sporogenes, fluorescent labelled, 233
C. tetani, toxins and indicators, 258
C. welchii,
    antitoxic antibodies, 215, 216
    auto-agglutination of, 196, 197
    fluorescent labelled, 234, 235
Colour sensitivity curves, 84
Colour temperature, of microscope lamps, 40
Coloured filters, in microscopy, 82–85
Coma, in microscopy, 44–45
Compensating eyepieces, 50–51
Compensation filters, in microscopy, 84
Competition curves,
    DNA homology by, 319
    RNA hybridization by, 355–358
Competition hybridization,
    direct hybridization and, 324
    DNA hybridization, for, 319–321
Complement, cell lysis and, 202–204
Complement—deviation, *see* Complement fixation.
Complement—fixation reactions, 202–207
    technique of, 204–207
    Wassermann reaction and, 207
Complement—fixation test, procedure, 204–207
Complement—fixing antibodies, antiserum production for, 214–215
Complementary colours, wavebands of, 83
Condensers, *see* Microscope condensers.
Conidia, fungal, fluorescence brightening of, 136, 143
Conjugates of antisera with labels,
    ferritin, 239–240
    fluorochromes, 222–228
Contrast filters, in microscopy, 82–85
Conversion filters, in microscopy, 84
Coombs antiglobulin test, basis of, 175
Corning filters, in microscopy, 87
Correction collar, for microscope objectives, 59

Correction filters, in microscopy, 85–86
Corynebacterium spp., "metachromatic granule" staining, 120–121
C. diphtheria,
  identification with fluorescent antibody, 237
  toxin and indicator effect, 258
Coulter counter, in micrometry, 99–100
Cover glasses,
  cleaning of, 106–108
  disposal of contaminated, 108
  gauge, 60
  mount, 59
  staining uses of, 106–111
  storage of, 107–108
  thickness of, 59–60
Coxiella burnetti, disintegration, 366
m-Cresol, in DNA preparation, 334
Critical angle, in reflectance, 16
Critical illumination system, in microscopy, 35–36, 56–57
Cross-linked DNA gels, 341–342
Crown glass lens, 42, 43
Crystal Violet, 115, 116, 117
Culture-tube method, for clonal motility studies, 149–150
Cushion distortion, 45–46
Cytological stains, for micro-organisms, 106, 120–134
Cytophaga sp., reference DNA from, 327
Cytophilic antibodies, 201, 202
Cytoplasm, removal from cell-wall preparations, 368–370
Cytoplasmic membrane, 368–370

D

Danysz phenomenon, in toxin-antitoxin combination, 263
Dark-field illumination, 71, 75–81
  annular diaphragm for, 75
  auxiliary telescope in, 79
  condensers for, 71, 75–77
  funnel stop in, 75–76
  immersion (mountant) media for, 76–78
  light source for, 80
  minimum effect aperture for, 75
  objectives for, 75, 77

objective iris diaphragm for, 75, 79–80
  oil immersion, and, 76, 80
  setting up system for, 78–81
  slide thickness in, 77
  uses of, 80
DEAE—cellulose, in preparation of t-RNA, 338
Denatured DNA, see also under Single-stranded,
  competition hybridization, in, 317, 319, 355–358
  DNA–RNA hybrids and, 332–333
  gels, in, 315–324, 339–342
  labelled with ^{14}C, 317
  u.v. absorbance of, 343
Denaturation temperature, of DNA, 317
Density gradient centrifugation, 369–372
Deoxyribonuclease, see also DNAase, 335, 337
Deoxyribonucleic acid, see DNA.
Depilation, of experimental animals, 288
Detergents, for DNA release, 312, 314
Deuterium oxide, (D_2O), labelled DNA hybrids in, 326
Dialister pneumosintes, size of genome, 302
Diaminobutyric acid,
  identification of, 382–383
  mucopeptide, from, 382–383
Diaminopimelic acid,
  cell wall hydrolysate, in, 396–397
  estimation of, 396–397
  identification of, 382–383
  isomers of, 399–400
  mucopeptide, from, 382–383
Diaminopimelic acid decarboxylase, 400, 404–406
Diaminopimelic acid epimerase, 400, 404–406
4,4'-Diamino-2, 2'-stilbene disulphonic acid derivatives,
  as fluorescent brighteners, 137
Diamino-trideoxy hexose, in teichoic acids, 384
Didymium filters, 84
Diffraction, of light waves, 6, 11–16
Diffuse reflection, 17, 18

Diffusion box, for antigen-antibody reaction analysis, 184, 186
Diffusion columns, for antigen-antibody reaction analysis, 187, 188, 189
Diffusion methods, for antigen-antibody precipitation studies,
box, 184, 186
column, 187, 188, 189
plates, 183, 184, 185
p,p′-Difluoro-m, m′-dimitrophenylsulphone, 240, 251
Dilute Carbol Fuchsin, preparation and use, 115
1-Dimethylaminonaphthalene-5-sulphonic acid, 221
Diphenylamine test, for bound DNA, 345
Diphenyloxazole, (PPO), 344
Diphtheria antitoxin,
hyperimmunization of horses for, 279
production of, 215
units of, 256, 259
Diptheria toxin, 257, 258
Direct hybridization, 322, 324
Distortion, in microscopy, 45–46
DNA, (deoxyribonucleic acid),
binding to filters, 342–343, 345
cistron, 348–349, 350, 355–358
filter-bound, assay of, 345
homology,
competition curves for, 319
percentage calculation, 232–234
labelling methods for, 313–314, 333–335
length measurement by e.m., 306–307
molecular weight per nucleoid, 301–309
chemical method, 303–306
electron microscopy for, 306–307
autoradiographic method, 308
preparation,
Marmur's method of, 314–315, 333–335
reference for, 319
yield from labelled cells, 314
DNA–agar,
DNA–DNA hybridization in, 318–324
DNA–RNA hybridization in, 339–341
preparation of, 316, 318, 339–340

DNAase, uses, 363, 368, 370
DNA–DNA hybridization, 311–328
bacterial taxonomy, in, 311
caesium chloride gradient technique in, 326–328
competition methods in, 319–322
DNA–agar method for, 315–324
DNA preparation in, 314–315
DNA–RNA hybridization and, 333
literature of, 328
membrane filter method for, 324–326
stock solutions for, 316, 317, 318
DNA-gel, 340, 341
DNA–RNA hybridization, 331–359
analysis of, 350–354
competition experiments in, 355–358
methods for, 339–348
agar gel, 339–341
cross-linked DNA gel, 341–342
Gillespie and Spiegelman, 342–348
membrane filter, 343–347
DNA–RNA hybrids,
assay of, 346–347
detection of, 332–333
equilibrium constant for, 350–351
melting temperture of, 351
non-specific filter absorption of, 346–347
ribonuclease, resistance to, 347
Dodecyl sulphate, sodium, 334
Dose-response curve, for toxins, 258
Dowex-50, use in DNA–RNA hybridization, 337, 341
Dry system objectives, 69
Duponol, 314

E

Eberthella typhosa, motility of, 161
Egg albumin, as immersion medium in microscopy, 75
Ehrlich's reagent, in hexosamine assays, 381
Einstein's theory of radiation, 7
Electromagnetic spectrum of radiation, 2, 4, 5
Electromagnetic (wave) theory of light, 6–7
DNA per nucleoid measured by, 306–307

Electron microscopy,
   ferritin labelled antibody technique
      for, 237–245
   locomotive organelle observation by,
      153–155
"Emission" theory, of vision, 5
Empty magnification, 32
Endospores, see also Spores, staining of,
   123–124
Endpoint, of antitoxin calibration,
   definition, 259
Enterobacter cloacae, fimbrial antigens,
   196
Enterobacteriaceae, growth for DNA
   hybridization, 312
Enterotoxin, 258
"Envelope" fraction, of cell wall, 361,
   364
Enzyme labelled antibodies, for antigen
   localization, 250–254
   background to use of, 250–251
   coupling methods for, 251
   staining with, 251–253
   results of uses of, 253–254
Escherichia coli,
   disintegration of, 366
   DNA–DNA hybridization in, 321,
      325
   DNA–RNA hybridization in,
      efficiency of, 348–349
      kinetics of, 344
      ribosomal RNA and, 355, 356
   fimbrial antigens, 196
   fluorescent antibody identification,
      237
   genome size, 302
   messenger RNA of, 350
   Mitomycin C and, 73
   motility mutants of, 155–156, 159,
      160, 161
   tactic responses, of, 161
   trypsin sensitivity of, 369
2-Ethoxyethanol, (cellusolve), 334,
   344
1-Ethyl-3(3-dimethylaminopropyl) carbodiimide, 251
Ethylene diamine tetra-acetic acid,
   (EDTA), 333–334
Euthanasia, of experimental animals,
   297–298

Evolution of bacteria, DNA hybridiza-
   tion and, 312
Experimental animals, see Animals.
Eye,
   colour sensitivity of, 82, 84
   resolution limits of, 22
Eye lens, 21–22
Eye pieces, see under Microscope.

F

Ferritin labelled antibodies, in antigen
   localization, 237–245
   conjugation procedure, 239–241
   ferritin preparation for, 237–239, 240
   results of application, 242–245
   uses of, 241–242
Ficoll, in DNA hybridization, 325
Field depth, in microscopy,
   determination of, 33
   numerical aperture and, 34
   resolving power and, 34
   values, for objectives, 34, 35
Field-of-view index, 51–52
Filament, flagellar, 165, 166
Filar micrometer,
   calibration of, 92–93
Filters, see also under Microscope,
   DNA binding to, 324–326, 342–343
Fimbriae, antigenic constitution, 195–
   196
Fimbrial antibodies, production of,
   213
Fimbrial antigens, 196
Flagella
   agglutination, 194–196
   bundles of, 151, 153, 165
   isolation, 166–167
   fixation, for staining, 106, 112, 230
   morphological features, 165–166
   motility and, 145–146
   mutations affecting, 157–159
   observation, 151–155, 165
   production, 165–172
   protein, 168–170
   purification, 167–168
   staining, 128–130, 151–153
Flagellar agglutinogens, (antigens), 194,
   195
Flagellar antibody, preparation, 213

Flagellar antigens, labelled antisera to, 222
Flagellates, flagella of, 146, 151
Flagellin, 166, 168–170
Flat field objective, 48
Fleming particle eye micrometer, 92, 95
Flint glass lens, 42–43
Flocculation, as antigen-antibody reaction, 177, 179–181
Flocculation tests, in toxin-antitoxin assay, 268–275
  antitoxin testing, 270–275
  apparatus for reading, 269
  haemolytic test, 271, 273–274
  intracutaneous test, 271, 274–275
  lecithinase test, 271–273
  materials required for, 268
  precipitin test compared with, 178
  toxin testing, 268–270
Fluolite C, 221
Fluorescein isocyanate, 223
Fluorescein isothiocyanate, for antibody labelling, 221, 223–225
Fluorescence microscopy, see also under Fluorescent, 143–144
Fluorescent antibodies, see also Fluorescent labelled antibodies,
Fluorescent brightening agents, see also under Optical,
  microbiological staining, 135–144
  plant and animal cell staining, 138–139
Fluorescent labelled antibodies,
  antigen localization with, 219–236
  methods of use, 219–221
  microscopy equipment for, 230–232
  preparation of, 222–228
  staining procedure with, 228–230
  uses of, 233–237
Fluorescent microscopy
  equipment for, 230–232, 249–250
  filters for, 232, 249
  optically brightened microbes, of, 143–144
Fluorescent staining,
  acid-fast bacteria, for, 119
  examination of specimens, 230–233
  procedure for, 228–230
  results of, 233–237
  scheme of methods for, 220

Fluorite lenses, 42, 43, 46
Fluorite objectives, 46
Fluorochromes, for antibody labelling,
  conjugation with antisera, 222–228
  examples of, 221
Fluors, vital, 136, 141–143
Flying-spot microscopy, for cellular motility studies, 148
Focusing condenser, in dark-field illumination, 77
Formaldehydogenic end groups,
  cell walls, 391–392
  estimation of, 391–392
Formalin, cell lysis prevention by, 363
Formamide—SSC, in DNA–RNA hybridization, 342
Formic acid, in labelled RNA estimation, 344
Fovea centralis, vision and, 22
Fraünhofer diffraction, 14–16, 24–25
Free amino groups, estimation in cell wall, 390–391, 411–412
French pressure cell,
  cell disintegration by, 365, 367–368
  DNA shearing in, 317
  labelled RNA extraction by, 336
Fresnel diffraction, 13–14
Fungi
  disintegration of, 363, 364
  insoluble polymers from, 373–375
  spore separation in, 363
Funnel stop, in dark-field illumination, 75–76, 80

G

Galactosamine,
  cell wall hydrolysates, in, 394
  estimation of, 394
  identification of, 381–382, 386
  mucopeptide, from, 381–382
  teichoic acid, in, 386
Galactose, in teichoic acids, 386
Gas chromatography, in cell wall analysis, 378, 388–389
Genetic relationship, DNA–RNA hybridization and, 359
Genome, bacterial,
  DNA hybridization and, 311
molecular weight of, 301–309
  size for specific organisms, 302

Geon 101 polyvinyl beads, 341
Gibberella fujikuroi, optical brightening, 139
Giesma stain, 132–133
Glass beads, for cell disintegration, 364
Gliding movement, microbial, 145
B-Glucanase, preparation and use, 374–375
Glucans, in cell wall, 373–374
Glucosamine,
    cell wall, estimation of, 394–396
    identification, 381–383, 386
    mucopeptide, 381–383
    teichoic acids, in, 384, 386
Glucosaminopeptide, *see under* Muco-peptide,
Glucose, in teichoic acid, 384, 386
Glucosidases, for teichoic acid degrada-tion, 387
Glutamate, isomer determination, 400, 403
Glutamate dehydrogenase, for gluta-matic assay, 400, 403
Glutamate-pyruvate transaminase, 400–403
Glutamic acid, *see also* Glutamate,
    identification from mucopeptide, 382–383
Glutaraldehyde, in labelled antibody preparation, 251
Glycerol, identification in teichoic acids, 386, 388
Glycine, identification in mucopeptide, 382–383
Glycine buffer, for DNA hybridization, 326–327
Glycopeptide, *see* Mucopeptide.
Glycosididase, estimation of, 408–410
Glycosidic bonds, in mucopeptide, 408
Goniometer eyepiece, in micrometry, 100
Gram stain,
    preparation and use, 115–117
    rapid methods for, 117
    type, 106
Gram-negative bacteria, cell wall poly-mers, 377
Gram-positive bacteria, mucopeptide content, 377
Grease removal, from slide, 107

Green filters, in microscopy, 83–86
Guanine, RNA labelled with ^{14}C or ^{3}H, 336–337
Guanine-cytosine percentage, hybridi-zation temperature and, 320–321
Guinea-pigs, handling techniques for,
    blood withdrawal, 294
    depilation for toxin testing, 277
    euthanasia, 298
    injections, 286, 293–294
    working conditions and, 283, 292–294
    varieties of, 292–293
Guru Swarmi micrometer eyepiece graticule, 95–96

                            H

Haemolytic test, in toxin-antitoxin assay, 271, 273–274
Haemophilus influenzae, genome size, 302
"Hanging drop",
    cellular motility studies with, 146–147
    preparation, 110
Hanks' solution, 201
HBO 200 (mercury) lamp, 39, 40
Heat filters, in microscopy, 87
Heat fixation, for staining, 112
Heavy labelled DNA, 326
Heerbrugg condenser, 70
Heptoses, identification in lipopoly-saccharides, 381
Heterologous DNA, in hybridization studies, 316, 320, 359
Hexosamines,
    estimation in cell wall hydrolysates, 393–396
    identification in mucopeptides, 381–382, 387
High eye-point eyepieces, 52
Hiss's capsule-stain, preparation and use, 123
Homoserine, in mucopeptide, 383–384
Hook, of flagellum, 165, 166, 168
Horses, hyperimmunization in, 278–279
Hughes press,
    cell disintegration by, 364
    DNA preparation, in, 314
    RNA extraction, in, 336

Human plasma albumin, as immersion medium, 75
Huygens negative eyepiece, 48–50
Huygen's principle, 9
Hybridization, see also under DNA–DNA and DNA–RNA,
   curves, 351–355
   efficiency of, 348–349, 354–355
   methods of, 315–328, 339–359
   temperature, 320, 350
   time, 321
   volume, 319
3-Hydroxy-diamino pimelic acid, identification, 383–384
8-Hydroxyquinoline, in DNA preparation, 334
Hyperimmunization, in horses, 278–279

I

Image formation, in microscopy, 24–26, 49, 60–64
   Abbe theory of, 24
   biconvex lens, by, 23–24
Image quality, in phase-contrast microscopy, 64–68
Image resolution, 27–32
Image-shearing eye piece systems, in micrometry, 92–95
Imaging errors, 40–46
Immersion media,
   dark-field illumination, for, 76–78
   phase-contrast microscopy, for, 73–75
   refractive index of, 78
Immersion objective, 48
Immersion refractometry, of living cells, 73–75
Immobilized DNA, 315–339
Immuno-diffusion, for antigen-antibody reaction analysis, 182–191
Immuno-electrophoresis, for antigen purification and analysis,182, 190, 191–193
Immunoglobulins, see also Antibodies,
   ferritin labelling of, 239–241
   fluorescent labelled, 229
   fluorochrome conjugation with, 223–228
India ink, capsule staining and, 121–122

Indicator effects, of toxins, 258
Indicators, for toxin-antitoxin assays, 263–268
   lecithovitellin, 264
   Ramon flocculation test, 265–268
   red cells, 264
   skins, 263–264
   tissue cultures, 265
   toxoid sensitized sheep red cells, 264–265
   whole animals, 263
Injection routes and techniques, 285–288
   guinea-pigs, 285, 293–294
   intravenous, 286–287
   mice, 285, 289–290
   rabbits, 285, 295–296
   rats, 285, 291–292
   scarification as, 287–288
   subcutaneous, 287
Inorganic phosphates, identification in teiochic acid, 386, 388
Interference filters, in microscopy, 84, 87
Interference fringes, 10–11
Internal reflection, 16, 17
Intercardiac injection,
   guinea-pigs, of, 294
   rats, of, 292
Intracutaneous test, in toxin-antitoxin assays, 271, 274–275
Intradermal injection techniques,
   general, 287
   guinea-pigs, of, 293
   mice, of, 289
   rats, of 291–292
Intramuscular injection techniques,
   general, 287
   guinea-pigs, of, 294
   mice, of, 289
   rabbits, of, 295–296
Intraperitoneal injection technique,
   general, 287
   guinea-pigs, of, 293
   mice, of, 289
   rabbits, of, 296
   rats, of, 291–292
Intravenous injection technique,
   general, 286–287
   guinea-pigs, of, 293

mice, of, 289–290
rabbits, of, 296
rats, of, 292
Ion exchange chromatography, of teichoic acid hydrolysates, 389
Iris diaphragm, in microscopy, 55–58, 68, 70
Iron, of ferritin labelled antibody, 237
Isopropyl alcohol, in DNA preparation, 334

J

Joyce–Loebl double beam recording densitometer, 327

K

Kel–F centrepiece, 327
Kellner eyepiece, 50, 90
Kieselguhr, as DNA–gel support, 341
Killing, of laboratory animals, techniques for, 297–298
Klebsiella aerogenes, fimbrial antigens of, 196
Klett colorimeter, 314
Kodak Wratten filters, in microscopy, 87
Kohler illumination system, 24–25, 36–38, 57–60, 70–72

L

Labelled antibody techniques, for antigen localization, 219–254
Labelled micro-organisms,
    DNA–DNA hybridization and, 313–314
    DNA–RNA hybridization and, 333–339
Laboratory animals, see Animals.
Lactate dehydrogenase,
    alanine estimation by, 400–403
    muramic acid determination by, 395–396
    preparation of, 395–396
Lactobacillus arabinosus teichoic acid from, 376
Lamps, see Microscope lamps.
Lecithinase test, in toxin-antitoxin assay, 264, 271–273

Lecithovitellin,
    preparation of, 278
    toxin-antitoxin assay indicator, 264
Leifson's flagella stain, 128–129
Leitz condensers, 54, 70–71, 77
Leitz Periplan eye-piece, 51–52
Leitz phase-contrast objectives, 69
Leitz phase plate, 67
Lenses, see also under Microscope,
    auxiliary, 23
    biconvex, image formation by, 23–24
    eye, vision in, 21–22
Leucocidin, 258
Leucocytes, optical brightening agents for, 139, 140
Leuconostoc mesenteroides, lactate dehydrogenase, 395–396
Light, theories of, 5–7
Light balancing filters, in microscopy, 82–85
Light waves, 2–21
    amplitude specimens, and, 18–20
    diffraction of, 11–16
        Airy disc in, 14–15
        Fraünhofer, 14–16
        Fresnel, 13–14
    filtration of, 18
    fronts of, 9
    interference of, 10–13
    phase relations of, 10–13
    phase specimens, and, 20–21
    plane, 8–10
    propagation of, 9
    reflection of, 16–18
    refraction of, 16–18
    spherical, 8–10
    trains of, 11–13. 19. 20
Lipid inclusion granules, stain for, 124
Lipids, cell wall, 384
Lipopolysaccharides,
    cell wall, 373, 377–380
    extraction of, 378
    hydrolysis, of, 379–380
    sugars from, 380–381
Lipoproteins, of cell wall, 377–378
Lissamine flavine FSS, 221
Lissamine rhodamine B 200, antibody labelling, 221, 225
Lister objective, 48

Living cells, refractive indices, 73–75
Locomotive organelles,
  electron microscopy of, 153–155
  observation of, 151–155
  optical microscopy of, 151–153
Loeffler's methylene blue, preparation
  and use, 114–115
Low-voltage lamps, physical characteristics, 39
Lyophilization, in DNA hybridization,
  313
Lysine,
  determination, 400, 406
  identification, in mucopeptide, 382–
  383
Lysine decarboxylase, analytical use,
  400, 406
Lysozyme,
  DNA preparation, in, 314, 333
  mucopeptide digestion by, 376
  RNA preparation, in, 337

M

Macroglobulin and microglobulin,
  separation, 182
Macroglobulin antibody, 174
Magnesium fluoride, in phase plates, 66
Magnification,
  chromatic difference of, 42–43
  compound microscope, of, 26–27
  eye–piece, by, 27
  maximum, 32–33
  minimum, 32–33
  objective, by, 27
Malachite green, as stain component,
  118, 120, 123
Mannans, in microbial cell walls, 373–
  374
Marmur's method, of DNA preparation, 314, 315
Mass action effect, on DNA–RNA
  hybridization, 360–361
Maxwells electromagnetic theory, 6
Melting temperature, in DNA–DNA
  hybridization, 317
Membrane filters, in nucleic acid
  hybridization methods, 324–
  325, 342–348
Messenger RNA, see also under RNA,

base composition of, 350
binding efficiency of, 348–349
detection of, 332
DNA hybrids, temperature effects
  on, 350
fractionation of, 349–350
preparation of, 336–337
ribosomal RNA, in, 337–338
"Metachromatic granules", staining for,
  120–121
Methylene Blue, see Loeffler's methylene
  blue.
Mice, techniques for handling, 288–
  291, 298
anaesthesia, 290
euthanansia, 298
handling, 289
injections, 285, 286, 289–290
varieties of, 288
withdrawal of body fluids, 290–291
working conditions and, 283
Mickle disintegrator, 364–365, 367
Microbial cells,
  autolysis of, 362–363
  cell walls, from, 362–372
  disintegration of, 363–368
  preparation of, 362–363
Micrococcus radiodurans, pigment localization, 372
Microelectrophoresis, in cellular
  motility studies, 149
Microglobulin antibody, 174
Micrometry,
  angle measurement by, 99–100
  area measurement by, 98–99
  depth measurement by, 99–100
  drawn images, and, 96–97
  eye-piece systems for, 88–100
    Filar, 91–93
    fixed, 89–91
    Fleming, 92, 95
    goniometer, 100
    graticules for, 95–96
    image-shearing, 92–95
    moveable, 91–93
    scale for, 89–91
    Vickers, 92–94
    Watson, 92, 94–95
  linear measurements by, 88–98
  macroscopic measurements by, 88–89

Micrometry—*cont.*
  Nikon microscope in, 98
  photomicrographs and, 97
  projected images and, 97
  stage micrometer in, 89–91
  stage vernier scales in, 88–89
  standard scale for, 89–90
  volume measurement by, 99–100
Micro-projection drawing mirror, in
      micrometry, 96
Microscope,
  bright field use of, 21–55
  condensers,
    Abbe, 53–54
    achromatic, 54–55
    aplanatic, 54
    dark-field, 75–77
    Leitz, 54, 70–71, 77
    lens systems for, 24–25
    numerical aperture of, 29
    phase contrast, 68, 70–71
    resolution by, 29
    Wild, 54–55
  dark field illumination, in, 75–81
  eye-pieces,
    adjustment of, 56, 58
    compensating, 50–51
    field-of-view index, 51–52
    high eye point, 52
    Huygens, 48–50
    image formation by, 48–49
    Kellner, 50
    Leitz, 51
    lens in, 48–53
    magnification by, 26–27
    negative projection, 53
    Ramsden, 50
    wide field, 51–52
  features of, 3
  filters, 18, 81–88
    barrier, 87–88
    coloured, 82–85
    contrast, 82–85
    correction, 85–86
    didymium, 84
    green, 83–86
    heat, 87
    interference, 84, 87
    light balancing, 82–85
    neutral density, 81–82

    polarizing, 82
    Rheinberg, 84
    selective transmission by, 18
    transmission curves of, 84
    transmission wavebands of, 83
  illumination systems for, 35–38, 56–
      60
  image,
    drawn, 96–97
    formation of, 24–26, 60–64
    photographed, 97
    projected, 97
    quality of, 64–68
  lamp, 37–40, 56–58
    adjustment, 56–58
    colour temperature, 40
    comparison, of, 39
  lens,
    achromatic, 42
    anastigmatic, 44
    apochromatic, 42
    astigmatism in, 43–44
    chromatic aberation in, 41–43
    coma in, 44–45
    correction of, 27
    crown glass, 42–43
    distortion by, 45–46
    flint glass, 42–43
    fluorite, 43
    image formation by, 24–26
    imaging errors in, 40–46
    semi-apochromatic, 42
    spherical aberation in, 41
  objectives,
    achromatic, 46
    apochromatic, 46–47
    changing of, 57
    correction collar for, 59
    cover glass thickness and, 59
    dark-field illumination and, 75, 77,
      80
    depth of field values for, 34–35
    dry system, 69
    flat field, 48
    fluorite, 46
    immersion, 48, 57, 59, 69, 76, 80
    Leitz, 69
    lens system in, 24–25, 46–48
    Lister, 48
    magnification by, 27

Microscope—*cont.*
    numerical aperture of, 29
    phase contrast, for, 66, 68–69
    plane, 47–48
    resolving power of, 28–29
    semi-apochromatic, 46
    use magnification, and, 32–33
    Zeiss, 47
    phase contrast systems, in, 60–75
    setting-up of, 55–60
    slides, 59, 77, 106–111
    Zeiss, 3
Milk, staining bacteria in, 119–120
Millipore filters, 324
Minimum effect aperture, in dark-field illumination, 75
Minimum effective dose, definition of, 257
Minimum haemolytic dose, 257
Minimum lethal dose, toxin potency expressed as, 256, 258
Mitomycin c, 73
Molar cytosine and guanine percentages, 317
Morgan–Elson reaction,
    N-acetyl-amino sugars, for, 409
    hexosamines, for, 393–395
Motility, 108–110, 145–160
    capillary tube method for, 147
    cellular, 146–149
    clonal, 149–151
    culture-tube method for, 149–150
    environmental factors affecting, 160–161
    flagellar agglutination and, 195
    flying-spot microscopy for, 148
    hanging-drop method for, 146–147
    individual bacteria, of, 146–149
    locomotive organelle observation, 151–155
    measurement and observation of, 146–151
    media for study of, 160
    microcinematography for, 148–149
    microelectrophoresis for, 149
    microscopic observation of, 146–149
    mutants, isolation and typing, 155–160
    plate method for, 150
    population of bacteria, of, 149–151
    special chamber method for, 147–148
    stroboscopy for, 148–149
    tactile responses and, 161
    types of, 145–146
    wet mounts for determining, 108–110
Motility mutants, 155–160
    isolation of, 155–157
        LP-type, 157, 158
    non-spreading types, 157–158
    paralysed type, 157–158
    pauci-flagellated, 158
    SD–type, 157, 158
    types, identification of, 157, 159
    weak-spreading types, 158–159
Motility-phages, 156
Mountant media, *see under* Immersion media.
Mounting, fluorescent antibody stained specimens, 230, 248
Mucopeptides,
    amino-acids from,
        estimation, 396–408
        identification, 382–384
    analysis of, 382–384
    extraction of, 363, 376–378
    hydrolysis of, 379
    muralytic enzyme effect on, 376–377, 408
    "soluble", 377, 413–414
Muralytic enzymes,
    cell wall lysis by, 377, 408
    estimation of, 408–413
    mucopeptide, degradation by, 408
Muramic acid,
    cell wall hydrolysates, in, 394–396
    estimation of, 394–396
    identification, 381–383
Murein, *see under* Mucopeptides,
Mutants, non-motile, *see also under* Motility, 155–160
Mycobacteria,
    disintegration of, 366
    lipid extraction from, 363
Mycoplasma gallisepticum, genome size, 302
M. hominis, genome size, 302
Mycoplasms, antisera to, 216
Myxobacteria, motility of, 145
Myxomycetes, motility of, 145

N

Nail varnish, as cover slip sealant, 109
Negative phase contrast, 66–68
Negative projection eyepiece, 53
Negative staining method, 154–155
Negative wedge window, 327
Neisseria gonorrhoeae, fluorescent antibody identification, 237
Neurospora crassa, genome size, 302
Neutral density filters, in microscopy, 81–82
"Newton's rings", 6
Nigrosin, 121
Nikon condenser, 70
Nikon microscope, use in micrometry, 98
Nikon phase plates, 67
Ninhydrin reagent, for hexosamine determination, 381–382
Nitrocellulose membranes, *see under* Cellulose nitrate.
N-terminal amino-acids, determination of, 411–412
Nuclear stain, Robinow's method, 130–131
Nucleic acids, *see also* DNA *and* RNA, in cell wall preparations, 373
Nucleotide sequences, 348, 355–358
Numerical aperture, of microscope condensers, 27–35, 75, 77
Nutrient gelatin agar, motility medium, 160

O

Oakley and Fulthorpe columns, *see* Diffusion columns.
Objective iris diaphragm, 75, 79, 80
Objectives, *see* Microscopic objectives.
Oil-immersion objectives,
  dark-field illumination, in, 76, 80
  phase contrast microscopy, for, 69
  setting up of, 57, 59
Optical brightening agents, *see also under* Fluorescent,
  development, 136
  fluorescence microscopy with, 143
  general uses, 137
  incorporation into media, 141–143
  nature and properties, 135–136
  sources of, 136
  staining with, 137–143
  techniques involving, 135–144
  toxicity to some bacteria, 141
  types "A" and "N", 137
Ornithine,
  identification, 382–383
  isomers of, 400, 406–408
  mucopeptide, in, 382–383
Ornithine transcarbamylase, analytical use, 400, 406–408
Orthophosphate, ^{32}P, nucleic acids labelled with 335, 336–337
Osmic acid solutions, storage, 113
Ouchterlony plates,
  antigen-antibody reaction analysis in, 183–184
  examples of, 183
  identity and non-identity reactions in, 185
Ouchterlony methods, for antigen-antibody reaction analysis, 182–191
Oxacyanine compounds, as fluorescent brighteners, 137
Oxoid No. 3 agar, for DNA-agar, 318

P

Paper chromatography,
  amino-acids, 382–383
  hexosamines, 381–382
  teichoic acid hydrolysis products, 386, 388
Paper powder, as DNA gel support, 341
Paraboloid condenser, in dark-field illumination, 77
Parasites, optical brightening agents for, 139–141
Pasteurella pseudotuberculosis, motility of, 161
Pauci-flagellated motility mutants, 158, 159
Penicillium chrysogenum, disintegration, 366
Pentose determination, 381
Peptidase,
  estimation of, 410–413
  mucopeptide degradation by, 408
  site of action of, 412
Peptidoglycan, *see* Mucopeptide.

Periodate oxidation, of cell walls, 391–392
Peritoneal fluid, withdrawal, 291
Phase contrast microscopy, 60–75
  annular diaphragm for, 61–62
  auxiliary telescope for, 62
  bacteria, of, 73–75
  basic conditions for, 60–64
  chick myoblast, of, 74
  condensers for, 68, 70–72
  image formation in, 60–64
  immersion media for, 73–75
  Kohler illumination system in, 70–72
  living cells, of, 73–75
  marker for, 72–73
  negative, 66–68
  objectives for, 66, 68–69, 70–74
  phase plate for, 61–68, 70–72
  phase ring in, 62–63, 65–66
  positive, 60–68
  setting up of systems for, 70–75
  specimens, for, 72–73, 111
  use of, 70–75
Phase plate, 61–68
  image quality, and, 64–68
  negative, 66–68
  phase retardation by, 64
  positive, 61–68
  preparation of, 66
  properties of, 67
  Phase ring, 62–63, 65–66, 71–72
Phase specimens, light absorption by, 20–21
Phenol, in DNA preparation, 334
Phenol-cresol mixture, 334, 336
Phosphatase, degradation of teichoic acid, 387
Phosphate buffer, preparation, 248
Phosphate esters, in teichoic acids, 386, 388
Phosphates, inorganic, in teichoic acids, 386, 388
Phosphocellulose acetate, as single-stranded DNA support, 332
Phosphodiesterase,
  DNA hybridization, in, 326
  teichoic acid degradation by, 387
Phospholipids, in cell wall, 378
Phosphomonoesterases, in teichoic acid degradation, 387

Photographic emulsions, colour sensitivity of, 82, 84
Photometric summation, law of, 11
Photomicrographs, 27, 97–98
Photon theory, of light, 6–7
Phycomycetes, disintegration of, 364
Phylogeny, DNA hybridization and, 312
Pili, see Fimbriae.
Plane objectives, in microscopy, 47–48
Plane waves, 8–10
Plank's constant, 7
Plant cells, optical brighteners for, 138–139
Plasmodium sp., detection in blood films, 139, 141
P. berghei, 141
Plate methods, for clonal motility studies, 150
Pneumococcus, teichoic acid, 384
Polar planimeter, in micrometry, 99
Polarizing filters, in microscopy, 82
Polydeoxyribothymidylic acids, 332
Polyglucose, as immersion medium, 75
Poly-β-hydroxybutyric acid, extraction of, 363
Polyols, identification in teichoic acid, 386, 388
Polyriboadenylic acid, 332
Polysaccharide inclusions, stains for, 124–126
Polysaccharides, see also under specific names,
  cell wall, in, 377–379
  extraction of, 378
Polyvinylpyrrolidone, in DNA hybridization, 325
Positive phase plates, preparation of, 66
Precipitating antibodies, production, 213–214
Precipitation reaction, in antigen-antibody reaction, 177–193
  diffusion techniques for, 182–191
  flocculation as, 177, 179–181
  immuno-electrophoretic techniques for, 191–193
  Ouchterlony methods for, 182–191
  ring test in, 177–179
  techniques of, 177–181
Precipitin test, flocculation and, 178

Precipitin reaction, extensions of, 182–193
Pronase, in DNA preparation, 315
Protein precipitation, in nucleic acid preparation, 334, 337
Proteins, in cell wall, 378, 379
Proteus sp., motility studies of, 146, 160, 161
P. hauseri, motility, 161
P. mirabilis, motility, 161
P. vulgaris, motility, 161
Protozoa, flagella of, 146, 151
Prozone, agglutination phenomenon, 197–198
Pseudomonas sp.,
    DNA hybridization, growth for, 312
    motility, factors affecting, 161
P. campestris, genome size, 302
P. fluorescens,
    genome size of, 302
    labelled DNA from, 314
P. putida, genome size, 302
P. viscosa, factors affecting motility of, 161

Q

Quartz-iodine lamp, characteristics of, 39

R

Rabbits, techniques for handling, 294–297, 298
    euthanasia, 298
    general handling, 295
    injections, 285, 286, 295–296
    varieties of, 294–295
    withdrawal of blood, 296–297
    working conditions and, 283
Radiation, electromagnetic spectrum of, 2, 4, 5
Ramon flocculation test, 265–268
Ramsden (positive) eyepiece, 50
Randomly labelled RNA, preparation, 337–339
Rapidly–labelled RNA, *see also* Messenger RNA,
    DNA–RNA hybridization of, 336–337, 339
    hybridization efficiency with, 348–349

Rats, techniques for handling,
    euthanasia, 298
    injections, 285, 286, 291–292
    working conditions and, 283
Rayleigh criterion, for optical resolution, 28, 30–32
Red cells, as toxin-antitoxin assay indicators, 264
Reducing sugars, determination of, 408–409
Reflection, of light waves 16–18
Refraction of light waves, 16–18
Refractive index, 16, 327
    immersion media, of, 31–32, 73–78
    living cells, of, 73–75
Rheinberg filters, in microscopy, 85
Rhizobium sp, DNA hybridization in, 312, 313
R. japonicum, DNA hybridization yield, 313
Ribbon filament lamp, characteristics, 39
Ribi cell fractionator, 366, 368,
Ribitol, identification, 386, 388
Ribitol teichoic acids, 384–386
Ribonuclease,
    DNA preparation with, 315, 334
    DNA–RNA hybrid resistance to, 347, 350
    non-specific hybridization and, 353
    non-specific RNA absorption and, 347–348
    RNA preparation, and, 337
Ribonucleic acid, *see* RNA.
Ribosomal RNA,
    bacterial transcription and, 332
    DNA cistrons and, 350
    hybridization efficiency of, 348
    non-specific binding and, 350
    preparation of, 337–338
Rickettsiae,
    cell wall separation, 369
    mucopeptide of, 373
Ring test, in antibody-antigen reaction, 177–179
RNA, (ribonucleic acid), *see also* DNA–RNA, Messenger RNA, Ribosomal RNA *and* Transfer RNA,
    characterization of, 331–359

RNA—*cont.*
  preparation of,
    messenger, 336–337
    randomly labelled, 337–339
    ribosomal, 337–338
    transfer and 5 S, 338–339
  purity of, 353, 358
RNA–DNA hybridization, *see* DNA–RNA hybridization.
Rod-shaped bacteria, area measurement of, 99
Romanowsky stains, for blood films and parasites, 132–133
Roux flasks, 312

S

Saccharomyces cerevisiae,
  disintegration of, 366
  genome size, 302
Salmonella sp.,
  antigenic classification, 195
  disintegration of, 366
  fimbrial antigen, 196
  flagellins of, 169
  fluorescence brightening of, 132, 142
  motility of,
    medium for, 160
    mutants, 155–156, 157–158, 159
  motility-phages, 156
S. gallinarum, antigenic reactions, 195
S. paratyphi C, temperature and motility of, 161
S. pullorum, antigenic reactions, 195
S. typhi, agglutination, 197
S. typhimurium, motility, 157, 158, 161
Sartorius membrane filters, 343
Saturation analysis, in DNA–RNA hybridization, 351–354
Scarification, as injection technique, 287–288
Schaeffer and Fulton spore stain, 123–124
Schiff's stain, preparation and use, 124–125
Schleicher and Schull filters, 324, 343
Schott filters, in microscopy, 87–88
Scintillation mixture, for hybridized RNA, 344, 346

Screw-micrometer, 91–93
S D S, *see* Dodecyl sulphate *and under* Sodium.
Selective transmission, by filters, 18
Semi-apochromatic lenses, 42
Semi-apochromatic objectives, 46
Sephadex, in RNA preparation, 337, 338
Septa, fluorescent brightening of, 138
Serine, identification from mucopeptide, 382–383
Serratia, motility-phages for, 156
Serum, preparation from blood, 298–299
Shockman head, for cell disintegration, 370
Shadowing methods, for e.m. of flagella, 154
Sheared DNA, 313, 314, 317, 319, 351
Sheath, flagellar, 166
Shigella flexneri, fimbrial antigens, 196
S. sonnei, fluorescent labelled, 232
Single-stranded DNA, supports for 315–326, 332, 339–342
Skin, as indicator in toxin-antitoxin assay, 263–264
Slides and cover glasses, use in bacterial staining, 106–111
Slime moulds, *see* Myxomycetes.
Slime-producing bacteria, DNA hybridization and, 312
Smears, preparation for staining, 111–112, 228, 229
Socket lamp P, characteristics of, 39
Sodium dodecyl sulphate, (S D S),
  cell lysis prevention by, 363–364, 368
  cell wall isolation, in, 369
Sodium salts, *see under name of parent compound.*
Somatic agglutinogens, 194
Specimen produced retardation, in phase-contrast microscopy, 64–65
Spectral colours, wavelengths of, 4, 83
Specular reflection, 17
Spherical aberration, in microscopy, 41
Spherical light wave formation, 9
Spheroplast lysis, for RNA extraction, 336–337
Spinco model E ultracentrifuge, 327
Spirillum sp., motility studies, 149, 161

Spirochaetes,
    motility of, 146
    stains for, 131–132, 133
Spores,
    antigen analysis by immuno-electro-
        phoresis, 190
    disintegration, 364, 366
    germination studies with fluorescent
        brighteners, 136
    stain for, preparation and use, 123–
        124
Spurious hybridization, 347, 350
S S C buffer, 315, 334, 337, 339, 342,
        348
Stage micrometer, 89–91
Staining bacteria, see also Stains,
    flagella, 151–153
    fluorescent antibodies, with, 228–230
    slides and cover glasses for, 106–111
    standard procedures for,
Stains, for bacteria,
    acid-fast, 106
    "cytological", 106, 120–134
    general, 113–120
    Gram, 106
    histochemical reaction type, 106
    labelled antibodies as, 228–230
    storage of, 112–113
    types of, 106
Standard indicating effect, of toxins,
        259
Standard sera, in microbiology, 176
Staphylococcal protein, trypsin sensi-
        tivity, 369
Staphylococci, area measurement of, 99
Staphylococcus sp., fluorescent labelled,
        235
S. aureus,
    alpha-toxin antibody, 215, 258
    cell wall separation, 369
S. lactis, teichoic acids, 384
Star test, in microscopy, 60
Streptococcal M protein, trypsin sensi-
        tivity, 369
Streptococcus lactis, disintegration, 366
S. pyogenes,
    disintegration, 366
    toxins and indicator effects, 258
Streptokinase, 258
Streptolysins, 258, 260

Subcutaneous injection technique,
    general, 287
    guinea-pigs, 293
    mice, 289
    rats, 291–292
Sucrose gradients, preparation of, 370,
        372
Sudan Black, lipid stain, 124
Sugars, in cell wall, 378, 393
Swimming movement, microbial, 145–
        146
Syphilis, 207

                             T

Tactile theory, of vision, 5
Tanned red cell agglutination, 200–201
Taxonomy, DNA hybridization and,
        311
"Tea-bag" method, in DNA hybridiza-
        tion, 322–323
Teepol, 322
Teichoic acids,
    bacterial cell walls, in, 375–377, 384–
        389
    constituents of,
        analysis, 384–389
        chromatography, 386, 388–389
        enzymic analysis, 387
    extraction of, 375–376
    hydrolysis of, 380, 385–389
    identification of, 376
    muralytic enzymes and, 376–377
    periodate oxidation of, 391–392
Teichuronic acids, 376
Test dose, of toxin, 259
Tetanolysin, 258
Tetanospasmin, 258
Tetanus antitoxin,
    hyperimmunization of horses for, 279
    units of, 259
Tetanus toxin, 257, 259, 278, 279
Threo-3-hydroxy-glutamic acid, muco-
        peptide, 384
Threonine, in mucopeptides, 382–383
Thymidine, 2-14C- and 6-3H-, for
        labelling DNA, 325, 335
Tinopal AN, 137
Tissue cultures, as toxin-antitoxin assay
        indicators, 265
Toluene 2,4-di-isocyanate, 234, 240

Toluene-liquefluor, for counting DNA, 325

Toluidine Blue, 120

Total combining power tests, in toxin-antitoxin assay, 275–277

Toxin-antitoxin assay, 255–280
accuracy and reproducibility of, 277
flocculation tests, 268–275
indicators for, 263–268
standard antitoxin for, selecting, 261–263
terminology of, 257–261
total combining power tests in, 275–277

Toxins, bacterial, see also Endo-toxins and Exo-toxins,
definition of, 257
discovery, 255–257
examples of, 258
flocculation test for, 268–270
indicator effects of, 258

Toxoid, definition of, 257

Toxoplasma, detection in blood smears, 140

Transfer RNA,
DNA–RNA ratios and, 358
preparation of, 338–339

Transmission curves, of microscope filters, 84

Transmission wavebands, of coloured filters, 83

Treponema,
fluorescent antibody detection, 211
syphilitic antibody test, 207

T. pallidum, identification, 237

Trinocular head,
dark-field illumination, in, 80
phase contrast microscopy, in, 72

Tris buffer, 325, 337, 338

Trypanosome vivax, optical brightening of, 142

Trypanosomes,
detection in blood smears, 140
optical brightening agents for, 136, 140
staining for, 133

Tubercle bacillus,
decontamination from glassware, 108
fluorescent stain for, 119
Ziehl–Neelson staining for, 117

Turret condenser, in phase contrast microscopy, 70–72

U

Ultracentrifugation, in nucleic acid hybridization, 326–328, 332

Ultrasonication, in DNA preparation, 314

Ultra-violet absorbance, of denatured DNA, 342

Ultra-violet light,
barrier filters in microscopy, 87
DNA gel formation, in, 341

Unit equivalents, of toxins or toxoids, 259

Universal condensers, 70

Uracil, 2-^{14}C-and ^{3}H-labelled, in nucleic acids, 313, 314, 335, 336–337

Useful magnification, of microscope, 32–33

V

Vibrio sp., factors affecting motility of, 161

Vickers condenser, 77

Vickers image-splitting micrometer eyepiece, 92–94

Vickers phase plates, properties of, 67

Virtis homogenizer, for cell disintegration, 364

Viruses, antibodies against, 208–210

Visible radiation, 5

Vision,
lens in, 21–22
theories of, 5

Vital fluorescent staining of micro-organisms, 141–143

W

Waring blender, for cell disintegration, 364

Wasserman reaction, 207

Water immersion objectives, 69

Watsons image-shearing micrometer eyepiece, 92, 94–95

Well slides, for motility studies, 108–109, 110

Wet mounts, for microscopic examina-
        tion, 108–111
    motility determination, 108–110
    phase contrast microscope, 111
Wide field eyepices, in microscopy, 51–
        52
Wild condenser, 54–55, 70
Wild phase plates, properties of, 67

                    X

Xanthomonas sp.,
    DNA hybridization growth for, 312
    heavy-labelled DNA from, 326
XBO 162 (Xenon) lamp, characteristics,
        39, 40
Xylans, in microbial cell wall, 373–374
m-Xylylene di-isocyanate, 239, 240

                    Y

Yaws, 207
Yeasts, *see also under specific names*,
    cell wall polymers from, 373–375
    disintegration of, 364

                    Z

Zeiss optical equipment,
    camera lucida, in micrometry, 96
    condenser, in dark-field illumination,
        77
    filters, in microscopy, 81
    microscope, 3
    objectives, data for, 47
    phase plates, properties of, 67
Zeocarb 225, as DNA-gel support, 341
Ziehl–Neelsen's carbol fuchsin, prep-
        aration and use, 115, 117–119